Advances in Additive Manufacturing and Their Applications (2nd Edition)

Advances in Additive Manufacturing and Their Applications (2nd Edition)

Guest Editors

Petru Berce
Răzvan Păcurar

MDPI

Basel • Beijing • Wuhan • Barcelona • Belgrade • Novi Sad • Cluj • Manchester

Guest Editors

Petru Berce
Department of
Manufacturing Engineering
Technical University of
Cluj-Napoca
Cluj-Napoca
Romania

Răzvan Păcurar
Department of
Manufacturing Engineering
Technical University of
Cluj-Napoca
Cluj-Napoca
Romania

Editorial Office
MDPI AG
Grosspeteranlage 5
4052 Basel, Switzerland

This is a reprint of the Special Issue, published open access by the journal *Metals* (ISSN 2075-4701), freely accessible at: https://www.mdpi.com/journal/metals/special_issues/P90Y08R4F3.

For citation purposes, cite each article independently as indicated on the article page online and as indicated below:

Lastname, A.A.; Lastname, B.B. Article Title. *Journal Name* **Year**, *Volume Number*, Page Range.

ISBN 978-3-7258-4229-2 (Hbk)
ISBN 978-3-7258-4230-8 (PDF)
https://doi.org/10.3390/books978-3-7258-4230-8

Contents

Preface

This Special Issue presents ten original research articles that explore key challenges and solutions in Additive Manufacturing for metallic applications, including powder reuse, hybrid modeling, microstructural control, and real-time monitoring. Addressed to professors, scientists, industrial engineers and experts in metal Additive Manufacturing, the Special Issue aims to highlight innovations that enhance process efficiency, component performance, and sustainability of these modern manufacturing processes. The contributions presented in the Special Issue reflect the ongoing shift toward intelligent and integrated Additive Manufacturing solutions for metallic applications.

Petru Berce and Răzvan Păcurar
Guest Editors

metals

Editorial

Challenges and Trends in Additive Manufacturing for Metallic Applications: Toward Optimized Processes and Performance

Petru Berce and Răzvan Păcurar *

Department of Manufacturing Engineering, Faculty of Industrial Engineering, Robotics and Production Management, Technical University of Cluj-Napoca, Blv. Muncii, No. 103-105, 400641 Cluj-Napoca, Romania; petru.berce@tcm.utcluj.ro
* Correspondence: razvan.pacurar@tcm.utcluj.ro

1. Introduction and Scope

Additive Manufacturing (AM) for metallic applications continues to redefine how complex, high-performance components are designed and fabricated across a wide range of sectors, including the aerospace, biomedical, and automotive fields, etc. [1]. Recent advancements in materials science, process optimization, and modeling have contributed to more reliable, efficient, and sustainable manufacturing practices in this area [2]. Despite this progress, several challenges persist. These include improving machinability and surface quality, optimizing the process parameters to minimize energy consumption, reusing materials without quality degradation, and reliably predicting performance outcomes through accurate modeling [3]. Furthermore, achieving consistent mechanical properties and reducing thermal residual stresses across different AM methods, such as Wire Arc Additive Manufacturing (WAAM), Laser Powder Bed Fusion (L-PBF), and Direct Energy Deposition (DED), remain areas that require significant research attention in the future [4–6].

This Special Issue aims to address these knowledge gaps by presenting a collection of ten original articles covering a spectrum of AM challenges and solutions for metallic applications—from powder reuse and hybrid modeling frameworks to microstructural control and other advanced monitoring techniques [7]. Collectively, the articles published in this Special Issue emphasize not only the performance of manufactured components, but also the efficiency and sustainability of the AM processes in a wide area of metallic applications. The contributions presented in this Special Issue emphasize a growing shift towards integrated, intelligent AM ecosystems for metallic applications that are combining materials science, machine learning, and process engineering. Looking forward, future research must focus on standardized benchmarking protocols, environmentally conscious process design, and advanced post-processing strategies that ensure component reliability in critical AM for metallic applications.

2. Contributions

This Special Issue presents 10 scientific articles that address key challenges in metal additive manufacturing, including process optimization, material reuse, stress modeling, surface enhancement, and real-time monitoring, highlighting current trends and future directions in the field of AM for metallic applications.

The first contribution of this Special Issue is a review article by Capasso, I. et al., which presents an extensive overview of the additive manufacturing (AM) technologies used for metallic applications, with a special focus on the field of construction engineering. The article thoroughly examines the evolution, principles, and classification of various

https://doi.org/10.3390/met15050525

AM processes for metallic applications, including powder bed fusion (SLM, EBM), direct energy deposition (WAAM, LMD), and binder jetting. It highlights the advantages and limitations of each method in terms of accuracy, surface finish, material compatibility, and build volume. Particular attention is paid to the process parameters influencing part quality, post-processing methods, and the challenges related to scale-up, standardization, and cost. The review also discusses the integration of topology optimization with AM for metallic applications and the potential for structural repair in civil infrastructure. By synthesizing a broad range of the current literature, the study provides valuable insights for both researchers and practitioners, reinforcing the strategic relevance of AM for metallic applications in the construction sector while outlining critical areas for future development.

The second contribution of this Special Issue is a research article by Zhang, H. et al., which presents a machine learning-based framework for optimizing process parameters in Wire Arc Additive Manufacturing (WAAM), with an emphasis on energy efficiency. The study employed three machine learning models—Support Vector Regression (SVR), Backpropagation Neural Network (BPNN), and XGBoost—to predict the wire feed speed to welding speed (WFS/WS) ratio based on geometric descriptors such as bead width (BW), height (BH), and cross-sectional area (BCSA), derived from 3D laser scanning and point cloud analysis. Among these, SVR demonstrated the highest predictive accuracy, which was further enhanced by using Particle Swarm Optimization (PSO). The optimized SVR model enabled the reverse prediction of input parameters from target bead shapes, reducing reliance on empirical trial-and-error. Experimental validations confirmed the method's effectiveness, achieving energy consumption reductions of up to 11.47%. This data-driven approach offers a practical path to process planning in WAAM, combining geometric fidelity with sustainable operation.

The third contribution of this Special Issue is a research article by Pereira, J.C. et al., which investigates the effects of powder reuse on the microstructural and mechanical integrity of directed energy deposition (DED) components produced with two different alloys: cobalt-based Stellite® 21 and super duplex stainless steel UNS S32750. The authors conducted a comprehensive three-cycle reuse study, evaluating powder morphology, oxygen uptake, and particle size evolution, along with the densification, chemical composition, and defect formation in the resulting bulk materials. The results revealed that Stellite® 21 can be reused up to three times without significant degradation, while the duplex steel showed marked declines in phase balance, hardness, and porosity control after a single reuse. Particularly in the super duplex case, the ferrite content dropped from 50.1% to 37.0% due to increased oxidation, causing microstructural instability and pore formation. These findings emphasize the critical role of alloy selection and controlled reuse strategies in maintaining material integrity and process efficiency for DED applications.

The fourth contribution of this Special Issue is a research article by Failla, D.P., Jr. et al., which investigates the predictive fidelity of material models for estimating residual stresses in Laser Powder Bed Fusion (L-PBF)-manufactured components using Inconel 718. The authors compare a standard elastic–perfectly plastic (EPP) model with a more advanced internal state variable-based model, the Evolving Microstructural Model of Inelasticity (EMMI), within a sequentially coupled thermo-mechanical finite element analysis. The L-shaped geometry used in the study, featuring curved edges and holes, was replicated from a neutron diffraction experiment to validate the simulation results. While both models captured the general residual stress distributions, EMMI provided more accurate predictions at critical free surfaces due to its ability to reflect microstructural evolution under thermal cycling. However, EPP outperformed EMMI in certain regions, highlighting the influence of material calibration—EMMI was tuned for wrought IN718 rather than for AM-specific microstructures. This work emphasizes the importance of selecting and calibrating material models appropriately for accurate stress prediction in geometrically complex AM parts made from metallic materials like Inconel 718.

The fifth contribution of this Special Issue is a research article by Horr, A.M., which presents a novel hybrid modeling framework for the real-time control and digital twin integration of additive manufacturing (AM) processes for metallic applications, with a focus on Wire Arc Additive Manufacturing (WAAM). The study combines reduced-order modeling (ROM) techniques with machine learning (ML) algorithms to create predictive and corrective tools for process simulation and optimization. Using a case study based on aluminum WAAM, the author demonstrates how the integration of singular value decomposition (SVD) and radial basis function (RBF) neural networks enables fast and accurate temperature predictions with dramatically reduced computational costs—less than one second per scenario compared to nearly an hour with full-scale finite element (FE) simulations. The study further evaluates the performance of several modeling methods (e.g., SVM, kriging, regression, clustering) and concludes that the SVD-RBF hybrid outperforms others, especially under high heating rate conditions. This research highlights the significant potential of ROM-ML frameworks to support adaptive, real-time decision-making in AM environments, promoting smarter, greener, and more agile manufacturing systems for metallic applications.

The sixth contribution of this Special Issue is a research article by Yang, J. et al., which presents a comprehensive strategy to reduce thermal residual stress in topologically optimized automotive brake calipers fabricated via Powder Bed Fusion (PBF). Recognizing the challenges posed by anisotropic thermal accumulation in irregularly shaped metallic AM parts, the authors proposed an optimized scan strategy using an island pattern with a 5 mm hatching length combined with vertical build orientation. Comparative experiments on residual stress and thermal deformation, supported by X-ray measurements and cantilever testing, showed that the island scan pattern reduced residual stress and deformation by up to 8.41% and 8.33%, respectively, compared to traditional patterns. The study also included finite element analysis and topology optimization of a brake caliper made from Ti-6Al-4V, leading to a 20% weight reduction while maintaining mechanical integrity under hydraulic pressure. The manufactured caliper was evaluated via a brake dynamometer using the JASO C406 procedure, demonstrating comparable or superior performance to commercial aluminum alloy counterparts. The findings validate the effectiveness of combining strategic scan planning with optimized geometry in PBF for automotive light weighting applications.

The seventh contribution of this Special Issue is a research article by Ibrahim, M. et al., which presents a preliminary evaluation of nickel silicide (NiSi12-wt%) laser cladding for enhancing the corrosion resistance and mechanical performance of S355 structural steel. Using laser metal deposition (LMD), the authors successfully applied multilayer cladding onto S355 substrates and subjected the coated and uncoated samples to accelerated corrosion testing in ferric chloride ($FeCl_3$) solution following the ASTM G48 protocol. Microstructural analysis via Scanning Electron Microscopy (SEM) and Light Optical Microscopy (LOM) revealed that the cladding formed a dense, dendritic microstructure with strong metallurgical bonding. The NiSi12-wt% cladding significantly outperformed the bare steel in terms of corrosion resistance at both room temperature and 50 °C, reducing pit formation and mass loss rates. Micro-hardness tests confirmed a substantial increase in surface hardness—from 842 HV in the base steel to 1258 HV in the cladded state—being maintained even after corrosion exposure. The results emphasized the potential of nickel silicide cladding as a protective surface engineering solution for steels used in aggressive marine and industrial environments, offering a promising solution for durability and the lifespan extension of structural components.

The eighth contribution of this Special Issue is a research article by Santos, L.J.E.B. et al., which investigates the integration of electric arc signal analysis with microstructural examination for anomaly detection in walls produced by the Gas Metal Arc (GMA)-based Wire Arc Additive Manufacturing (WAAM) process. The authors conducted controlled contamination experiments using sand, chalk, and oil during the deposition of 316L-Si stainless steel walls, aiming to evaluate how these contaminants affect both arc stability and material quality. Real-time voltage and current data were collected during deposition and metallographic samples were analyzed to detect solidification defects and microstructural irregularities. The study identified strong correlations between arc signal features—such as peak counts, average values, and signal deviations—and the presence of microscopic defects in contaminated regions. The findings suggest that arc signal anomalies can serve as effective indicators of hidden process instabilities, and that these signals could support the development of real-time monitoring tools or predictive models for defect detection in WAAM. This work demonstrates the potential of integrated arc–microstructure monitoring as a non-invasive approach to improve reliability in AM production environments of metallic applications.

The ninth contribution of this Special Issue is a research article by Liu, J. et al., which investigated the microstructure and wear resistance of laser-cladded Ni60/60%WC composite coatings applied to 45 steel substrates. The study aimed to fabricate crack-free, high-hardness coatings with a thickness exceeding 1 mm using optimized laser cladding parameters. The resulting coatings exhibited a well-bonded microstructure, with a uniform distribution of WC particles and the formation of hard phases such as W_2C, $Cr_{23}C_6$, and $Fe_{3.57}W_{9.43}C_{3.54}$. XRD and SEM analyses revealed the layered distribution of these phases and localized elemental diffusion near WC particles. The coatings achieved an average micro-hardness of 1416 $HV_{0.2}$—over five times greater than that of the base steel—and a friction coefficient reduced by 43.5%. Furthermore, the wear rate was lowered by 79.13% compared to the uncoated substrate. The wear mechanisms were characterized as predominantly abrasive, with WC acting both as a reinforcing phase and as a wear debris source over time. The results validate the effectiveness of using high-WC composite coatings to significantly enhance surface durability in high-load applications.

The tenth contribution of this Special Issue is a research article by Kunčická, L. et al., which investigated the thermomechanical behavior of AISI 316L stainless steel fabricated via Selective Laser Melting (SLM) under various strain rates and temperatures to determine the optimal post-processing parameters. The study combines experimental uniaxial hot compression tests with finite element modeling to assess the deformation response and microstructural evolution of AM-prepared 316L steel across four temperatures (900–1250 °C) and four strain rates (0.1–100 s^{-1}). The results show that high strain rates at 900 °C significantly increase the micro-hardness (up to 270 HV) and flow stress (~380 MPa) due to pronounced sub-structural development, while higher temperatures facilitate dynamic recrystallization and grain coarsening, reducing the mechanical strength. Finite Element Method (FEM) simulations, which have been validated through experimental trends in strain distribution and force predictions, have confirmed that lower temperatures and higher strain rates are optimal for strengthening via plastic deformation. This work provides valuable insights for selecting the post-processing conditions that enhance the performance of AM 316L components for demanding applications.

Acknowledgments: As Guest Editors of this Special Issue entitled "Advances in Additive Manufacturing and Their Applications (2nd Edition)" we would like to extend our sincere thanks to all the contributing authors for their high-quality and innovative research concerning AM for metallic applications. We are also thankful to the dedicated reviewers whose insightful feedback helped to ensure the scientific rigor of each contribution and also to the editorial team of Metals and MDPI for their continued professionalism and support throughout the editorial process. We trust that this Special Issue will serve as a valuable resource for researchers, engineers, and practitioners engaged in the evolving field of additive manufacturing for metallic applications and that it will help to guide future advancements in this challenging domain.

Conflicts of Interest: The authors declare no conflicts of interest.

List of Contributions:

1. Capasso, I.; Andreacola, F.R.; Brando, G. Additive Manufacturing of Metal Materials for Construction Engineering: An Overview on Technologies and Applications. *Metals* **2024**, *14*, 1033. https://doi.org/10.3390/met14091033.
2. Zhang, H.; Bai, X.; Dong, H.; Zhang, H. Modelling and Prediction of Process Parameters with Low Energy Consumption in Wire Arc Additive Manufacturing Based on Machine Learning. *Metals* **2024**, *14*, 567. https://doi.org/10.3390/met14050567.
3. Pereira, J.C.; Irastorza, U.; Solana, A.; Soriano, C.; García, D.; Ruiz, J.E.; Lamikiz, A. Effect of Powder Reuse on Powder Characteristics and Properties of DED Laser Beam Metal Additive Manufacturing Process with Stellite® 21 and UNS S32750. *Metals* **2024**, *14*, 1031. https://doi.org/10.3390/met14091031.
4. Failla, D.P., Jr.; Dantin, M.J.; Nguyen, C.J.; Priddy, M.W. Material Model Fidelity Comparison for the Efficacy of Predicting Residual Stresses in L-PBF Additively Manufactured IN718 Components. *Metals* **2024**, *14*, 1210. https://doi.org/10.3390/met14111210.
5. Horr, A.M. Real-Time Modeling for Design and Control of Material Additive Manufacturing Processes. *Metals* **2024**, *14*, 1273. https://doi.org/10.3390/met14111273.
6. Yang, J.; Jung, Y.; Jung, J.; Ock, J.D.; Cho, S.; Park, S.H.; Lee, T.H.; Park, J. Optimized Build Orientation and Laser Scanning Strategies for Reducing Thermal Residual Stress in Topology-Optimized Automotive Components. *Metals* **2024**, *14*, 1277. https://doi.org/10.3390/met14111277.
7. Ibrahim, M.; Hulme, C.; Grasmo, G.; Aune, R.E. Preliminary Evaluation of Nickel Silicide (NiSi12-wt%) Laser Cladding for Enhancing Microhardness and Corrosion Resistance of S355 Steel. *Metals* **2024**, *14*, 1389. https://doi.org/10.3390/met14121389.
8. Santos, L.J.E.B.; Souto, J.I.V.; Azevedo, I.J.S.; Castro, W.B.; Lima, J.S.; Delgado, J.M.P.Q.; Santana, R.A.C.; Gomez, R.S.; Bezerra, A.L.D.; Lima, A.G.B. Integration of Arc and Microstructural Analysis for Anomaly Detection in Walls Manufactured by GMA-Based WAAM. *Metals* **2025**, *15*, 110. https://doi.org/10.3390/met15020110.
9. Liu, J.; Liu, C.; Zhang, X.; Yin, X.; Meng, F.; Liu, C. The Microstructure and Wear Resistance of Laser Cladding Ni60/60%WC Composite Coatings. *Metals* **2025**, *15*, 166. https://doi.org/10.3390/met15020166.
10. Kunčická, L.; Kocich, R.; Pagáč, M. Experimental and Numerical Study of Behavior of Additively Manufactured 316L Steel Under Challenging Conditions. *Metals* **2025**, *15*, 169. https://doi.org/10.3390/met15020169.

References

1. Tebianian, M.; Aghaie, S.; Razavi Jafari, N.S.; Elmi Hosseini, S.R.; Pereira, A.B.; Fernandes, F.A.O.; Farbakhti, M.; Chen, C.; Huo, Y. A Review of the Metal Additive Manufacturing Processes. *Materials* **2023**, *16*, 7514. [CrossRef] [PubMed]
2. Gopal, M.; Lemu, H.G.; Gutema, E.M. Sustainable Additive Manufacturing and Environmental Implications: Literature Review. *Sustainability* **2023**, *15*, 504. [CrossRef]
3. Fonfría, Í.; Holgado, I.; Ortega, N.; Castrillo, A.; Plaza, S. Improving the Efficiency and Quality of Sustainable Industrial CT by Optimizing Scanning Parameters. *Sensors* **2025**, *25*, 2440. [CrossRef] [PubMed]
4. Gurmesa, F.D.; Lemu, H.G.; Adugna, Y.W.; Harsibo, M.D. Residual Stresses in Wire Arc Additive Manufacturing Products and Their Measurement Techniques: A Systematic Review. *Appl. Mech.* **2024**, *5*, 420–449. [CrossRef]
5. Kaess, M.; Werz, M.; Weihe, S. Residual Stress Formation Mechanisms in Laser Powder Bed Fusion—A Numerical Evaluation. *Materials* **2023**, *16*, 2321. [CrossRef] [PubMed]

6. Yan, Z.; Guo, J.; Zou, X.; Wang, S. Integrated Control of Thermal Residual Stress and Mechanical Properties by Adjusting Pulse-Wave Direct Energy Deposition. *Materials* **2024**, *17*, 5231. [CrossRef] [PubMed]
7. Lanzutti, A.; Marin, E. The Challenges and Advances in Recycling/Re-Using Powder for Metal 3D Printing: A Comprehensive Review. *Metals* **2024**, *14*, 886. [CrossRef]

metals

MDPI

Review

Additive Manufacturing of Metal Materials for Construction Engineering: An Overview on Technologies and Applications

Ilaria Capasso [1,2,*], Francesca Romana Andreacola [1] and Giuseppe Brando [1,2]

1 Department of Engineering and Geology, University "G. d'Annunzio" of Chieti-Pescara, 65127 Pescara, Italy; francesca.andreacola@unich.it (F.R.A.); gbrando@unich.it (G.B.)
2 UdA-TechLab, Research Center, University "G. d'Annunzio" of Chieti-Pescara, 66100 Chieti, Italy
* Correspondence: ilaria.capasso@unich.it

Abstract: Additive manufacturing, better known as 3D printing, is an innovative manufacturing technique which allows the production of parts, with complex and challenging shapes, layer by layer mainly through melting powder particles (metallic, polymeric, or composite) or extruding material in the form of wire, depending on the specific technique. Three-dimensional printing is already widely employed in several sectors, especially aerospace and automotive, although its large-scale use still requires the gain of know-how and to overcome certain limitations related to the production process and high costs. In particular, this innovative technology aims to overtake some of the shortcomings of conventional production methods and to obtain many additional advantages, such as reduction in material consumption and waste production, high level of customisation and automation, environmental sustainability, great design freedom, and reduction in stockpiles. This article aims to give a detailed review of the state of scientific research and progress in the industrial field of metal additive manufacturing, with a detailed view to its potential use in civil engineering and construction. After a comprehensive overview of the current most adopted additive manufacturing techniques, the fundamental printing process parameters to achieve successful results in terms of quality, precision, and strength are debated. Then, the already existing applications of metal 3D printing in the field of construction and civil engineering are widely discussed. Moreover, the strategic potentiality of the use of additive manufacturing both combined with topological optimisation and for the eventual repair of existing structures is presented. It can be stated that the discussed findings led us to conclude that the use of metal additive manufacturing in the building sector is very promising because of the several benefits that this technology is able to offer.

Keywords: metal additive manufacturing; 3D printing process parameters; 3D-printed metals in construction; topology optimisation in additive manufacturing

1. Introduction

Metal 3D printing is already widely adopted in a variety of engineering fields, including biomedical, aerospace, and automotive. However, there are still very few applications in the fields of structural and construction engineering, despite the numerous benefits that this technology can offer, such as raw materials and waste savings, ability to produce complex and optimised shapes and geometries impossible to achieve with conventional methods, possibility of producing multiple parts in the same batch reducing time to market, and minimization of warehouse stocks [1].

Large-scale application is not yet possible due to a lack of in-depth knowledge about, e.g., the mechanical properties of the materials and printed parts, short- and long-term performance, geometric accuracy and quality, lead times and costs, and a gap in certification and standards. Safety, cost-effectiveness, and environmental sustainability of metal additive methods represent other key aspects to be analysed.

Despite their wide diffusion in the last decades, the first application of printing methodologies can be dated back to 1979, when the first patent concerning a technology ancestor of modern 3D printing (US4247508A) was filed by Ross Housholder [2]. The patent proposed, for the first time, the creation of objects using the approach of overlapping consecutive layers, anticipating the methodology typical of the modern three-dimensional printing process and rapid prototyping (RP) [2]. In fact, RP, which was developed in the 1980s to create models and prototype parts, can be considered the first form of creating a 3D object using computer-aided design (CAD) layer by layer, and the origins of the current additive manufacturing (AM) processes can be dated back to it [3,4].

The first machines for 3D printing processes appeared in the late 1970s, when Alan Herbert and Charles Hull in America and Hideo Kodama in Japan developed, at the same time, a system for selective solidification of a photopolymer, capable of creating an object by successive layers.

The fundamental historical milestones for the development of 3D printing machines are listed below and schematically summarized in Figure 1 [5–9]:

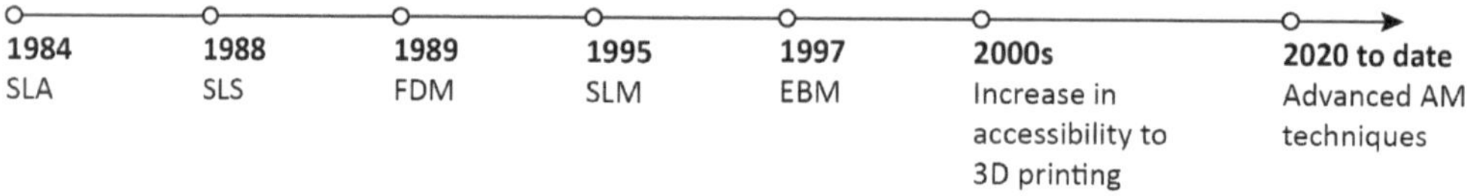

Figure 1. Historical evolution of additive manufacturing technology.

1984: Stereolithography (SLA) was invented and patented.

1987: The world's first 3D printing machine for stereolithography was produced.

1988: Selective laser sintering (SLS) was patented.

1989: The patent for fused deposition modelling (FDM) technique was filed.

1992: Printers for SLS and FDM were developed.

1995: Selective laser melting (SLM) machines were marketed as an alternative technologies to stereolithography.

1997: Electron beam melting (EBM) was firstly introduced.

2000: A growing number of companies and specialists became interested in the possibilities and benefits of 3D printing, leading to a considerable development of printers and technologies.

2011: The world's first additive manufactured aircraft and prototype car were released.

2020 to date: Worldwide spread of advanced 3D printing techniques due to significantly reduced printer equipment costs and production times combined with the wide range of currently available filaments and materials [10,11].

It is possible to identify a general sequence of eight fundamental steps, schematically summarized in Figure 2 [12], for the production of a component using AM technology from design to final part, regardless of the additive manufacturing technique selected.

The production of a digital model of the object is the starting point for any AM process. The digital model can be created through CAD or by 3D scanning of a real object, and its quality directly affects the final product, so an accurate virtual representation is essential for a good result. Many parameters must be taken into account, including geometrical limitations, support material, and escape hole requirements. Once the CAD file is available, the following step is to convert the file into a printer-readable format. The first step is to transform the CAD model into an STL (the acronym can stand for stereo lithography interface format, standard triangle language or standard tessellation language) file, that describes the surface of an object using triangles, simplifying the CAD model. The STL file just created is imported into a slicer software, which converts it into a G-Code file. G-Code is a numerical control programming language to monitor and manage CNC (computer numerical control) machines and 3D printers [13]. The slicer programme splits the design into the several layers that will be required to manufacture the object. It defines also the building parameters of the 3D printer by specifying the support layout, layer height, and orientation of the workpiece.

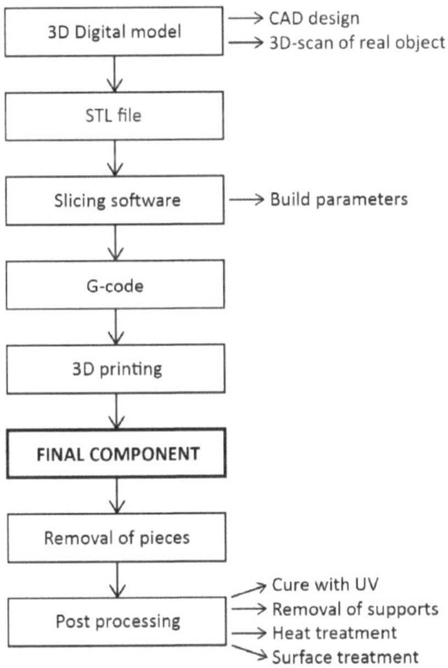

Figure 2. Process sequence for producing an AM component.

At this point, the three-dimensional object is ready to be printed, according to the specific printing technology selected. When the additive manufacturing process is complete, there is the stage of supports removal, which for some methods (mainly the polymer-based ones) consists only of "detaching" the printed part from the building plate, while for some others can involve high complex processes for extracting the supports from the printed part [14].

Several post-processes may be necessary at the end of additive manufacturing, and they mainly concern the improvement of surface characteristics or the modification of the mechanical properties of the material [15].

With regard to 3D printing techniques, there are many different types of 3D printers available, with unique features and capabilities that characterise the main additive manufacturing technologies. In 2015, ISO/ASTM Standard 52900 was introduced to define a technical terminology and to qualify the various 3D printing methods [13,16]. Seven macro-categories of production processes were established, each of them characterised by the possibility to use different materials and to obtain different final features of the printed element [17,18]. The seven macro-categories include vat polymerization, material extrusion, material jetting, binder jetting, powder bed fusion, sheet lamination and direct energy deposition. Each of them includes one or more production techniques [16,19].

Additive manufacturing methods can be classified using several criteria [20]. All the printing processes presented and analysed below are the most widespread in the industry, even if there are many other techniques that are still in the development phase or have limited presence on the market [6,7,15,21,22].

Vat polymerisation comprises all resin-based 3D printing techniques, in which a liquid photopolymer contained in a vat is selectively cured by a heat source, instead of being injected by a nozzle. The main technologies belonging to this category are stereolithography (SLA) and digital light processing (DLP) [22].

SLA was the first 3D printing method to be developed and marketed. This technology exploits the photo-polymerisation of light-sensitive resins using a low-power laser [3].

The SLA system consists of a vat, containing the liquid resin, and the printing surface at the base of it (see Figure 3) [23]. The part of liquid hit by the light polymerizes and solidifies and, when one layer has solidified, the plate is lowered and the process continues until the part is completed [24]. The laser is moved over the printing area by a system of mirrors controlled by galvanometers.

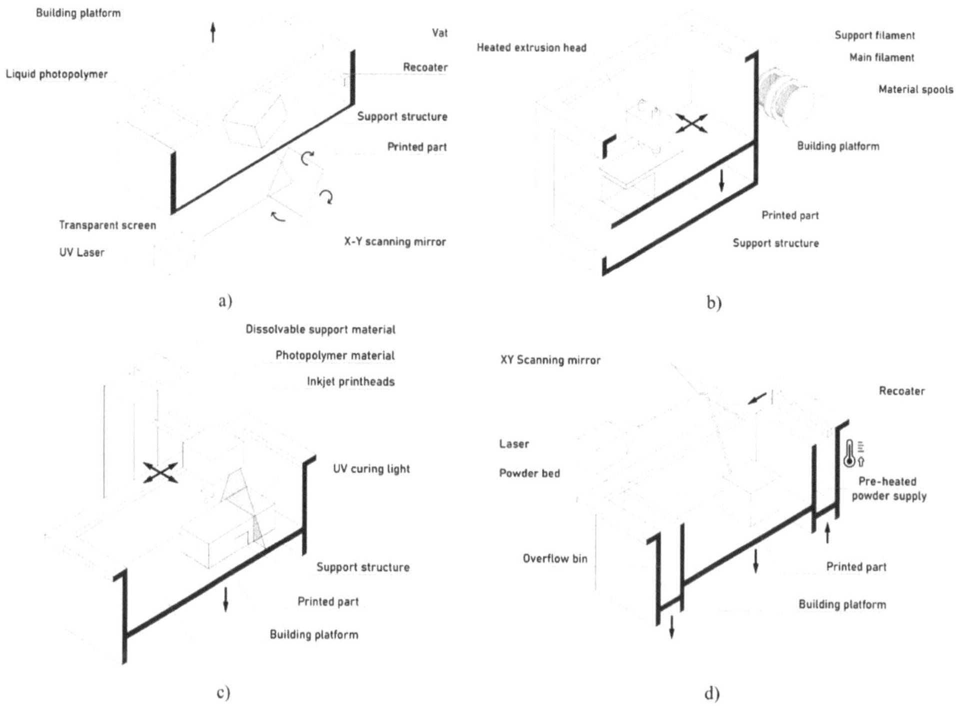

Figure 3. Schematic representations of different types of printers: (**a**) SLA printer; (**b**) FDM printer; (**c**) MJ printer; (**d**) SLS printer.

Three-dimensional printing technology based on material extrusion uses a continuous filament of thermoplastic polymer as starting material, which acts as a tool for rapid and cost-effective prototyping.

The most widespread material extrusion method is the fused filament fabrication (FFF), more commonly known as fused deposition modelling, a term coined in the late 1980s and registered in 1990 by the company Stratasys [22,25].

FDM is a process based on the extrusion of remelted material to generate the subsequent layers of the three-dimensional object [22,26]. The materials used in this type of technology are mainly polymers, but there are also some variants that use other materials such as ceramic pastes. The polymers are supplied as wire spools, continuously bringing material to the extruder (moved by computer control), which allows the extrusion of filaments through the nozzle, heating the material to melting. The nozzle moves according to the layout of the final object to be produced, releasing melted material that cools and solidifies generating each layer. After finishing a layer, the build plate is lowered and the process is continued layer by layer until the 3D part is built [27,28].

The 3D printing production technique of material jetting (MJ) is often compared to the 2D inkjet process that print ink on paper.

Using photopolymers, metals, or wax that solidify when exposed to light or heat (similar to stereolithography), MJ releases material in the form of very small droplets from hundreds of small nozzles in the printhead to build the part layer by layer. When the drops are deposited on the building platform, they are directly polymerised and solidified using UV light [29]. After a layer is produced, the printing platform is lowered by one layer of thickness and the process is repeated until the three-dimensional part is built [22]. MJ processes require support structures, which are often 3D-printed simultaneously during construction from a dissolvable material and then removed during post-processing. This manufacturing process allows different materials to be 3D-printed within the same component.

MJ is one of the most refined additive manufacturing techniques, which allows the production of very smooth surfaces and printed objects with layer thicknesses of about 16 microns [28]. Moreover, it is considered the most precise method of 3D printing because deformation and shrinkage are rare due to the absence of heat treatments during the manufacturing process.

The most popular techniques are material jetting (MJ) and drop on demand (DOD) [25].

Among the powder bed fusion technologies, the technique most commonly used is SLS, which produces parts starting from polymeric powders [30]. The process begins heating the container of polymer powder to a temperature just below the melting point of the material, to minimise both the energy required by the laser and the effects of phase change. The recoater then deposits a layer of powder on the building plate and the laser beam, moving through a system of mirrors, begins to scan the working surface according to the shape of the section, resulting in the solidification of a cross-section of the object through selective sintering. Following the completion of a layer, the printing platform lowers to allow the deposition of a new layer of powder and the sintering of the next section. In this way, the process is repeated until the workpiece is finished [31]. In order to ensure better bonding between the layers, an initial heating of the powder takes place before the exposure to laser sintering. In SLS machines, 50% of the non-sintered powder can be recovered and reused.

Figure 3 shows the schematic representation of the production process of the above-mentioned non-metallic additive manufacturing methods.

After describing non-metallic additive manufacturing methods, Figure 4 gives an overview of metallic 3D printing techniques, which will be discussed in detail in the following sections.

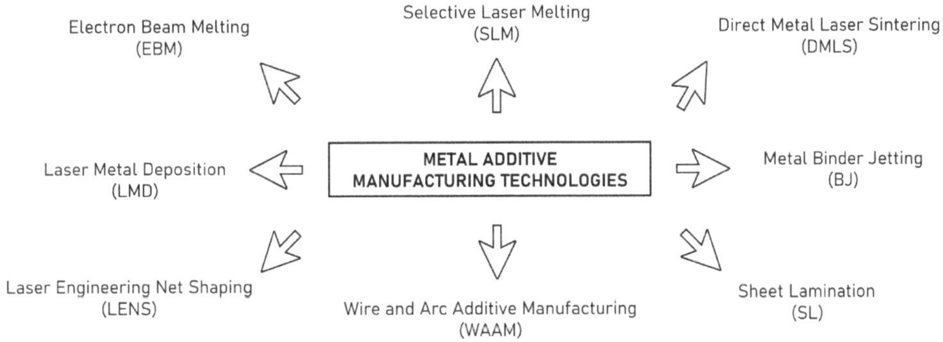

Figure 4. Overview of the metal AM methods.

It is important to emphasise, however, that 3D printing in the civil engineering and construction sector is revolutionizing the way buildings are designed and built [32]. By using additive manufacturing techniques, architects and engineers can create complex and customized structures with greater efficiency [33]. Common methods include extrusion-based 3D printing, where material is deposited layer by layer, and binder jetting, which uses a binder to solidify powdered materials. A wide range of materials can be used, including concrete, mortar, and even specialized geopolymers [34]. Applications span from small-scale components like architectural models to full-scale buildings and infrastructure projects. For instance, 3D-printed houses have been constructed in various parts of the world, demonstrating the potential of this technology to address housing shortages and create sustainable communities [35].

In this paper, several of the abovementioned challenges will be investigated, with a specific focus on metal 3d printing methods and more in detail on selective laser melting (SLM).

In particular, a detailed overview of metal additive manufacturing technologies is given in Section 2, focusing on a description of the currently most widely used processes, especially in the construction engineering sector. Then, the main process parameters inherent to metal 3D printing methods are discussed in Section 3. Real examples of metal 3D printing in the construction sector and the potential of metal additive technologies, with a focus on topological optimisation procedures, are presented, respectively, in Sections 4 and 5.

The aim of this review paper was to collect together the works related to 3D printing in order to provide readers, both experts and nonexperts, with a comprehensive overview, divided by topic, that is able to offer a complete knowledge on the topic of metal additive manufacturing. Of course, the results of the cited studies are also discussed in a general approach, as they are not the results of the authors' own research.

2. Metal Additive Manufacturing Technologies

This section will discuss more in detail the metal 3D printing technologies, with a special focus on selective laser melting.

Powder bed fusion (PBF) 3D printing technologies result in geometrically complex products with an excellent level of precision, using a heat source to melt the powder particles layer by layer, producing a solid part. The main difference between the various processes, lies in the use of different energy sources, mainly laser or electron beams [36].

Within this category, the most popular methods are selective laser melting (SLM) and electron beam melting (EBM) [7].

The selective laser melting process is very similar to SLS, but it is used for metal parts fabrication. The range of metals available includes aluminium alloys, steel, titanium, cobalt, chrome, and nickel [37].

An alternative method to SLM is direct metal laser sintering (DMLS), which sinters metal powders instead of completely melting them. This method is limited to metal alloys only and cannot be used for pure metals [38].

SLM involves the complete melting of metal powder, layer after layer, within an inert environment. This process produces a much more compact and homogenous element than sintering. A high-power laser beam melts the deposited powder layer according to the desired geometry. Once a layer is completed, new powder is deposited and the process continues until the 3D part is created [37,39].

Support structures are essential to limit the effects of possible distortions and the occurrence of warping and buckling caused by residual stresses developed because of the high operating temperatures. In addition, supports are required to allow the printing of projecting parts and to dissipate heat.

The performance of SLM technology printers is strongly influenced by a wide number of factors, which must be considered during production [40]. Specifically, the quality of the finished part depends on the diameter of the laser beam, the geometry of the particles, and the thickness of the layers. For these reasons, SLM printers need to be properly configured, as they need severe operating, calibration, material management, post-processing, and maintenance protocols. These specifications result in the demand for highly qualified operators [41].

The quality of the surfaces largely depends on the orientation of the part; in fact, the upper surfaces have a better quality than the lower ones, which are directly in contact with the printing plate. Anyway, without any surface treatment, a "stepped" appearance due to overlapping layers is always observable, even in components produced with SLM where the quality is very high, as can be seen in Figure 5.

Figure 5. Specimen manufactured with SLM, in which the stepped surface due to the different overlapping layers is visible (layer thickness: 50 microns).

Despite the considerable thermal gradients due to the high temperatures developed during printing, which generate residual stresses in the parts, the dimensional accuracy achieved is still high.

After the printing of the part, there are several post-processing possibilities. Among them, there are the removal of supports and residual powder from the build chamber and heat treatments. The supports, which increase the cost of the component, require mechanical removal, and the contact surfaces need treatment with files or grinders to smooth out imperfections (see Figure 6). Heat treatments are essential to relax internal stresses developed during production or to modify specific mechanical characteristics. Some heat treatments are useful to change the microscopic structure [40].

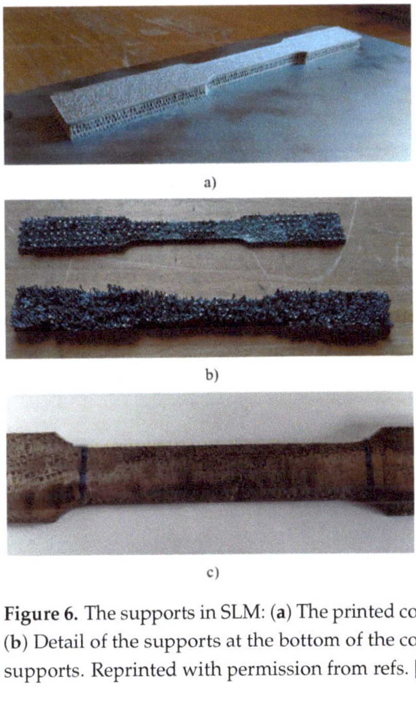

Figure 6. The supports in SLM: (**a**) The printed component on the construction plate with its supports. (**b**) Detail of the supports at the bottom of the component. (**c**) Defects due to nonperfect removal of supports. Reprinted with permission from refs. [42,43]. Copyright 2023 Elsevier Ltd.

In addition, further post treatments can be applied to upgrade the surface finish, including machining, media blasting, polishing, and micromachining. Finally, metal plating can be applied to enhance part performance (corrosion, strength, and hardness).

The main applications of SLM involve the production of parts where high accuracy and high customisation are required. SLM is widely employed in the medical and dental industry, as well as in the aerospace and automotive sectors [44]. A schematic representation of the SLM process is shown in Figure 7.

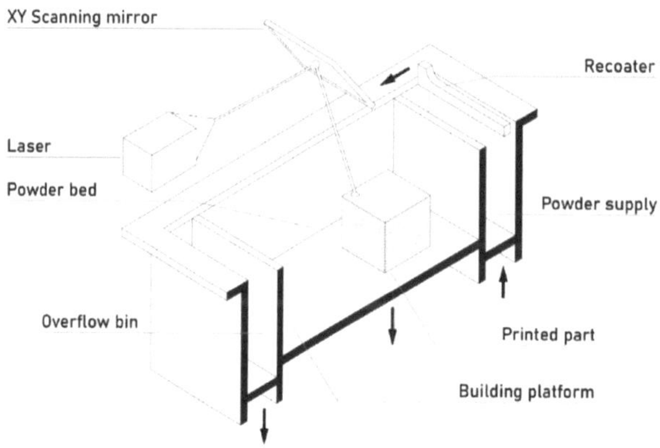

Figure 7. Schematic representation of an SLM printer.

The operating principle of the electron beam melting is similar to that of SLM. In fact, EBM is based on the fusion of powder metallic material by means of an electron beam instead of a laser beam. The electron beam scans a layer of powder, producing local melting and solidification of a cross-section of the workpiece. Compared to the other powder bed fusion methods, EBM can guarantee a higher building rate due to the higher energy density involved. In addition, the minimum elements and particles sizes, the layer thickness, and the surface finish quality are usually better. Most of the specifications of the SLM method are equally valid for EBM [44,45]. For the operating scheme, please also refer to Figure 7, which is indicative of the SLM technique.

Binder jetting (BJ) can be considered as a merger of SLS and MJ as it involves powdered material and a nozzle that deposits a binding agent to produce three-dimensional objects. The BJ process operates with metals but also with other materials, including sands and ceramics [46].

This technology uses nozzles, moving over the building plate, which deposit drops (approximately 80 μm of diameter) of a binding agent in liquid form on top of the pre-deposited thin layer of powder. In this way, the powder particles are bonded together to obtain every layer of the part [47]. When the layer is complete, the powder bed moves downwards and a new layer of powder is placed on top of the one already formed to restart the process. The process is repeated until the whole workpiece is completed [48,49]. After printing, the part remains in the powder to cure and strengthen. Then, the component is cleared from the powder bed and the unbound excess powder is removed with compressed air. In BJ, 100% of the unbonded powder can be reused. After printing, parts are in a "green", or unfinished, state with weak mechanical properties and may require two further post-processing sessions before they are ready for use [48].

The printed parts must then be placed in a furnace where the binder is burnt out, creating voids in the piece. Subsequently, the voids are filled by capillary effect with bronze, producing parts with both high density and good strength. The last step consists of sintering in an oven until a high density is reached [50,51]. However, the mechanical performance offered by metal parts produced with powder bed fusion cannot be achieved [44]. The main post processes consist of the elimination of excess powder from the workpiece. Metal components can be post-processed in the same way as traditionally produced metals, and heat treatments can be applied to improve mechanical properties. The secondary processes of infiltration and sintering enable the manufacturing of functional metal parts, thanks to the wide variety of materials available and the potential to design challenging geometries that are extremely expensive and difficult to produce conventionally [25].

The quality of the final product is mainly affected by the thickness of the layers, the size of the droplets dispensed, and the size and geometry of the powder particles. As with SLS, BJ binder jetting does not require support structures, because the part is sustained during printing by the surrounding powder. This helps to save both time needed for post-processing and material waste [50]. Unlike SLS, however, the parts are manufactured in the absence of heat, which reduces the problems related to temperature differences that can cause distortions and deformations. Shrinkage may only occur during the sintering phase [49].

The limited cost and production rate make BJ ideal for the development of casting models that would be hard to achieve with traditional methods. The principles of the binder jetting BJ printing process can be seen in Figure 8.

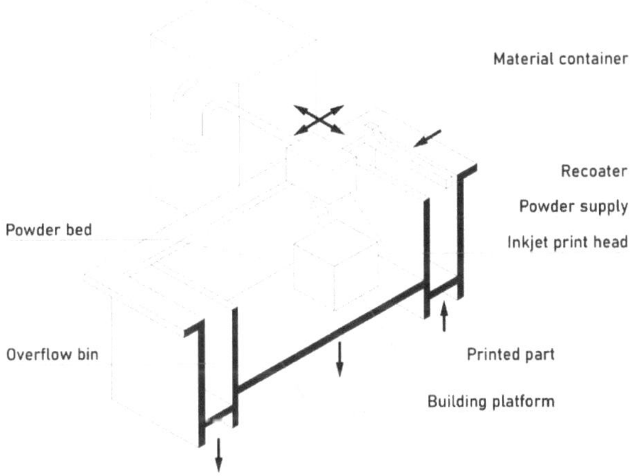

Figure 8. Schematic representation of a BJ printer.

Sheet lamination (SL) and direct energy deposition (DED) are reported in order to provide a complete overview of additive manufacturing methods, even if they are methods not widely adopted yet and of minor relevance to the 3D printing industry [19].

The SL manufacturing technique, also known as laminated object manufacturing (LOM), is based on the deposition of sheets of solid-state material, mainly aluminium foil, supplied in the form of rollers, which are bonded to each other. They are then cut by a laser to fit the cross-section of the object and form the object layer after layer. This technique allows one to obtain very small thicknesses for each individual layer, thus increasing the resolution of the final product. The most commonly used material is paper sheets with a thermoplastic coating on one of the two sides [21,52].

Laminated objects are frequently employed for both aesthetic and visual purposes and are not intended for structural application [21,52].

DED printing technology creates parts by directly melting materials and depositing them on the part, layer by layer. The operating principle is that of metal welding, so any material that can be welded could be employed. This additive manufacturing technique is mostly used with metal powders or wire source materials. In addition to the ability to build parts from scratch, DED is also capable of repairing complex damaged parts, such as turbine blades or propellers. The term direct energy deposition can cover several technologies which can be categorized according to the way the material is fused. The most popular are laser engineering net shaping (LENS), laser metal deposition (LMD), and wire and arc additive manufacturing (WAAM) [21,22].

Among these, the most widely adopted technique is WAAM, in which most printers are very large industrial machines that require a closed, controlled environment to operate. The typical WAAM equipment consists of a nozzle mounted on a multiaxis arm (theoretically without size limits) within a closed frame, which deposits the molten material onto the surface of the part, where it solidifies [53]. The process is similar in principle to the material extrusion printing technique, but with DED, a nozzle can move in multiple directions, with up to five different axes compared to only three on most FDM printers [54].

LENS is a type of additive manufacturing process specifically designed for metals. It is a highly precise and versatile technique in which a fine metal powder is continuously fed into the build chamber and a high-power laser beam is focused onto the powder bed. The laser beam melts the powder, creating a small molten pool, and a nozzle directs a stream of inert gas onto the molten pool, preventing oxidation and promoting rapid solidification. As the building platform moves, the laser beam melts the next layer of powder, fusing it to the previous one. As with other 3D printing technologies, this process is repeated layer by layer until the entire 3D part is built [55]. Briefly, the key advantages of LENS are the possibility to produce parts with very fine feature sizes and excellent mechanical properties, such as high strength and durability. Also, a wide range of metals can be processed using LENS, including titanium, stainless steel, and nickel alloys. However, it must be considered that there may be some drawbacks, such as the need for post-processing, poor surface finish of the components, and distortion of the components due to residual stresses [48].

LMD is an additive manufacturing process, closely related to LENS, that utilizes a high-power laser to melt and deposit a metal wire onto a substrate, for the creation of complex metal components layer by layer. Unlike LENS, which uses metal powder, LMD employs a continuous metal wire as its source material, with a high-power laser beam focused on the metal wire. The laser melts the wire, creating a molten droplet that is deposited on the surface. The molten droplet quickly solidifies, bonding to the previous layer. The metal wire and laser beam move in a coordinated way to create the desired component shape. Moreover, a broad range of metal materials can be used, both in wire and powder form [56].

LMD typically offers a faster deposition rate compared to LENS, making it suitable for producing larger components, and the surface finish of components produced with LMD is generally smoother than those produced with LENS. In summary, LMD is another powerful additive manufacturing technique for metals, offering specific advantages in terms of deposition rate and surface finish [57,58].

A little-known variant of the LMD is the shaped metal deposition (SMD) method. It differs from LMD only in the use of a metal wire of circular or profiled cross-section as the base material. In addition, it offers a higher deposition rate and better component surface quality than LMD [59,60].

2.1. Materials

When selecting a metal for 3D printing, it is crucial to consider a number of interrelated factors, including the physical and mechanical properties of the material, the peculiarities of the AM process, the final application, and the cost of the material, which is an important economic factor, especially for mass production [61].

Among the material characteristics, it is important to consider:

- The mechanical strength, as yield stress, ductility, wear, and fatigue resistance are key properties in determining whether the metal can withstand the expected stresses.
- The density of the material, which affects the weight of the finished component and can be a critical factor in many applications.

- Thermal and electrical conductivity, important properties for electronic and thermal applications.
- Corrosion resistance, since if the component will be exposed to corrosive environments, a resistant metal must be chosen.
- The microstructure of the 3D-printed metal, considering its features, imperfections, and defects, which can influence the properties of the finished part.
- Biocompatibility, since, for example, for medical applications, the material must be biocompatible and nontoxic.

As far as the 3D printing process is concerned, it is important to consider compatibility with the technology, as not all metals are suitable for all metal 3D printing technologies, layer thickness, which affects the resolution and surface quality of the component, and the melting speed of the metal, which affects the printing rate and quality of the component.

Then another important aspect is the final application. Assessments must be made on the type of load the component will have to withstand during its life, the environment in which the component will be used, and aesthetic requirements, as surface finish and appearance are important in some areas of application. It must also be considered that specific certifications and compliance with particular standards are required in certain sectors or for certain applications.

There are also other factors that must be taken into account, including material availability, as not all metals are readily available in powder or wire form for 3D printing, and the type and cost of post-processing, as some metals require post-print heat or surface treatment. Finally, the environmental impact of the production process and material use must be considered, as this is an increasingly important factor nowadays [62].

In light of these general remarks, the most commonly used metals in 3D printing are stainless steels, due to their high corrosion resistance and good mechanical strength; titanium, because of its high specific strength, biocompatibility, and corrosion resistance; aluminium for its lightness and good thermal and electrical conductivity; nickel, due to its high strength and good corrosion resistance; and cobalt-based alloys because of their high resistance to high temperatures [61].

The world of metallic 3D printing is continuously growing, and with it, also the range of metallic materials used. Therefore, in addition to the traditional materials, innovative materials are becoming increasingly attractive and popular. Nickel-based superalloys offer excellent resistance to high temperatures and corrosion, making them ideal for aerospace applications, gas turbines, and engine components [63,64]. One of the most widely adopted is Inconel [65], which is known for its resistance to fatigue and oxidation; it is used in sectors such as aerospace and automotive. Aluminium alloys include high-strength aluminium, which offers an excellent strength-to-weight ratio, making them ideal for lightweight structural components in sectors such as automotive and aerospace [66,67]. Then there are shape memory alloys (SMAs), among them Nitinol, which are able to change their shape in response to thermal or mechanical inputs, finding applications in sensors, actuators, and medical devices, and conductive metals such as copper, silver, and gold, which are used to create electronic circuits and sensors directly through 3D printing [68–70]. Finally, there would be high-entropy alloys (HEAs), the use of which in additive manufacturing, despite their potential, still presents some challenges related to cost, optimisation of printing parameters, and characterisation of the materials themselves [71,72].

2.2. Overview Remarks

In order to compare the operating principles and the production features of the different metal printing techniques, the benefits and drawbacks of the main AM methods (SLM, BJ, and WAAM), widely examined in the previous section, are summarized in Table 1.

Table 1. Main advantages and disadvantages of the different AM methods.

3D Printing Technique	Advantages	Disadvantages
SLM-EBM-DMLS	• Production of complex parts with high level of customization • Production of geometries impossible to achieve with traditional methods	• Very high costs of materials and printers • Build size • No large-scale production • High degree of expertise of operators
BJ	• Absence of residual stresses • Low operating costs • Ability to print large parts • Inexpensive binder agents • Cheaper than PBF methods	• Costs of metal powder • Poor mechanical properties • Secondary process always required for functional parts • Grainy surface finish
WAAM	• Possibility to print large size components • Wide range of materials • Design flexibility • Stand-alone or integrated solution	• Wide tolerances and low accuracy • Surface roughness • Requirement of skilled operators

In terms of surface finish and level of accuracy, it emerges that, among the printing methods analysed, good results can be obtained through SLM, allowing the production of smooth surfaces. Surface roughness is a key parameter in assessing the quality of a 3D-printed component, as it affects aspects such as aesthetics, functionality, and corrosion resistance. The average roughness range can vary significantly depending on the metal 3D printing technology used. SLM generally offers a lower surface roughness than EBM and WAAM due to the use of a focused laser that precisely fuses the metal powders [73]. However, the roughness can vary significantly depending on the process parameters. For EBM, the surface roughness is slightly higher than for SLM, mainly due to the higher heat diffusion of the electron beam [74]. Finally, the surface roughness of WAAM is typically the highest of the three technologies, due to the nature of the deposition process and the larger size of the traces left by the wire [75]. However, there are some factors influencing roughness. These include process parameters, such as laser power/electron beam, scanning speed, layer thickness, and scanning pattern; material properties, such as the fluidity of the molten metal; and post-processing, as treatments such as sandblasting or sanding can significantly reduce roughness. The average roughness of parts produced with metal BJ technology is a parameter that depends on several factors and can vary significantly from one component to another. Roughness is highly dependent on the particle size of the metal powder and the density of the binder, as the metal powder particles are joined by a liquid binder, forming a porous structure that requires a subsequent sintering process to achieve the desired density. Therefore, it is hard to give a precise range for the average roughness of metal BJ-printed parts, as it depends on the factors mentioned above. However, in general, it can be stated that the average roughness of parts produced with metal BJ is usually higher than that achievable with technologies such as SLM or EBM [76].

In terms of mechanical properties, the only processes capable of assuring good performances are the powder bed fusion methods, i.e., SLM. However, the great disadvantage of these technologies is the high costs of machines and materials, as well as the need for highly specialised operators. The WAAM method, which can produce very large components, can also provide excellent mechanical strength, but at the cost of poor quality. Also, BJ is a very expensive method due to the elevated costs of raw materials. The widest variety of materials is available for SLM for metallic materials, although the WAAM method can ideally use any metal suitable for welding. Finally, considering that the size of the workpieces that can be printed is one of the most limiting factors for large-scale use of additive manufacturing, the only method that allows the production of larger parts are BJ and WAAM.

Furthermore, a comparison of the specifications of the main metal additive technologies under investigation is shown in Table 2. It displays the main characteristics in terms of operating principles, energy source employed, available materials, print sizes, resolution, and layer thickness, and fields of application of the most important additive technologies reviewed.

Table 2. Comparison of the features of the main 3D printing methods.

	Solid Based	Powder Based		
Technology	**WAAM**	**SLM-DMLS**	**EBM**	**BJ**
Working Principles	Material extrusion + welding	Melting	Melting	Binding
Source	Electric arc	Laser beam	Electron beam	Bonding agent
Material Group	Metal wires	Metal powders	Metal powders	Metal powders
Main Available Materials	Titanium Steel Nickel Aluminium (or any weldable metal)	Stainless steel Aluminium alloys Titanium alloys Nickel alloys	Stainless steel Titanium alloys Nickel alloys Cobalt chrome	Stainless steel Bronze
Supports Requirement	No	Yes	Yes	No
Build Volume	Unlimited build volume	From 100 × 100 × 100 mm^3 (small sizes) to 800 × 500 × 400 mm^3 (large sizes)	350 × 350 × 450 mm^3	Up to 800 × 500 × 400 mm^3
Resolution	1 mm	0.1 mm	0.1 mm	0.2 mm
Roughness	50–250 μm	10–50 μm	15–75 μm	variable
Layer Thickness	min 1–2 mm	30–50 μm	30–50 μm	100 μm
Applications	Aerospace, energy sector, research and development, cladding and repair components	Medical and dental industry, aerospace and automotive sectors		Realistic models, coloured components, casting models with complex shapes

From Table 2, it is possible to deduce that there is a further criterion to classify the printing techniques, the physical state of the raw materials, so there are three categories, namely, solid-, liquid-, and powder-based. Metal 3D printing techniques all involve material in powder form. The powder-based methods are SLM, EBM, and BJ. The only solid-based process is the WAAM, which involves the deposition layer after layer of metallic wires in the form of weld bead combined with an electric arc used as the heat source, for the production of components that do not demand any particular aesthetic properties. Wire and arc additive manufacturing, as we will discuss in Section 4, is the most widely diffused technology in the field of construction engineering due to the possibility of producing large elements with excellent mechanical properties.

In addition to Tables 1 and 2, all the characteristics discussed in this section are graphically summarised in Figure 9.

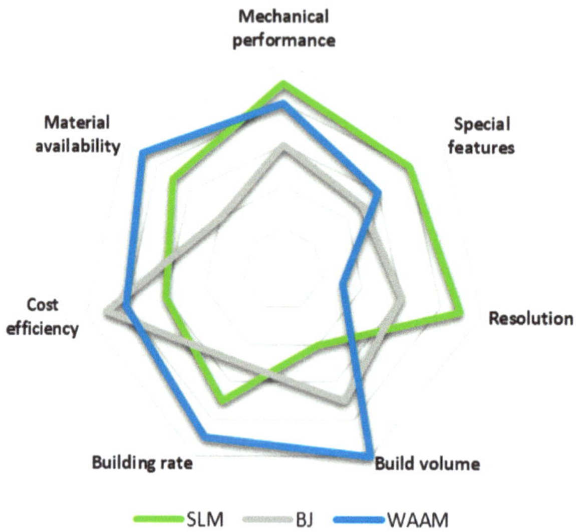

Figure 9. Graphical outline of the main features of metal 3D printing methods.

2.3. Features of 3D Printing Methods

Figure 10 aims to provide a useful tool to select the most adequate technique, according to the requirements for a particular application, classifying the additive technologies as a function of several parameters, such as the target performance, the raw materials employed, the specific production process, and the visual appearance of the final product. An overview of all additive technologies is given, with a focus on metal-based methods.

The selection of the production process may be influenced by the processing principles or the energy source involved. However, the different technologies have already been extensively discussed in the previous sections.

Considering the raw materials, the possible options may be polymers, metals, or, more rarely, sands. Polymers can be found in the form of filaments, as with FDM, powder, as with SLS, or resin, as with SLA and MJ. Metals can be found in the form of powder, as in the case of SLM, BJ, LMD, and LENS, or wire, as in the case of WAAM [19]. The only method that allows the manufacture of materials different from polymers and metals is BJ, which also allows the 3D printing of sand and silica [21].

Moreover, if the final performance is the main goal of the specific application, it is worth noting that if precision of the shape and a high level of detail are demanded, SLA for polymer-based methods, and MJ and SLM for metal-based technologies, have to be selected. If, on the other hand, no special features in terms of accuracy are required, metallic technologies should not be considered, except for WAAM, where mechanical performance is not matched by adequate performance in terms of detail precision, and, together with SLS and FDM, could be a proper choice. SLM and WAAM are the only appropriate techniques if excellent mechanical behaviour is desired, even if good strength can be also guaranteed by SLA and SLS printed components, always bearing in mind, however, that this involves plastic materials. If mechanical performance is the main target, in fact, FDM, MJ, and BJ should not be taken into account [6]. With regard to the size limits of printable objects, only the WAAM allows for no constraints, as there are no build chambers, but only robotic arms with potentially any dimension required for the purpose.

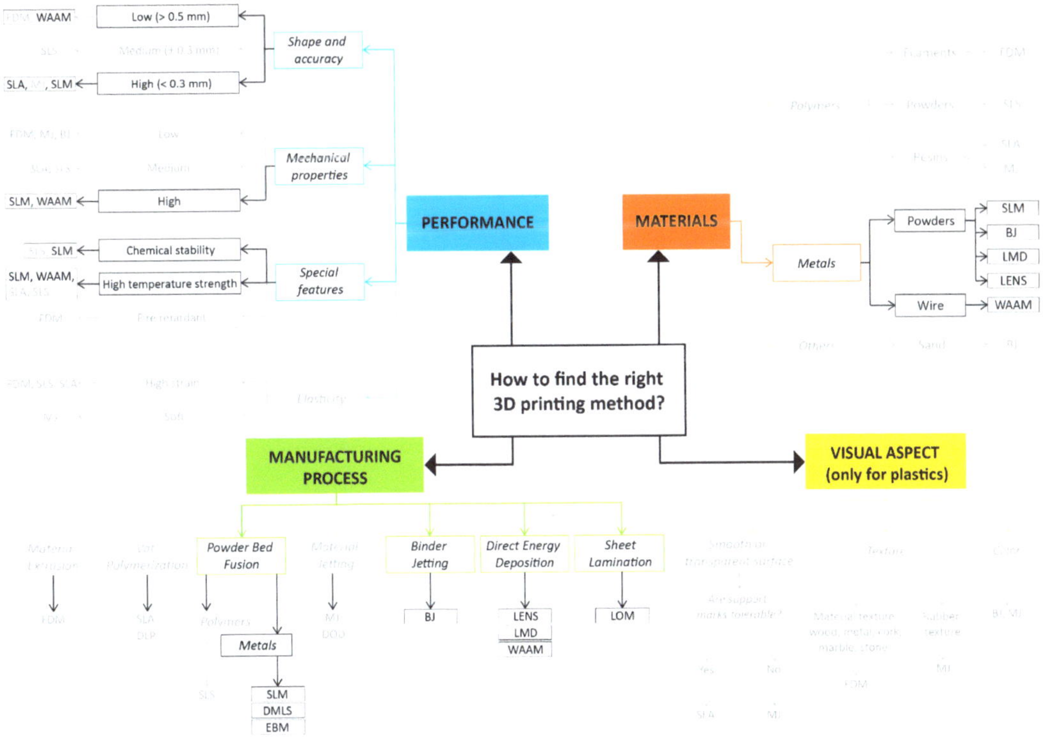

Figure 10. Outline of additive technologies according to different parameters.

Finally, referring to more specific features, products deriving from SLS and SLM are considered chemically stable, parts from FDM as fire-retardant, and components from SLS, SLM, WAAM, and SLA as resistant to high temperatures.

It can, therefore, be summarised that in terms of mechanical and structural performance, and of accuracy and precision in finishing, metal 3D printing methods are unrivalled.

The last parameter that can be analysed is the visual appearance of the printed product. Considering only plastic materials, smooth or transparent parts can be obtained, with textured or coloured effects and, depending on whether the marks left by the supports removal on the surface are acceptable or not, SLA or MJ can be selected, respectively. If components with a texture similar to other materials are desired, FDM results as the best choice. A rubber-like finish is offered by MJ products. Finally, if fully coloured parts are required, MJ and BJ are the leading techniques [15].

3. Printing Process Parameters for Metal AM

The quality of the printed part, both from the surface finish and mechanical points of view, can depend on several factors, known as process parameters. The presence of a very large number of process parameters in AM technologies makes it very complex to find the perfect combination to obtain the optimal result. In fact, specific properties can be achieved by proper modifications of these parameters, which can be printer-related or raw-material-related, and they can introduce changes in microstructure, porosity, and mechanical features. Some process parameters are manageable by the operator, while others are predefined by the printer manufacturer or powder supplier but, in general, they mainly concern the laser beam, the scanning process, the feedstock in powder form, and the operating temperature, as illustrated in Figure 11. Figure 10 shows the process parameters of the PBF methods, the focus of this review, which are discussed in detail in the following sections.

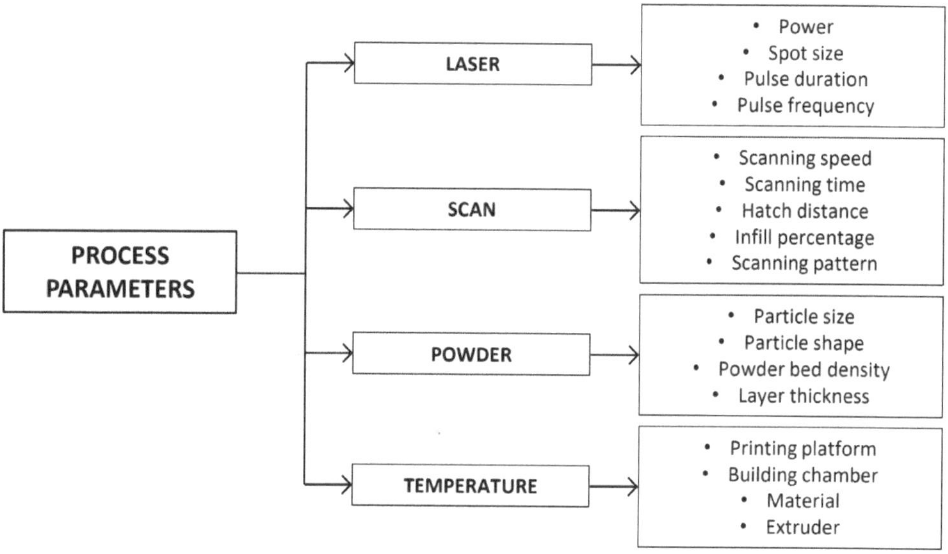

Figure 11. The main process parameters involved in AM processes.

In addition, there are different directions and orientations that can be adopted during the printing process, and they can affect the mechanical behaviour of the printed part, due to the anisotropy generated by the additive manufacturing process. All the parameters involved in the printing process contribute to define the so-called "scanning strategy" [77,78].

As far as the WAAM method is concerned, the main process parameters to be considered for a successful result are current, arc voltage, and welding speed as far as the welding process is involved, and deposition rate and wire feed rate as far as the material side is considered [79].

3.1. Laser-Related Parameters

The main parameters related to the laser beam are the output power, the diameter of the laser spot, the duration of the laser pulse, and its frequency. More specifically, the power of the laser (see Figure 12) represents the power of the beam directed onto the powder bed required to melt it to obtain the desired geometry. It indicates the amount of energy per unit of time and it is, therefore, essential to set the correct speed value in order to assess the amount of energy affecting a certain area. Thus, a right power value must always be coupled with a proper scanning speed [80], which depends on the type of printer. It is worth noting that, when setting the laser power, it should also be taken into consideration that insufficient power does not ensure the proper heating and fusion of the powder bed, resulting in a failed remelting of the previous layer and, consequently, in a nonadhesion of the layers. In addition, higher power increases the degree of particle fusion, reducing porosity [81,82]. The laser power contributes, together with the spot diameter, scanning speed and distance between scan track spacing to define the laser energy [83].

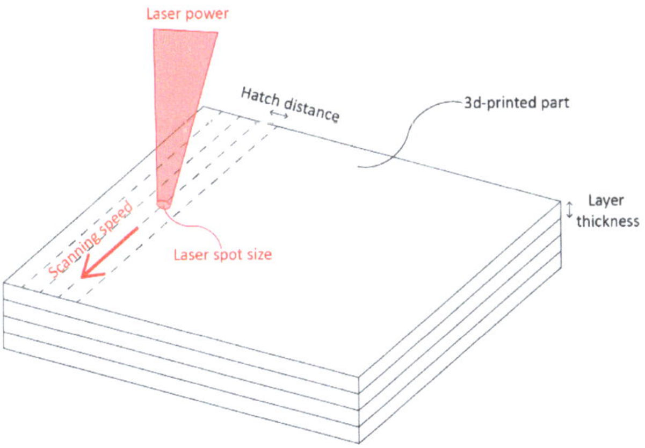

Figure 12. The printing process parameters.

As 3D printers produce parts in three dimensions, it is necessary to consider the minimum element size of the XY-plane and the Z-axis resolution. The reason for the best printing quality lies in the resolution of the XY plane, which is defined by the diameter of the laser beam. The smaller the diameter, the greater the reproducible detail and accuracy.

The laser spot size contributes (see Figure 12), together with the thickness of the layers, to define the resolution of the printed part. When the laser beam hits the construction platform, it projects a surface, called a spot, where the powder melting phenomena occur. The geometric shape is idealised as a perfect circle, although in practice it turns out to be an ellipse of varying size. In fact, during the printing process, the angle of incidence between the laser beam and the surface of the powder bed changes continuously. More generally, it is possible to state that larger spot sizes ensure higher build speeds but result in lower accuracy and dimensional tolerance of the components produced [78].

Finally, the duration and frequency of the laser pulse are quantities that contribute to define the laser output energy.

3.2. Scan-Related Parameters

The parameters related to the scanning process of the single two-dimensional element layer are scanning speed, scanning time, hatch distance, and scanning pattern (see Figure 12).

The scanning speed expresses the speed of the laser beam movement and it changes depending on the material used. It is an important factor as it influences the heating and cooling rate of each layer and, consequently, the microstructure of the printed product [80].

The scanning time defines the minimum time for scanning a single layer, measured in seconds. When the scanning of the desired geometry is completed earlier, the machine waits for the set time to produce the next layer. It can significantly influence the mechanical properties as it represents the time to ensure cooling of the layer manufactured.

The hatch distance, also called hatch spacing or scan spacing, is the distance between two consecutive passes of the laser beam (see Figure 12). It is also known as infill percentage, as it represents the scan density of the geometry to be produced. It is a parameter directly proportional to the production rate. Obviously, the shorter the hatch distance, the higher the level of detail and the time required to build the component. On the contrary, if the scan tracks are too far apart, the powder between them will not be melted and, at the end of the process, high-porosity zones will result due to lack of fusion. Bremen et al. [84] and Sefene et al. [78] highlighted that the hatch distance is defined by the diameter of the laser and proposed an optimal hatch spacing value of 70% of the diameter.

Finally, the scanning pattern defines the path where the laser selectively fuses the section on the powder bed and plays a significant role in limiting the presence of defects [78]. The selection of the appropriate scanning pattern influences not only the final density of the parts, but also the residual stresses occurring [77,85,86]. There are three main possible configurations of scanning patterns, namely, stripes, chessboard or islands, as illustrated in Figure 13. In the stripe pattern (Figure 13a), the laser builds the product by scanning the geometry through horizontal or vertical bands, progressively fused. The chessboard pattern (Figure 13b) divides the geometry into a series of squares and the white squares are printed first and then the black ones, as in the scheme of a chessboard. Finally, the island pattern (Figure 13c) is a random form of chessboard pattern, in which each square is printed randomly over the whole level, without any particular sequence, until there is no more not-melted powder remaining. Miao et al. [87] and Sefene et al. [78] showed that the use of the chessboard pattern, compared to the striped pattern, develops lower thermal gradients and reduces the number of defects and amount of surface roughness. Zai et al. [77] confirmed that the scanning pattern can change the thermal history of each layer, also modifying its porosity and microstructure.

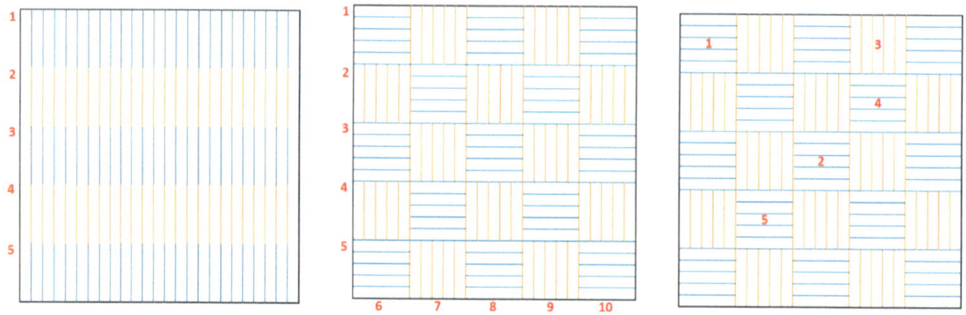

(a) (b) (c)

Figure 13. The three most widely used scanning patterns: (**a**) stripes, (**b**) chessboard, (**c**) islands.

3.3. Powder-Related Parameters

The parameters concerning the characteristics of the powder are the size and shape of the particles, the powder bed density, and the thickness of the layers. The particle size represents the size of a single grain of metal powder (in the range of microns), while the shape represents the final shape of the single grain as a function of the production technique of the metal powder (most of the powders used in the SLM process are mainly obtained from gas or water atomisation methods) [88]. It is important to bear in mind that powders produced by gas atomisation have a higher sphericity. Both size and shape are considered two key material properties and depend on the nature of the metal alloy. In fact, smaller particles perform better from the point of view of microstructure compactness but are prone to have a low flowability; thus, the deposition of a homogenous layer is harder. Granulometry and morphology of the particles affect the density of the powder bed, which represents the packing density of the individual layer [81,89–91]. Furthermore, these characteristics have a significant influence on the final component density, on the mechanical properties, on the microstructure, and on the surface roughness [78,92,93].

Finally, the layer thickness (Figure 12) represents the thickness of the individual layer deposited in the z-direction, which can strongly influence the quality and accuracy of the printed part. Optimal and recommended values are provided by the metal powder manufacturers. Lower thicknesses guarantee better part quality, but at a lower production speed. Furthermore, layer thickness has an effect on the mechanical properties of additively manufactured components [94].

3.4. Temperature-Related Parameters

The temperature-related parameters include the temperature of the powder bed and the temperature of the feeding system. The temperature of the powder bed is a function of the temperature of the building plate, which can vary from room temperature up to 200 °C, to reduce the thermal gradient as much as possible in the post-melting cooling phase; while the build chamber temperature can reach very high values depending on the material, the final intended application of the printed component, and the type of printer. Indeed, the high temperature gradients that can develop during the SLM process can lead to high levels of residual stress within the additively manufactured metal structure [86]. For this reason, the temperature within the feeder can also be modified to preheat the powder [95].

3.5. Printing Directions and Orientations

One of the most important aspects to consider in the 3D printing process, often underestimated by designers, is the orientation of the printing parts (or build orientation). The way the components are positioned on the build plate, in fact, plays a significant role in achieving the desired final quality. Part orientation can affect accuracy, production time, strength, and even surface finish [96,97].

To comprehend how part orientation impacts the accuracy of 3D-printed parts, a hollow cylinder is examined (see Figure 14). Figure 14 shows the example of a part produced using FDM, as the layers are thicker and the critical aspects are more visible. When the workpiece is oriented in the vertical direction, the final cylinder will have a fairly smooth external surface as the part will consist of a series of overlapping concentric circles. If the cylinder is oriented horizontally, the part will be built from a series of overlapping rectangles (of slightly different widths) and the surface of the cylinder that contacts the build platform will be flat. Therefore, it is evident how different production directions lead to significant differences in quality of printed products.

Figure 14. Two identical pieces printed in vertical (**left**) and horizontal (**right**) direction. Reprinted from ref. [98].

Building direction can also impact strongly on production speed, in particular in case of parts of large dimensions. The horizontally printed components usually require shorter printing time than those produced vertically, as the number of layers is reduced [78,95].

Several studies demonstrated the anisotropy of elements produced through 3D printing in terms of mechanical strengths. In fact, as highlighted in Figure 15, parts are usually stronger in the XY-direction than in the Z-direction, due to the production process of adding layers, resulting as being much more vulnerable to loads applied perpendicularly to the production direction than to the longitudinal. Therefore, for functional parts, it is important to consider the application and direction of the loads [25].

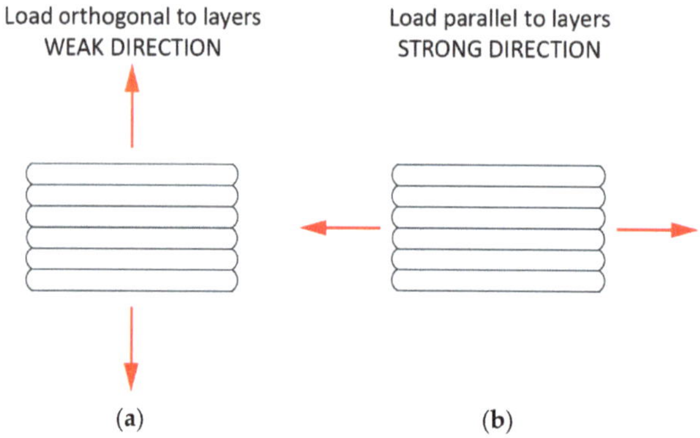

Figure 15. Effects of a tensile load applied in the perpendicular (**a**) and parallel direction of layer deposition (**b**).

Figure 16 shows the most commonly applied printing directions, namely, a vertical layout with the longitudinal axis perpendicular to the construction plane, and two horizontal layouts with the longitudinal axis parallel to the building plane, flat (xy) and on edge (xz).

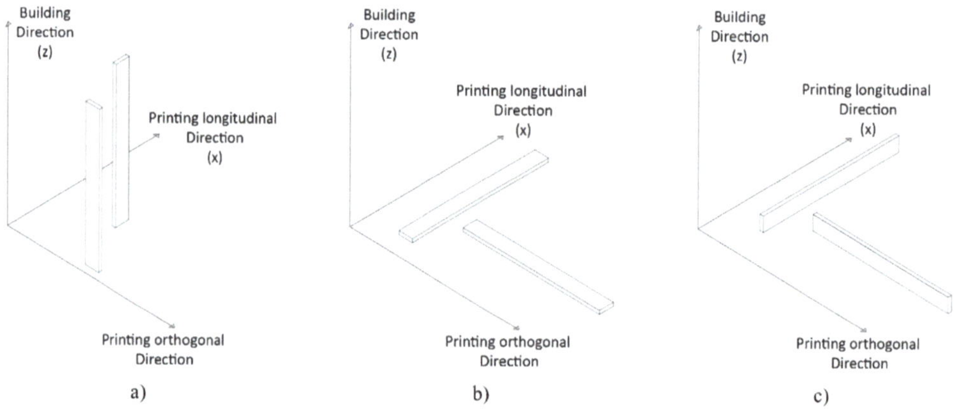

Figure 16. The main printing directions and orientations of the parts: (**a**) vertical, (**b**) horizontal flat, (**c**) horizontal on edge.

The orientation of the parts affects the thermal history, as it influences the final microstructure and, consequently, the mechanical behaviour [77,99].

3.6. Effects of Process Parameters on the Properties of 3D-Printed Metals

There are a number of studies in the literature that have examined the effects of the printing process parameters on the mechanical performance of different metal alloys produced with SLM. Among the most frequently investigated parameters, there are printing direction and orientation [43,100,101], scanning time [42], layer thickness [102], and scanning strategy [82,103,104].

However, there are still some aspects that need to be further explored in order to have a complete knowledge of additively manufactured metals, especially if the material is intended for structural purposes. Among these aspects is certainly the sensitivity of the material to certain ranges of strain rates, as knowledge of the mechanical performance of the material under dynamic conditions may be very important when designing specific devices (for example, dampers or in general seismic energy dissipation devices) or structural components (for example, joints or structural members).

Indeed, the influence of strain rate on 3D-printed metals has been little studied: as far as we know, the only research works on the effects of strain rate on metals produced by additive manufacturing are by Mazzucato et al. [65] and Forni et al. [105] concerning the nickel-based alloy Inconel 718 produced via LMD, and by Brando et al. [106] regarding the 17-4PH alloy steel produced through selective laser melting.

The selection and use of adequate process parameters strongly reduce the formation of defects such as excessive porosity, presence of incompletely fused areas, and cracks. In fact, it is important to consider that many problems in PBF techniques in general, and SLM in particular, are related to the choice of wrong printing parameters.

The number of defects increases as the scanning speed increases. This can be attributed to the effect of the scanning speed on the energy of the laser beam, as the higher the speed, the shorter the time for energy transfer. Melting failure occurs because the energy emitted is not sufficient to completely melt the powder. This leads to the formation of not-melted particles trapped in the pores (lack of fusion). Other negative consequences can be delamination (separation of adjacent layers due to incomplete fusion of neighbouring layers) and balling (loss of continuity of the fusion bed). Therefore, the presence of these defects can be controlled by reducing the scanning speed and increasing the laser power [83].

Residual stresses are another factor closely related to the production process. As pointed out by Michla et al. [107] and Haghdadi et al. [95], the preheating of the powder bed or the building plate and, in smaller measure, the type of scanning pattern, can help to mitigate the temperature gradient and, consequently, the residual stresses. According to Zhu et al. [108], an energy density increase, which can occur by either enhancing the power of the laser or reducing the scanning speed, raises the possibility of part shrinkage, resulting in higher residual stresses in the printed part [82,108]. Furthermore, according to the studies of Ramos et al. [109], the adjustment of laser power and speed can also influence residual stresses. In fact, in the spot where the laser melts the powder, a strong temperature gradient is produced, leading to the development of a localised melt pool in the powder bed. When the laser moves, the instantaneous heating results in the thermal expansion of the heated area, which is confined by the already cooled layer below. This phenomena has been referred to as the "temperature gradient mechanism" by Mercelis and Kruth [85] and is responsible for residual stresses that can compromise the quality of the finished part and its mechanical properties. The achievement of fully dense and functional parts is one of the main challenges in additive manufacturing, especially for applications where adequate resistance to actions and loads is required. In this regard, Zai et al. [77] point out that the process parameters to be monitored are laser energy density, scanning pattern, and powder bed preheating. The volumetric laser energy density E (J/mm^3), as also reported by other studies [78,81], is determined through the following Equation (1):

$$E = \frac{P}{v \cdot h \cdot t} \tag{1}$$

It depends on the laser power P (W), the scanning speed v (mm/s), the hatch spacing h (mm), and the layer thickness t (mm).

Enneti et al. [110] confirmed that density is a parameter inversely proportional to scanning speed and hatch distance. According to studies by Klocke et al. [111] and Larimian et al. [82], a higher energy density is able to increase the density of the final element. Furthermore, Sefene [78] found that also laser power and scanning speed, which are parameters useful for achieving complete melting of the powder and high density of the printed part, as well as a good surface finish, are always inversely proportional. In conclusion, laser power is the principal energy parameter that strongly influences the final density of the component. Density increases as the power used in the process rises if no defects occur during manufacture [82,112].

4. Metal Additive Manufacturing in Construction

Small metal components are usually produced by AM. Due to their high precision, they are widely used, especially in the aerospace, automotive, and healthcare industries [21,113]. Up to now, there are few examples in the context of large-scale applications, either in the literature or in industrial applications [21,33,114,115]. The reason is that, when extending the scale of the component to be created, there is a loss in the quality of the final parts (which forces an additional post-processing step), and both costs and production times increase considerably [6].

The main advantages of using metal AM are the possibility to create elements with complex geometries and optimised performance related to a more efficient use of material [116,117]. This section will review the main examples and applications of metal AM in the construction field.

4.1. Optimized Structural Node by Arup

An interesting application of metal AM in construction [114,118–120] was realised by ARUP, a global engineering and design company. To explore the benefits of the combined use of AM techniques and topological optimisation (TO) processes, the ARUP team redesigned the structural node of a tensegrity structure used for street lighting.

Considering the variability of the inclination and cable attachment points of the original structure, the integrated use of AM and TO proved extremely efficient in rationalising the original geometry.

Starting from the original geometry, a first topological optimization process was conducted (AM Node 1.0). The optimization analysis was carried out by defining as an objective function the minimization of the structural weight. Then, the maximum Von Mises stress and geometric restrictions were imposed as constraints. The result of the optimization process is shown in detail in the paper [120]. The element was produced by DMLS with ultra-high-strength steel powders (Maraging steel grade 1.2709). During the fabrication phase, in order to respect production constraints, changes were applied to the optimised node (AM Node 1.0) by introducing appropriate self-supporting elements.

After the first TO process, the AM 1.0 node showed a significant weight reduction, 30% less than the original node. The resulting shape fulfils all functional requirements, providing a better flow of internal forces and a more rational material distribution.

A second version of the optimised structural node (AM 2.0 node) [118] was created to obtain a more compact and lighter element, maintaining the same functional requirements.

The objective function of the optimisation process was to minimise structural weight. While Von Mises stresses were set as a constraint and limited to a maximum value of 80% of the material maximum tensile strength, the AM 2.0 node was produced with the DMLS technique using 316L stainless steel powder. Although the structural behaviour remained similar to previous versions, the AM 2.0 node showed a weight reduction of 75% compared to the initial node. This result highlights how significant savings are possible, both in terms of material uses and production time, keeping the functional requirements constant.

4.2. MX3D Pedestrian Bridge

The first large-scale application of AM in the field of structural engineering was the completely functional pedestrian bridge by MX3D (Figure 17), a Dutch company specialising in robotic metal 3D printing [79,114,116].

For the production of the bridge, manufactured from 308 L grade austenitic stainless-steel wire, 6-axis robotic welding arms using the WAAM technique were employed. This demonstrates the enormous potential of the application of multiaxis 3D printing technology. In the project, the MX3D company worked with various partners such as Arup, Autodesk, ArcelorMittal, the University of Twente, and Imperial College London.

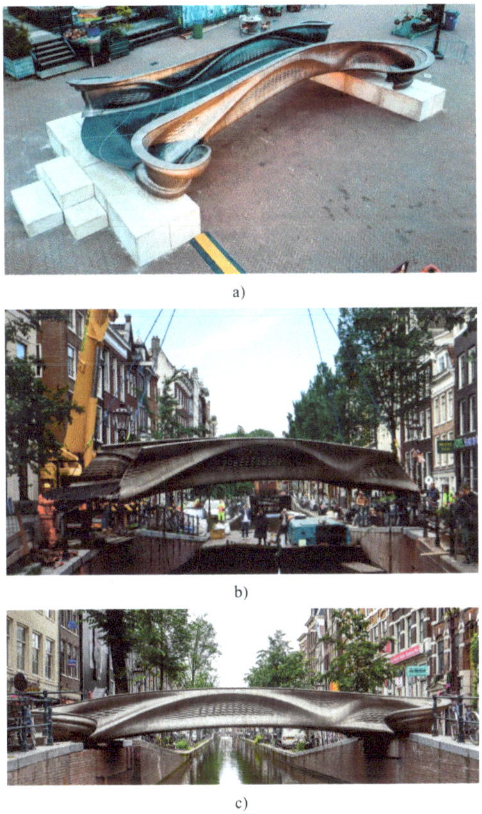

Figure 17. The MX3D bridge: (**a**) The finished 3D-printed bridge, before installation; (**b**) the bridge during installation; (**c**) the bridge after the inauguration on the Oudezijds Achterburgwal canal in Amsterdam. Reprinted with permission from refs. [79,121]. Copyright 2020 Elsevier Ltd.

The bridge, which measures 12 m in length and was built from a single metal part weighing 4.5 tonnes, required six years from the design to the final construction. It was inaugurated in Amsterdam on 21 July 2021, after capacity tests with a 20-tonne load had been successfully completed (Figure 18). In addition, the partners were involved in the design, development, and testing of a smart sensor network to monitor the bridge conditions in real time.

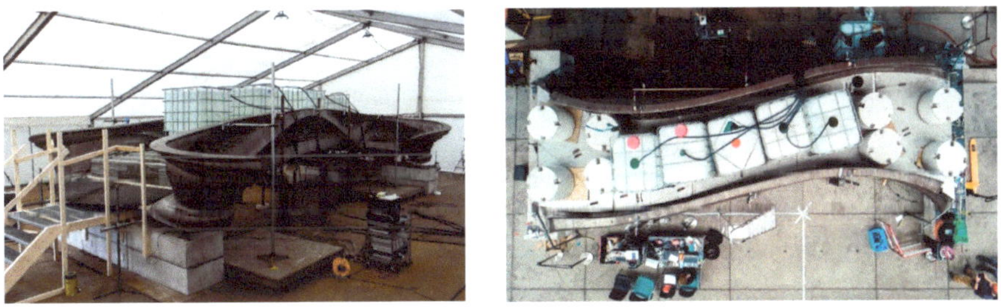

Figure 18. Load tests of MX3D bridge at the University of Twente. Reprinted with permission from ref. [79]. Copyright 2020 Elsevier Ltd. Reprinted from ref. [122].

4.3. Takenaka Connector

The Takenaka connector was recently created by the collaboration between the Takenaka corporation and MX3D. The structural steel connector (Figure 19) was designed by MX3D and Takenaka to overcome the limitations of conventional processes and produce complex components for large structures in the construction industry. The element has a hollow structure, which is filled with cast mortar through the use of robotic arms (Figure 19b). The metal part is produced from Duplex stainless steel using the WAAM technique. The solution of filling the inner core with mortar is to prevent local buckling of the steel, while the outer steel offers high resistance to bending and tensile forces. The net weight of the structure is 40 kg, which increases to 45 kg when the inner cavities are filled with mortar.

a)

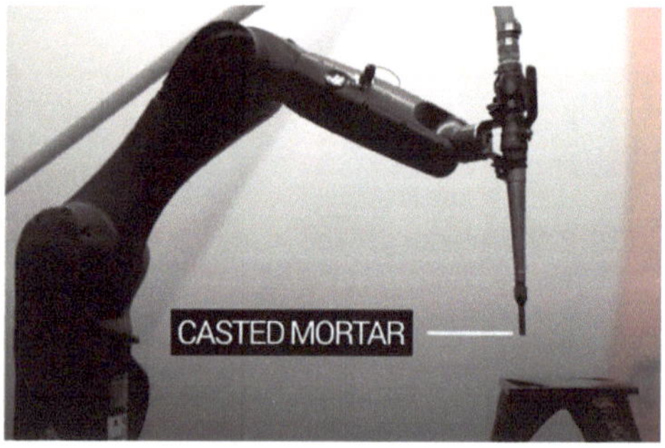

b)

Figure 19. The Takenaka connector: (**a**) Geometry; (**b**) robotic arm filling the connector with casted mortar. Reprinted from ref. [123].

The final shape is the result of an optimisation process that led to a very efficient configuration (Figure 20). The entire process was guided by design constraints as the position of the element in the structure was known in advance and was taken into account in the optimisation process.

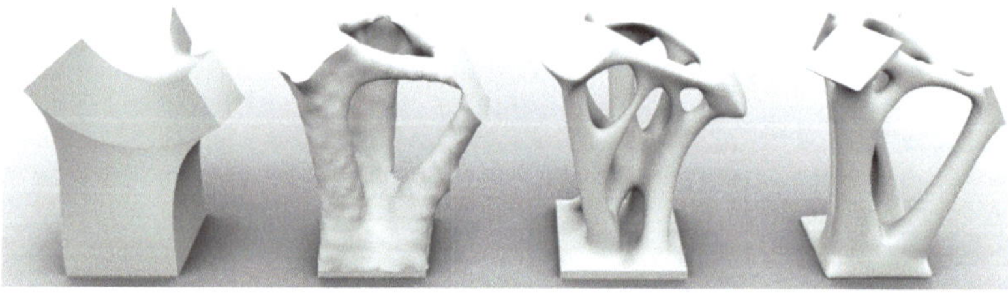

Figure 20. Optimization processes of the Takenaka connector. Reprinted from ref. [123].

4.4. AM Steel Reinforcement for Concrete

Early studies on the production of AM steel reinforcements for concrete [124–126] show encouraging results. However, their application in the construction industry is still limited. At this stage, it is essential to continue the research, in particular by investigating the quality of bars produced with AM, their fatigue behaviour, and their durability in concrete environments.

In fact, AM research involving the construction sector has mainly focused on the production techniques of cement-based materials. Despite the growing interest in the topic, few indications on the use of reinforcements with concrete 3D-printed elements have been provided [127].

Recently, some studies [124,125] have explored the potential of AM steel reinforcements as a supplement to 3D-printed concrete. Mechtcherine et al. [124] studied the mechanical behaviour of bars made using the GMAW (gas metal arc welding) technique. The samples were subjected to a series of experimental tests to compare their performance with that of conventional bars. In particular, uniaxial tensile tests were carried out to determine the tensile behaviour of bars with a diameter of 8 mm at the loaded ends and 7.5 mm in the middle. Pull-out tests were performed with two different bonded lengths (16 and 32 mm) to determine the interface adhesion forces between 8 mm diameter bars and fine-grained concrete. Both tests were also conducted on conventional B500B steel bars with a diameter of 8 mm.

The results obtained from the tensile tests showed that the performance of 3D-printed bars was significantly lower in terms of yield stress and tensile strength. In terms of strain capacity, the 3D-printed reinforcement showed significantly higher values than conventional bars. This result highlights the extremely ductile behaviour of 3D-printed bars, which can also be observed from the visual aspect of the fracture surfaces.

The results of the pull-out tests showed lower shear stress values for the 3D-printed bars for both bond lengths than for conventional steel bars. However, in the case of the bond length of 32 mm, the shear stress–displacement curve exhibited, after the peak, an approximately constant trend for the 3D-printed samples. A more pronounced softening phase is observed for conventional bars. For a bond length of 16 mm, no significant differences in the behaviour after the peak were visible.

The pull-out results did not provide a complete description of the behaviour of the 3D-printed samples, so further investigations are necessary. The low shear stress values resulting from the pull-out tests for the 3D-printed bars can find partial justification in the lack of "ribs" on the outer surface. In fact, solutions capable of improving the adhesion between bars and concrete matrix, such as ribs that prevent mutual sliding between the two materials, were not considered in this study. Please refer to the paper for details of all the results described [124].

4.5. Joining Aluminium Profiles

The variety of production allowed by AM also lies in the possibility of using many different types of materials. In sectors where the use of metal AM is widespread, several applications use aluminium alloys [128], titanium alloys [129,130], and nickel-based superalloys [131]. Regarding the construction field, Baptista et al. [132] proposed the combined use of WAAM and joining by forming to connect hollow aluminium profiles and fix them to composite sheets. The connection is realised through a "mortise and tenon" joint (Figure 21).

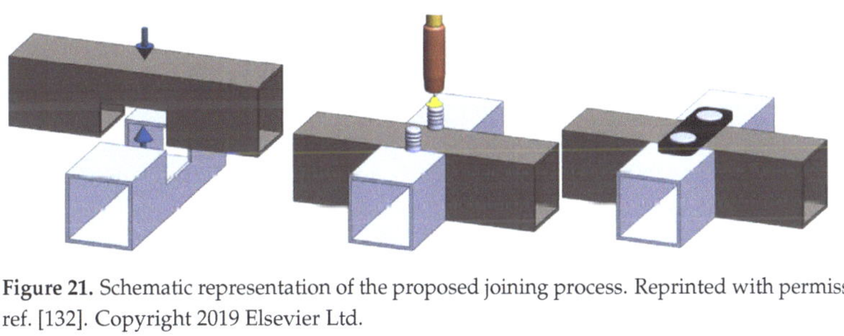

Figure 21. Schematic representation of the proposed joining process. Reprinted with permission from ref. [132]. Copyright 2019 Elsevier Ltd.

Concerning the connection assembly, initially, the hollow aluminium profiles were coupled, then tenons were made directly on the profiles using WAAM. Finally, through the compression of the tenons against the mortises, a mechanical joint was generated. To determine the mechanical performance of the connection system, two destructive tests were performed: pull-out and shear tests (Figure 22). The results showed the failure due to a pull-out force of 2 kN recorded after the detachment of the individual components for a displacement of approximately 0.6 mm. With regard to shear strength, a value of approximately 3.2 kN was recorded.

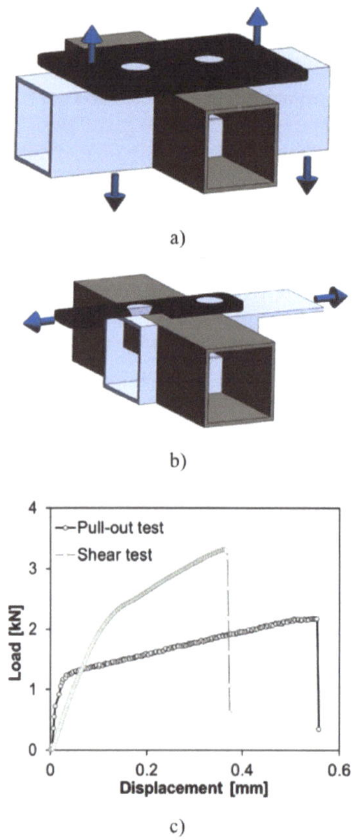

Figure 22. The mechanical tests on the connection: (**a**) Scheme of the pull-out test; (**b**) scheme of the shear test; (**c**) results. Reprinted with permission from ref. [132]. Copyright 2019 Elsevier Ltd.

4.6. Future Applications

One of the main innovations introduced with additive manufacturing, which overcomes most of the limits imposed by traditional production methods, concerns the possibility of creating innovative types of structures and components [133–135]. Future applications in the field of structural engineering could involve steel connections and stiffening elements for steelwork; some interesting examples are shown below [136,137].

A beam-to-column connection, where the suggested solution involves welding a hook obtained by topological optimisation, printed directly onto the column and connected to a bolt, is showed in the works of Feucht and Lange [136] and Lange et al. [137]. The main advantage of this system is the simplicity of construction and the elimination of bolted connections. The first tests carried out in the laboratory on small components showed that direct welding on metal elements, using the WAAM technique, is possible without observing significant distortions in the welds.

With the aim of ensuring better load transfer within a double-T profile and to avoid the occurrence of flange buckling phenomena, topologically optimised and additively manufactured reinforcing elements can be introduced to replace stiffeners. In this case, the main advantages consist of (1) reduction in material used, (2) reduction in printing time, as unstressed areas are removed, and (3) possibility of printing directly on the element to be reinforced, in contrast to conventional production, where unloaded regions are hard to eliminate, as this requires extra effort and the production of nonreusable waste.

Destructive compressive tests, performed after the production of complete stiffeners on a double-T beam, showed that, despite the evident distortions of the weld, the stiffeners produced with 3D printing can withstand stress levels equal to those of stiffeners manufactured using traditional techniques.

When rigid connections with an end plate are employed, the moment and tensile force are transmitted with an eccentricity, as the joint lies outside the flange. The tensile force is transferred from the flange to the bolt through the bending of the end plate, which often has large thicknesses to transfer this bending moment. Considering the potential of additive manufacturing, the geometry can be modified in a way that improves its efficiency: the use of topology optimisation to find the shape led to a first hypothesis which was also produced using WAAM. Furthermore, it was proven that only 40% of the mass of the initial element is required to achieve the same load capacity: this is due to the formation of the lever arms, for which the greater the length, the higher the eccentric stress transferred. Further details are provided in the papers [136,137].

In light of the applications just reviewed, it is clear that 3D printing technologies can be applied directly on metal components and can potentially provide numerous benefits in the fields of structural engineering and construction. Further developments could concern the introduction of automated processes for monitoring and maintenance, as the use of additive manufacturing in the repair of existing structures could introduce several advantages, which, however, require further research to assess their potential in real applications. In fact, in the field of construction, one of the greatest opportunities of additive technology may lie in the repair of existing structures: in some cases, considering the production costs of high-performance components, it may be cost-effective to repair worn parts rather than replace them with new ones.

However, studies on the subject are still in their beginnings and are still not significant, so further investigations are needed to confirm the encouraging results obtained so far.

5. The Potentials of Metal AM in Topological Optimization

5.1. Nonconventional Geometries

Conventional manufacturing techniques have encouraged the development of simple geometries. In contrast, as was already said in the previous sections, AM allows higher freedom in designing complex geometries with unique features that difficult or impossible to achieve with conventional manufacturing methods. Although AM offers greater flexibility in the design of new shapes and geometries, an adequate level of performance and safety must be ensured in each application. Several studies in the aerospace [129] and automotive [138] sectors demonstrated that topological optimisation (TO) processes can guarantee high levels of performance for projects realised with additive techniques.

In Seabra et al. [129], the complete process (from optimisation to production and testing) for an aircraft bracket was described (Figure 23). SLM was used to produce the component, and to improve performance, the printed part underwent hot-isostatic pressing (HIP) heat treatment and surface treatment.

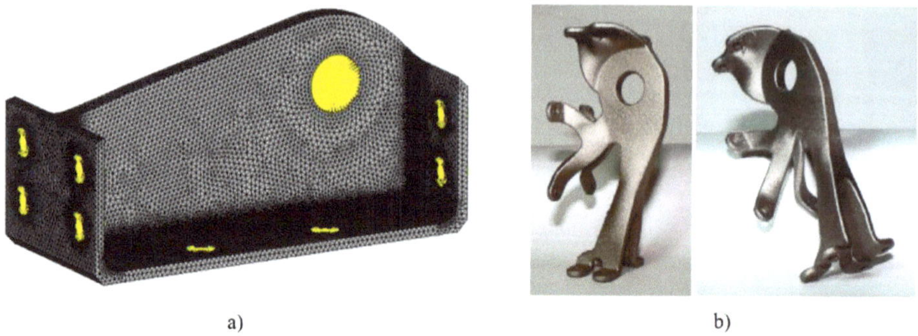

Figure 23. Aircraft bracket: (**a**) Mesh of the original component; (**b**) 3D-printed optimized shape. Reprinted with permission from ref. [129]. Copyright 2015 Elsevier Ltd.

Thanks to its manufacturing skills, SLM is adopted for the production of high-precision end-use parts. Topology optimisation is used to achieve a final design with high stiffness against an initial volume reduction. In addition, the optimised design is analysed with FEM to verify the compliance with requirements. Once the previous steps are completed, the component is created using SLM. However, before construction, an intermediate step is necessary to consider the limits of the production process adopted (overhangs, support structures, dimensional limits) to prevent possible problems in the parts destined for final use.

The final optimised aircraft bracket reduces the material volume by 50% compared to the original part. Furthermore, tests show a good match between the FE model and the manufactured part.

In the automotive sector, an interesting example was provided by Walton et al. [138]. The study proposed to redesign a pair of suspension uprights with a view to reducing mass. Indeed, the final component, manufactured in Ti6Al4V ELI by EBM without additional surface finishing or heat treatments, exhibited a significant decrease in structural weight with an improvement in the safety factor (Figure 24).

Figure 24. Optimized suspension uprights. Reprinted with permission from ref. [138]. Copyright 2017 Elsevier Ltd.

Critical considerations about manufacturing cost and raw material use suggested that the use of TO components via AM is limited to high-performance designs. Indeed, in these applications, function prevails over production costs. Despite its great potential, the ongoing growth of AM requires the evolution of design practices to take into account production limitations and cost considerations [139].

Design methods for AM can be divided into process-driven shape and designer-driven shape [139]. The former allows the creation of customised parts using TO processes, while the designer-guided design method uses lattice structures to achieve the required performance and reduce support structures. TO and lattice structures are the main strategies for exploring the potential of AM [140]. Further details are provided in the following sections.

5.1.1. Topology Optimization

In general, optimisation problems can be mainly divided into the following:

- Size optimisation, which considers the variation in size of the elements. In order to find the optimal solution (weight, stress, etc.), the cross-sectional areas of the beams, etc., are adjusted (Figure 25a).
- Shape optimisation, which concerns the changing of the structural form. It allows one to remodel holes in the model, but not to eliminate them (Figure 25b).
- Topological optimisation, which is the general form of structural optimisation. It allows specific parts to be added or removed in the design domain (Figure 25c).

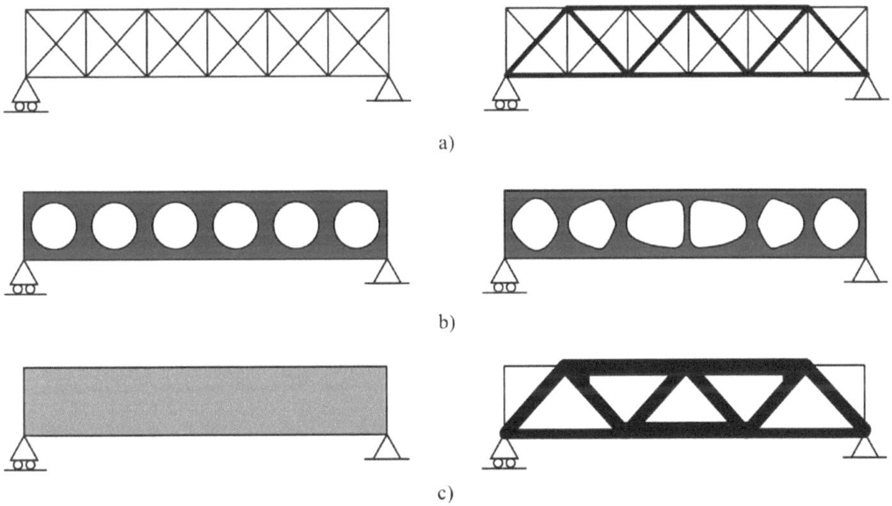

Figure 25. Types of optimization processes: (**a**) Size optimization; (**b**) shape optimization; (**c**) topology optimization. Reprinted with permission from ref. [131]. Copyright 2017 Elsevier Ltd.

Topological optimisation (TO) is an iterative process to find the best distribution of material in a design domain subject to a specific set of constraints. The definition of "best distribution" takes on different meanings depending on the target of the optimisation. Usually, with optimisation processes, the lightest or stiffest solution is found by subjecting the design domain to constraints such as stresses, displacements, or frozen areas.

Mathematically, the problem can be expressed like this [141–143]:

- Minimise/maximise f(x,y).

The function f(x,y) is subject to

- Behaviour constraints on y.
- Design constraints on x.
- Equilibrium constraints.

In which

- f is the objective function of the optimisation problem, i.e., the target of the process; typically, it represents the parameter subject to constraints in the design domain.
- x represents the design variables, which describe the geometry and material. Constraints can also be set as geometric restrictions on parts of the domain to be constrained or on dimensions.
- y represents the state variables, which describe the structural response. They can be expressed in terms of stresses, displacements or forces.

SIMP method

The most popular numerical FE-based topology optimization method is the solid isotropic material with penalization (SIMP) [144]. The SIMP method is a density-based TO method which can be expressed by the following Equation (2):

$$E_h = \rho^p \cdot E_0 \quad (2)$$

where

- E_h is the Young's modulus of the optimized element.
- ρ is the pseudo-density.
- p is the penalization factor ($p > 1$).
- E_0 is the initial Young's modulus.

Through an appropriate choice of p, the SIMP approach allows intermediate densities to be penalised. Therefore, a binary solid/void solution is promoted, which guarantees high efficiency and easy integration with production processes. The choice of $p > 1$ ensures that intermediate densities are penalised. In fact, values of ≥ 3 provide good results in both the 2D and 3D cases [145].

Initially, in the SIMP method, densities are uniformly distributed over all elements in the design domain. Once the iterative analysis is started, the equilibrium equations are solved in the first step, and then, with a sensitivity analysis, the derivatives of the design variables are calculated. Before updating the element densities in the domain, filtering techniques are employed to ensure numerical stability. Subsequently, a new FE analysis is performed until convergence.

Level-Set method

Recently, the use of level-set method (LSM) has been increasing for the topological optimisation of components and structures. The main advantages of LSM lie both in leading to efficient computational schemes and in managing topological changes such as the union and division of connected components. Well-defined boundaries enable the creation of geometries that also consider production and digitalisation issues. Interesting examples of using the LSM in optimisation problems can be found in Allaire et al. [146] and Wang et al. [147].

The application of topological optimization

Thanks to its considerable advantages in many industrial areas, topological optimisation is applied in the production of high-performance components. Nowadays, several commercial software packages are able to solve specific problems common to a wide range of industrial applications [140].

Figure 26 shows the distribution of the most common topological optimisation methods used in commercial software. The density-based method (50% of the total) is the most widespread. Furthermore, if hybrid approaches (which consider the density-based approach with the level-set method) are also evaluated, the coverage reaches over 80%.

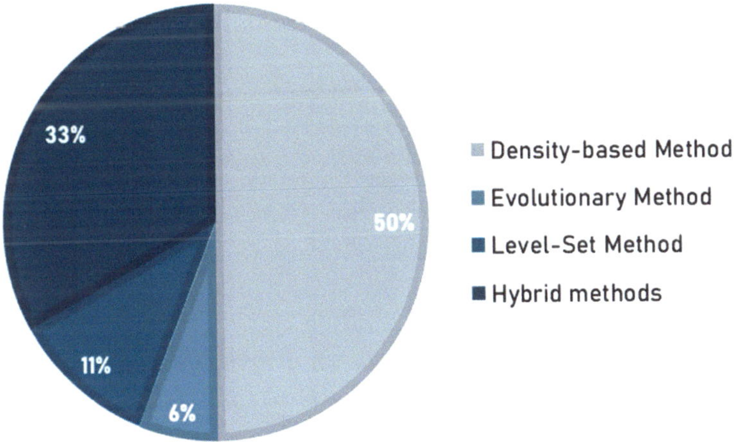

Figure 26. The most common methods adopted by topological optimisation software. Reprinted from ref. [140].

To evaluate the advantages of the combined use of AM and topological optimisation, Saadlaoui et al. [131] compared the performance provided by different formulations of the optimisation problem. The investigations were carried out by assessing the numerical and experimental performance of a cube subjected to uniaxial compression. The commercial software used in the study were Abaqus and Optistruct.

In detail, the formulations used in [131] are the following:

- Stress-constrained optimisation (SCO). According to this approach, the goal is to minimise the structural weight for a stress constraint. In this case, stresses are required to be less than the elastic limit of the material (Inconel 718) divided by a factor of safety
- Continuous compliance optimisation (CCO). In most optimisation problems, the structural compliance parameter (to be understood as the inverse of stiffness) is used as the objective function to be minimised. Usually, in this formulation, the constraint is represented by an arbitrary value on the volume of the material.
- Discrete compliance optimisation (DCO). The terms of the optimisation problem follow those of the CCO approach. However, it considers discrete variables that return the problem to a binary solution: 1 (solid), 0 (void).

No significant discrepancies in terms of displacement and stress were found in the three formulations examined. However, noteworthy differences were obtained in terms of reduction in the original volume. The SCO approach resulted in a reduction of approximately 74%, while CCO was 69% and DCO 61%.

Once the numerical analyses were completed, the optimised geometries were produced using the SLM technique (Figure 27).

The stress-constrained optimisation (SCO) approach is the formulation with the highest computational cost, but the results show the best mechanical performance for the greatest reduction in structural weight.

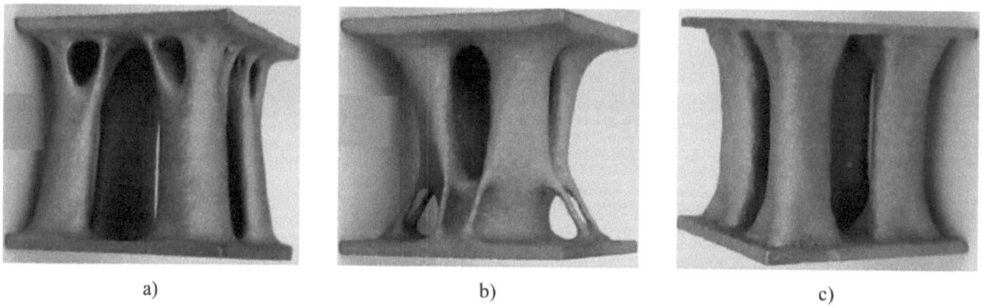

Figure 27. Optimized structures: (a) Sample based on DCO; (b) sample based on SCO; (c) sample based on CCO. Reprinted with permission from ref. [131]. Copyright 2017 Elsevier Ltd.

5.1.2. Lightweight Components

AM offers great opportunities in the production of lightweight components such as lattice structures. The term "lattice structures" refers to 3D open-cell structures resulting from the repetition of a unit cell.

Due to their versatility and mechanical properties, these structures are widely used, particularly in the aerospace and biomedical sectors [148].

Based on their mechanical behaviour [149], lattice structures are classified into the following:

- Bending-dominated structures: The design elements are mainly subject to bending moment. Therefore, these structures exhibit compliant behaviour.
- Stretch-dominated structures: The structures are mainly subject to axial loads. Generally, this type is stronger and stiffer than the previous one.

The unit cell can be generated from strut-based elements or from surface-based elements. The former is characterised by constructive simplicity, and several examples are shown in Figure 28a. The latter have several advantages, especially with regard to production problems (Figure 28b).

For more details on lattice structures, the contribution by Maconachie et al. [148] is recommended.

In order to also provide insight on manufacturing issues, Panesar et al. [150] compared different design strategies. In their paper, lattice solutions are generated using discrete solid/void or greyscale topology optimisation results (Figure 29).

The design strategies considered in [150] were the following:

- Solid: This strategy shows a topologically optimised result using the SIMP method (Figure 29a).
- Intersected lattice: According to this strategy, the solution is obtained by intersecting a topologically optimised discrete solid/vacuum result with a uniform lattice structure consisting of unit cells with constant volume fraction (Figure 29b).
- Graded lattice: The greyscale TO solution is the basis for mapping a lattice with variable volume fraction (Figure 29c).
- Scaled lattice: The rescaled greyscale TO solution is the basis for mapping a lattice with variable volume fraction (Figure 29d).
- Uniform lattice: The design domain is filled with a uniform lattice (Figure 29e).

To evaluate the performance of the proposed design strategies, the cantilever beam scheme shown in Figure 30 was used for the numerical investigation. The strategies presented were compared according to their mechanical performance in terms of total strain energy (SE). In addition, the study evaluated the designs, taking into account different manufacturing issues. The criteria analysed were support structure requirements, processing efforts, and design-to-production discrepancy.

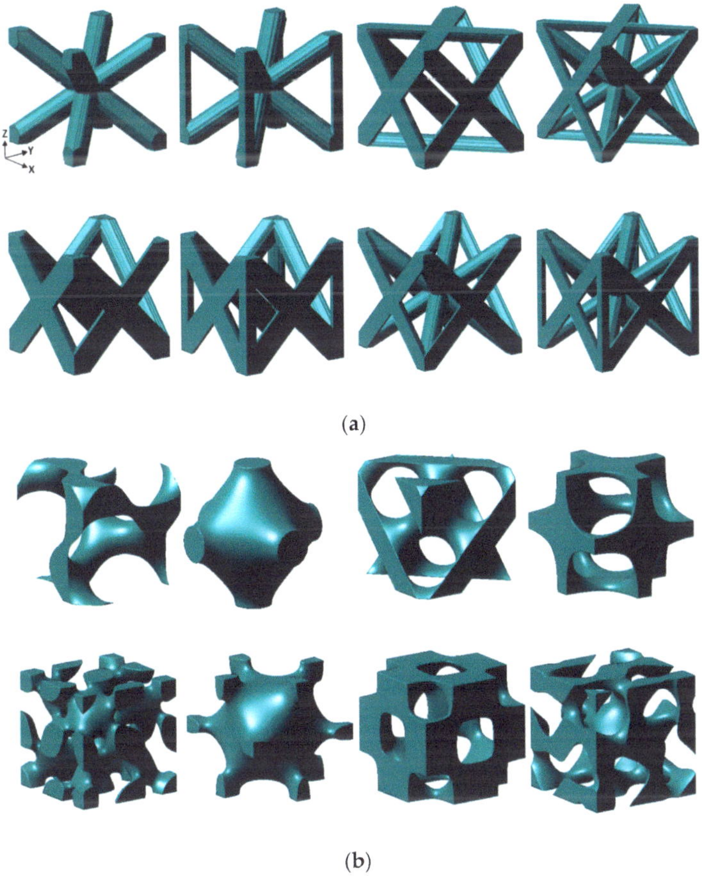

Figure 28. Examples of lattice structures: (**a**) Strut-based unit cells; (**b**) surface-based unit cells. Reprinted from ref. [150].

The numerical results, in terms of total SE, for the different design strategies applied to the cantilever beam are shown in Figure 31. The highest value of strain energy is attributed to the uniform lattice structure, which shows low stiffness.

The solid (SIMP) solution has the lowest SE. While the intersected and graded structures exhibit the same behaviour and the SE values are about 50% lower than the uniform lattice structure. The scaled lattice has a lower SE value and a better solution than the uniform lattice structures.

Regarding the production issues, Figure 32 shows a comparison of different design strategies. One of the main manufacturing considerations is related to support structures. These are essential elements to reduce the possibility of failures during the production process. However, the use of support structures affects production time, material consumption, and production efforts.

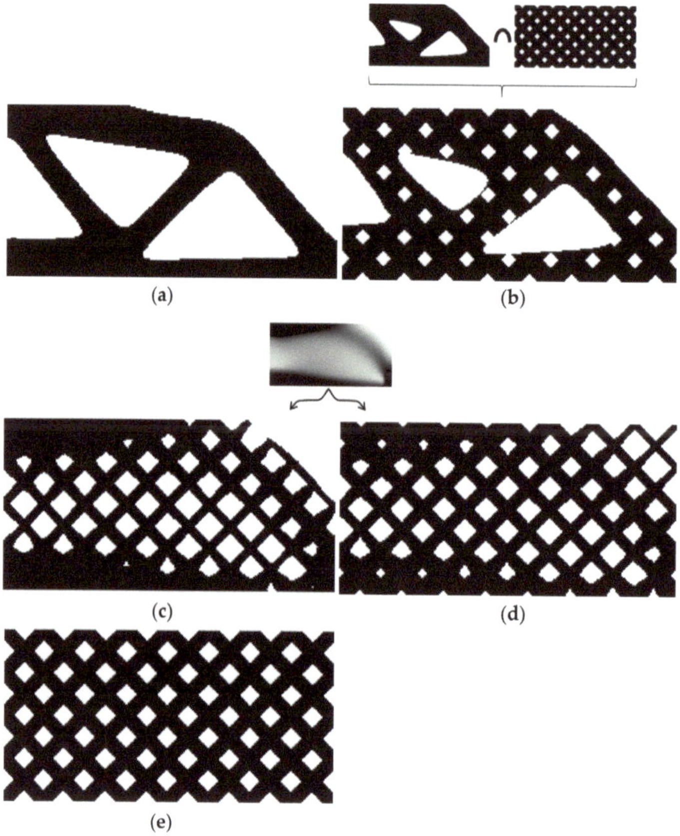

Figure 29. Overview of different lattice strategies considered in [150]: (a) Solid (SIMP) solution; (b) intersected lattice; (c) graded lattice; (d) scaled lattice; (e) uniform lattice. Reprinted from ref. [150].

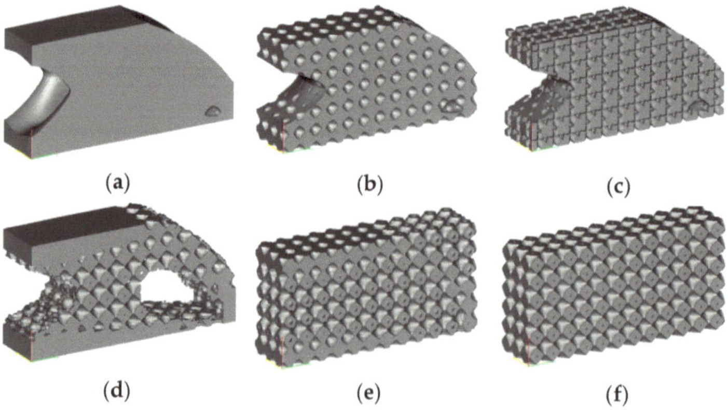

Figure 30. The models employed for the numerical investigation: (a) Solid (SIMP) solution; (b) intersected lattice of D-P; (c) intersected lattice of BCC; (d) graded lattice of D-P; (e) scaled lattice of D-P; (f) uniform lattice of D-P. Reprinted from ref. [150].

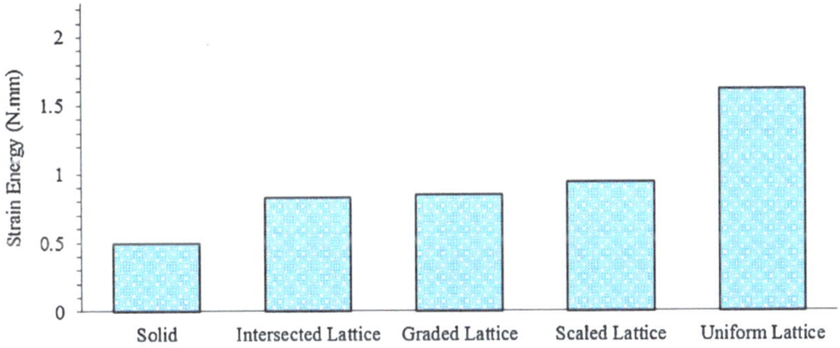

Figure 31. The total strain energy values for different design strategies. Reprinted from ref. [150].

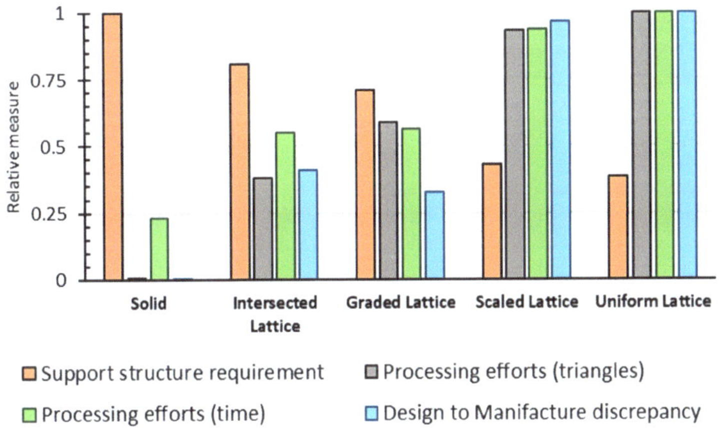

Figure 32. Manufacturing issues: comparison between different design strategies. Reprinted from ref. [150].

The solid solution is the strategy that requires the widest use of support structures. In contrast, lattice structures require less support because they make the structure self-supporting. In particular, both the uniform and scaled strategies require about 40% less support than the solid solution.

With regard to production efforts, in terms of time and geometric complexity, lattice structures are less cost-effective than the solid strategy.

Further considerations must be made about the discrepancy between design and production. For AM components, these deviations may relate to distortion due to residual stresses, and surface roughness. In particular, the latter is more common in lattice structures. This defect can lead to a decline in mechanical properties.

5.2. Use of AM in Repair of Existing Structures

One of the greatest potential of AM in construction lies in its use for the repair of existing structures [121]. In the aerospace industry, the repair of worn parts is of great interest [151–153]. Indeed, because of the production cost of high-performance components, it is cost-effective to repair worn parts rather than replace them with new ones.

The repair process consists of three phases [151]:

- Preparation stage. In the first step, considerations concern the cost-effectiveness of the repair/reinforcement of the degraded component. Next, a geometric check is performed between the worn elements and the nominal model. This comparison generates an error map, which highlights the errors between the two models. Finally, the repair area can be identified and judgements made on the extent of the damage.
- Production stage. In this stage, the previously identified area is repaired/reinforced through AM or hybrid manufacturing processes.
- Post-repair stage. In the final step, a geometric inspection is performed to verify the correct execution. In addition, the restored element can be mechanically characterised by means of material strength tests.

The same methodology can also be used in construction. Particular attention must be given to the preparatory stage. Both the choice of the appropriate process for acquiring geometric data and the judgement of the effectiveness of the repair/reinforcement are fundamental aspects. The use of objective data makes it possible to highlight the real demand for intervention. Visual inspections, which are strongly influenced by the subjective judgements of the technician, should, therefore, be avoided.

The use of noncontact digitising systems allows accurate 3D scan data to be obtained. These can be compared with the nominal model to generate an error map [152]. Therefore, through the geometry reconstruction method based on reverse engineering, structural deformations and material defects can be monitored. Subsequently, on the basis of the measurements, repair or reinforcement processes can be carried out through AM techniques.

AM technologies allow products to be printed directly onto metal components [136]. Therefore, the development of in situ stiffeners will be attractive. For example, stiffeners and components can be printed in areas where the need arises during the preparation stage. Furthermore, additional developments could concern the implementation of automated processes for monitoring and repair.

The use of AM in the repair of existing structures can have several advantages. However, to date, this is an unexplored topic. Therefore, further studies are needed to assess its potential in real applications.

In fact, compared to other sectors, construction is still lagging in the field of AM. A fundamental problem lies in the excessive fragmentation of the sector and its production processes. In the future, through the digitalisation of information related to the entire process, it will be important to aim at a design oriented towards complex components that combine several functions. One of the great potentials of AM lies in its ability to produce customised components and complex geometries without increasing costs and production efforts. Therefore, simplification consists of the production of complex components that are optimised to meet multiple functional requirements in order to enable an overall improvement in performance while reducing manufacturing errors, time, and costs.

6. Conclusions

As already discussed in the introduction, the idea of the authors was to provide a tool suitable for everyone in which all the studies were divided by topic, in order to find their way in the world of 3D printing, specifically metal.

Additive manufacturing is an innovative production technique that has been significantly developed in recent years, especially in specific sectors of engineering. Starting with rapid prototyping, the advancement of techniques led to encouraging results in terms of accuracy, precision, and mechanical performance. However, there are still limits to the large-scale application of 3D printing. They concern the production of large parts, the time and cost of making parts, the still very high costs of materials and printers, and the lack of in-depth expertise and knowledge.

Numerous studies have been recently added to the literature concerning metal additive manufacturing, which aims to provide a valid alternative to traditional manufacturing techniques. In this review article, a state-of-the-art of research and literature review on metal additive manufacturing and an overview of the main examples in the field of structural engineering were presented. Furthermore, the currently most widespread techniques were analysed in detail, as well as the process parameters and the factors influencing the quality and characteristics of the printed components. In this regard, this review referred to articles dealing with technology from a qualitative and quantitative point of view, investigating the aforementioned aspects. The performance of printed metal parts, which can have highly articulate and complex shapes, thanks also to the advanced optimisation algorithms available to date, also allows for their favourable employment in the construction sector. Several examples can already be found around the world, including bridges, joints, and nodes. However, there are still many aspects to be investigated, such as durability in the long term, defects due to the particular production process, and the real cost-effectiveness and environmental sustainability of the process. The printing process parameters to be set in the design and manufacturing stages are also both numerous and complex. The correct settings ensure the success of the printing process and, consequently, the achievement of the desired performance.

In conclusion, although there are still many aspects to be studied and explored for a large-scale application of additive manufacturing of metals in construction and beyond, the results obtained so far are remarkable and the advantages are clear. Only time and experience will allow existing limitations to be overcome, making 3D printing a commonly adopted technology in industry.

Author Contributions: Conceptualization, I.C. and G.B.; methodology, I.C. and G.B.; validation, I.C. and F.R.A.; data curation, F.R.A.; writing—original draft preparation, I.C. and F.R.A.; writing—review and editing, I.C., F.R.A. and G.B.; visualization, I.C., F.R.A. and G.B.; supervision, G.B. All authors have read and agreed to the published version of the manuscript.

Funding: This research received no external funding.

Data Availability Statement: Data sharing is not applicable.

Conflicts of Interest: The authors declare no conflict of interest.

References

1. Xiang, H.; Zhou, Y.; Zhang, X.; Li, J.; Huang, Y.; Mou, G.; Wu, C. Supportfree Printing in Laser Powder Bed Fusion: Formation Mechanisms of Discontinuity, Dross and Surface Roughness. *Opt. Laser Technol.* **2024**, *177*, 111201. [CrossRef]
2. Housholder, R.F. Molding Process. U.S. Patent 4247508, 27 January 1981.
3. Wong, K.V.; Hernandez, A. A Review of Additive Manufacturing. *ISRN Mech. Eng.* **2012**, *2012*, 208760. [CrossRef]
4. Chen, L.-Y.; Qin, P.; Zhang, L.; Zhang, L.-C. An Overview of Additively Manufactured Metal Matrix Composites: Preparation, Performance, and Challenge. *Int. J. Extrem. Manuf.* **2024**, *6*, 052006. [CrossRef]
5. Coon, C.; Pretzel, B.; Lomax, T.; Strlič, M. Preserving Rapid Prototypes: A Review. *Herit. Sci.* **2016**, *4*, 40. [CrossRef]
6. Abdulhameed, O.; Al-Ahmari, A.; Ameen, W.; Mian, S.H. Additive Manufacturing: Challenges, Trends, and Applications. *Adv. Mech. Eng.* **2019**, *11*, 1–27. [CrossRef]
7. Singh, R.; Gupta, A.; Tripathi, O.; Srivastava, S.; Singh, B.; Awasthi, A.; Rajput, S.K.; Sonia, P.; Singhal, P.; Saxena, K.K. Powder Bed Fusion Process in Additive Manufacturing: An Overview. *Mater. Today Proc.* **2019**, *26*, 3058–3070. [CrossRef]
8. Ligon, S.C.; Liska, R.; Stampfl, J.; Gurr, M.; Mülhaupt, R. Polymers for 3D Printing and Customized Additive Manufacturing. *Chem. Rev.* **2017**, *117*, 10212–10290. [CrossRef]
9. Herzog, D.; Seyda, V.; Wycisk, E.; Emmelmann, C. Additive Manufacturing of Metals. *Acta Mater.* **2016**, *117*, 371–392. [CrossRef]
10. Bhatia, A.; Sehgal, A.K. Additive Manufacturing Materials, Methods and Applications: A Review. *Mater. Today Proc.* **2021**, *81*, 1060–1067. [CrossRef]
11. Pratheesh Kumar, S.; Elangovan, S.; Mohanraj, R.; Ramakrishna, J.R. Review on the Evolution and Technology of State-of-the-Art Metal Additive Manufacturing Processes. *Mater. Today Proc.* **2021**, *46*, 7907–7920. [CrossRef]
12. Wang, Y.; Zhou, Y.; Lin, L.; Corker, J.; Fan, M. Overview of 3D Additive Manufacturing (AM) and Corresponding AM Composites. *Compos. Part A Appl. Sci. Manuf.* **2020**, *139*, 106114. [CrossRef]

13. Fijoł, N.; Aguilar-Sánchez, A.; Mathew, A.P. 3D-Printable Biopolymer-Based Materials for Water Treatment: A Review. *Chem. Eng. J.* **2022**, *430*, 132964. [CrossRef]
14. Jiang, J.; Xu, X.; Stringer, J. Support Structures for Additive Manufacturing: A Review. *J. Manuf. Mater. Process.* **2018**, *2*, 64. [CrossRef]
15. Karar, G.C.; Kumar, R.; Chattopadhyaya, S. An Analysis on the Advanced Research in Additive Manufacturing. In *Advances in Production and Industrial Engineering. Part of the Lecture Notes in Mechanical Engineering Book Series*; Springer: Singapore, 2021; pp. 229–277, ISBN 9789811555183.
16. *ISO/ASTM 52900*; Additive Manufacturing—General Principles—Terminology. ASTM—American Society for Testing and Materials: West Conshohocken, PA, USA, 2015.
17. Durai Murugan, P.; Vijayananth, S.; Natarajan, M.P.; Jayabalakrishnan, D.; Arul, K.; Jayaseelan, V.; Elanchezhian, J. A Current State of Metal Additive Manufacturing Methods: A Review. *Mater. Today Proc.* **2021**, *59*, 1277–1283. [CrossRef]
18. Shrinivas Mahale, R.; Shamanth, V.; Hemanth, K.; Nithin, S.K.; Sharath, P.C.; Shashanka, R.; Patil, A.; Shetty, D. Processes and Applications of Metal Additive Manufacturing. *Mater. Today Proc.* **2021**, *54*, 228–233. [CrossRef]
19. Badoniya, P.; Srivastava, M.; Jain, P.K.; Rathee, S. *A State-of-the-Art Review on Metal Additive Manufacturing: Milestones, Trends, Challenges and Perspectives*; Springer: Berlin/Heidelberg, Germany, 2024; Volume 46, ISBN 0123456789.
20. Lee, J.Y.; An, J.; Chua, C.K. Fundamentals and Applications of 3D Printing for Novel Materials. *Appl. Mater. Today* **2017**, *7*, 120–133. [CrossRef]
21. Ngo, T.D.; Kashani, A.; Imbalzano, G.; Nguyen, K.T.Q.; Hui, D. Additive Manufacturing (3D Printing): A Review of Materials, Methods, Applications and Challenges. *Compos. Part B Eng.* **2018**, *143*, 172–196. [CrossRef]
22. Wang, J.C.; Dommati, H.; Hsieh, S.J. Review of Additive Manufacturing Methods for High-Performance Ceramic Materials. *Int. J. Adv. Manuf. Technol.* **2019**, *103*, 2627–2647. [CrossRef]
23. Wang, Y.; Chen, R.; Liu, Y. A Double Mask Projection Exposure Method for Stereolithography. *Sens. Actuators A Phys.* **2020**, *314*, 112228. [CrossRef]
24. Mazzoli, A. Selective Laser Sintering in Biomedical Engineering. *Med. Biol. Eng. Comput.* **2013**, *51*, 245–256. [CrossRef]
25. Redwood, B.; Schoffer, F.; Garret, B. *The 3D Printing Handbook. Technologies, Design and Applications*; 3D Hubs B.V.: Amsterdam, The Netherlands, 2020.
26. Giri, J.; Sunheriya, N.; Sathish, T.; Kadu, Y.; Chadge, R.; Giri, P.; Parthiban, A.; Mahatme, C. Optimization of Process Parameters to Improve Mechanical Properties of Fused Deposition Method Using Taguchi Method. *Interactions* **2024**, *245*, 87. [CrossRef]
27. Alafaghani, A.; Qattawi, A.; Alrawi, B.; Guzman, A. Experimental Optimization of Fused Deposition Modelling Processing Parameters: A Design-for-Manufacturing Approach. *Procedia Manuf.* **2017**, *10*, 791–803. [CrossRef]
28. Hafsa, M.N.; Kassim, N.; Ismail, S.; Kamaruddin, S.A.; Hafeez, T.M.; Ibrahim, M.; Samsudin, Z.H. Study on Surface Roughness Quality of FDM and MJM Additive Manufacturing Model for Implementation as Investment Casting Sacrificial Pattern. *J. Mech. Eng.* **2018**, *5*, 25–34.
29. Emiliani, N.; Porcaro, R.; Pisaneschi, G.; Bortolani, B.; Ferretti, F.; Fontana, F.; Campana, G.; Fiorini, M.; Marcelli, E.; Cercenelli, L. Post-Printing Processing and Aging Effects on Polyjet Materials Intended for the Fabrication of Advanced Surgical Simulators. *J. Mech. Behav. Biomed. Mater.* **2024**, *156*, 106598. [CrossRef]
30. Yang, Y.; Bharech, S.; Finger, N.; Zhou, X.; Schröder, J.; Xu, B.X. Elasto-Plastic Residual Stress Analysis of Selective Laser Sintered Porous Materials Based on 3D-Multilayer Thermo-Structural Phase-Field Simulations. *npj Comput. Mater.* **2024**, *10*, 117. [CrossRef]
31. Rajesh, R.; Sudheer, S.; Kulkarni, M.V. Selective Laser Sintering Process—A Review. *Int. J. Curr. Eng. Sci. Res. (IJCESR)* **2015**, *2*, 91–100.
32. Paolini, A.; Kollmannsberger, S.; Rank, E. Additive Manufacturing in Construction: A Review on Processes, Applications, and Digital Planning Methods. *Addit. Manuf.* **2019**, *30*, 100894. [CrossRef]
33. Volpe, S.; Sangiorgio, V.; Fiorito, F.; Varum, H. Overview of 3D Construction Printing and Future Perspectives: A Review of Technology, Companies and Research Progression. *Archit. Sci. Rev.* **2022**, *67*, 1–22. [CrossRef]
34. Placzek, G.; Schwerdtner, P. Concrete Additive Manufacturing in Construction: Integration Based on Component-Related Fabrication Strategies. *Buildings* **2023**, *13*, 1769. [CrossRef]
35. Pacillo, G.A.; Ranocchiai, G.; Loccarini, F.; Fagone, M. Additive Manufacturing in Construction: A Review on Technologies, Processes, Materials, and Their Applications of 3D and 4D Printing. *Mater. Des. Process. Commun.* **2021**, *3*, e253. [CrossRef]
36. Scheel, P.; Wrobel, R.; Rheingans, B.; Mayer, T.; Leinenbach, C.; Mazza, E.; Hosseini, E. Advancing Efficiency and Reliability in Thermal Analysis of Laser Powder-Bed Fusion. *Int. J. Mech. Sci.* **2023**, *260*, 108583. [CrossRef]
37. Yap, C.Y.; Chua, C.K.; Dong, Z.L.; Liu, Z.H.; Zhang, D.Q.; Loh, L.E.; Sing, S.L. Review of Selective Laser Melting: Materials and Applications. *Appl. Phys. Rev.* **2015**, *2*, 041101. [CrossRef]
38. Song, B.; Zhao, X.; Li, S.; Han, C.; Wei, Q.; Wen, S.; Liu, J.; Shi, Y. Differences in Microstructure and Properties between Selective Laser Melting and Traditional Manufacturing for Fabrication of Metal Parts: A Review. *Front. Mech. Eng.* **2015**, *10*, 111–125. [CrossRef]
39. Du, X.; Chen, J.; She, Y.; Liu, Y.; Yang, Y.; Yang, J.; Dong, S. Effect of Process Parameter Optimization on Morphology and Mechanical Properties of Ti_6Al_4V Alloy Produced by Selective Laser Melting. *Prog. Nat. Sci. Mater. Int.* **2023**, *33*, 911–917. [CrossRef]

40. Fette, M.; Sander, P.; Wulfsberg, J.; Zierk, H.; Herrmann, A.; Stoess, N. Optimized and Cost-Efficient Compression Molds Manufactured by Selective Laser Melting for the Production of Thermoset Fiber Reinforced Plastic Aircraft Components. *Procedia CIRP* **2015**, *35*, 25–30. [CrossRef]
41. Liu, B.; Bai, P.; Li, Y. Post Treatment Process and Selective Laser Sintering Mechanism of Polymer-Coated Mo Powder. *Open Mater. Sci. J.* **2011**, *5*, 194–198. [CrossRef]
42. Andreacola, F.R.; Capasso, I.; Pilotti, L.; Brando, G. Influence of 3D-Printing Parameters on the Mechanical Properties of 17-4PH Stainless Steel Produced through Selective Laser Melting. *Frat. Intregrità Strutt.* **2021**, *58*, 282–295. [CrossRef]
43. Andreacola, F.R.; Capasso, I.; Langella, A.; Brando, G. 3D-Printed Metals: Process Parameters Effects on Mechanical Properties of 17-4PH Stainless Steel. *Heliyon* **2023**, *9*, e17698. [CrossRef]
44. Gokuldoss, P.K.; Kolla, S.; Eckert, J. Additive Manufacturing Processes: Selective Laser Melting, Electron Beam Melting and Binder Jetting-Selection Guidelines. *Materials* **2017**, *10*, 672. [CrossRef]
45. Murr, L.E.; Gaytan, S.M.; Ramirez, D.A.; Martinez, E.; Hernandez, J.; Amato, K.N.; Shindo, P.W.; Medina, F.R.; Wicker, R.B. Metal Fabrication by Additive Manufacturing Using Laser and Electron Beam Melting Technologies. *J. Mater. Sci. Technol.* **2012**, *28*, 167–177. [CrossRef]
46. Huang, N.; Cook, O.J.; Argüelles, A.P.; Beese, A.M. Review of Process–Structure–Property Relationships in Metals Fabricated Using Binder Jet Additive Manufacturing. *Metallogr. Microstruct. Anal.* **2023**, *12*, 883–905. [CrossRef]
47. Tischel, F.; Reineke, L.; Alrashdan, J.; Ploshikhin, V. Experimental Investigation and Modeling of Densification during Sintering of Binder Jetted Ti–6Al–4V. *Powder Technol.* **2024**, *444*, 119958. [CrossRef]
48. Aramian, A.; Razavi, S.M.J.; Sadeghian, Z.; Berto, F. A Review of Additive Manufacturing of Cermets. *Addit. Manuf.* **2020**, *33*, 101130. [CrossRef]
49. Mostafaei, A.; Elliott, A.M.; Barnes, J.E.; Li, F.; Tan, W.; Cramer, C.L.; Nandwana, P.; Chmielus, M. Binder Jet 3D Printing-Process Parameters, Materials, Properties, Modeling, and Challenges. *Prog. Mater. Sci.* **2021**, *119*, 100707. [CrossRef]
50. Blunk, H.; Seibel, A. Design Guidelines for Metal Binder Jetting. *Prog. Addit. Manuf.* **2024**, *9*, 725–732. [CrossRef]
51. Dwivedi, S.; Dixit, A.R.; Das, A.K.; Nag, A. A Novel Additive Texturing of Stainless Steel 316L Through Binder Jetting Additive Manufacturing. *Int. J. Precis. Eng. Manuf. Green Technol.* **2023**, *10*, 1605–1613. [CrossRef]
52. Mueller, B.; Kochan, D. Laminated Object Manufacturing for Rapid Tooling and Patternmaking in Foundry Industry. *Comput. Ind.* **1999**, *39*, 47–53. [CrossRef]
53. Kan, C.; Zhao, L.; Cao, Y.; Ma, C.; Peng, Y.; Tian, Z. Microstructure Evolution and Strengthening Behavior of Maraging Steel Fabricated by Wire Arc Additive Manufacturing at Different Heat Treatment Processes. *Mater. Sci. Eng. A* **2024**, *909*, 146804. [CrossRef]
54. Srivastava, M.; Rathee, S.; Tiwari, A.; Dongre, M. Wire Arc Additive Manufacturing of Metals: A Review on Processes, Materials and Their Behaviour. *Mater. Chem. Phys.* **2023**, *294*, 126988. [CrossRef]
55. Ahmed, N. Direct Metal Fabrication in Rapid Prototyping: A Review. *J. Manuf. Process.* **2019**, *42*, 167–191. [CrossRef]
56. Alami, A.H.; Ghani Olabi, A.; Alashkar, A.; Alasad, S.; Aljaghoub, H.; Rezk, H.; Abdelkareem, M.A. Additive Manufacturing in the Aerospace and Automotive Industries: Recent Trends and Role in Achieving Sustainable Development Goals. *Ain Shams Eng. J.* **2023**, *14*, 102516. [CrossRef]
57. Arrizubieta, J.I.; Martínez, S.; Lamikiz, A.; Ukar, E.; Arntz, K.; Klocke, F. Instantaneous Powder Flux Regulation System for Laser Metal Deposition. *J. Manuf. Process.* **2017**, *29*, 242–251. [CrossRef]
58. Pirch, N.; Linnenbrink, S.; Gasser, A.; Schleifenbaum, H. Laser-Aided Directed Energy Deposition of Metal Powder along Edges. *Int. J. Heat Mass Transf.* **2019**, *143*, 118464. [CrossRef]
59. Yilmaz, O.; Ugla, A.A. Shaped Metal Deposition Technique in Additive Manufacturing: A Review. *Proc. Inst. Mech. Eng. Part B J. Eng. Manuf.* **2016**, *230*, 1781–1798. [CrossRef]
60. Agelet de Saracibar, C.; Lundbäck, A.; Chiumenti, M.; Cervera, M. Shaped Metal Deposition Processes. In *Encyclopedia of Thermal Stresses*; Springer: Dordrecht, The Netherlands, 2014; pp. 4346–4355. [CrossRef]
61. Cooke, S.; Ahmadi, K.; Willerth, S.; Herring, R. Metal Additive Manufacturing: Technology, Metallurgy and Modelling. *J. Manuf. Process.* **2020**, *57*, 978–1003. [CrossRef]
62. Zhang, D.; Sun, S.; Qiu, D.; Gibson, M.A.; Dargusch, M.S.; Brandt, M.; Qian, M.; Easton, M. Metal Alloys for Fusion-Based Additive Manufacturing. *Adv. Eng. Mater.* **2018**, *20*, 1700952. [CrossRef]
63. Li, Y.; Liang, X.; Yu, Y.; Wang, D.; Lin, F. Review on Additive Manufacturing of Single-Crystal Nickel-Based Superalloys. *Chin. J. Mech. Eng. Addit. Manuf. Front.* **2022**, *1*, 100019. [CrossRef]
64. Mostafaei, A.; Ghiaasiaan, R.; Ho, I.T.; Strayer, S.; Chang, K.C.; Shamsaei, N.; Shao, S.; Paul, S.; Yeh, A.C.; Tin, S.; et al. Additive Manufacturing of Nickel-Based Superalloys: A State-of-the-Art Review on Process-Structure-Defect-Property Relationship. *Prog. Mater. Sci.* **2023**, *136*, 101108. [CrossRef]
65. Mazzucato, F.; Forni, D.; Valente, A.; Cadoni, E. Laser Metal Deposition of Inconel 718 Alloy and As-Built Mechanical Properties Compared to Casting. *Materials* **2021**, *14*, 437. [CrossRef]
66. Dixit, S.; Liu, S. Laser Additive Manufacturing of High-Strength Aluminum Alloys: Challenges and Strategies. *J. Manuf. Mater. Process.* **2022**, *6*, 156. [CrossRef]
67. Rometsch, P.A.; Zhu, Y.; Wu, X.; Huang, A. Review of High-Strength Aluminium Alloys for Additive Manufacturing by Laser Powder Bed Fusion. *Mater. Des.* **2022**, *219*, 110779. [CrossRef]

68. Farber, E.; Zhu, J.N.; Popovich, A.; Popovich, V. A Review of NiTi Shape Memory Alloy as a Smart Material Produced by Additive Manufacturing. *Mater. Today Proc.* **2019**, *30*, 761–767. [CrossRef]
69. Felice, I.O.; Shen, J.; Barragan, A.F.C.; Moura, I.A.B.; Li, B.; Wang, B.; Khodaverdi, H.; Mohri, M.; Schell, N.; Ghafoori, E.; et al. Wire and Arc Additive Manufacturing of Fe-Based Shape Memory Alloys: Microstructure, Mechanical and Functional Behavior. *Mater. Des.* **2023**, *231*, 112004. [CrossRef]
70. Wei, S.; Zhang, J.; Zhang, L.; Zhang, Y.; Song, B.; Wang, X.; Fan, J.; Liu, Q.; Shi, Y. Laser Powder Bed Fusion Additive Manufacturing of NiTi Shape Memory Alloys: A Review. *Int. J. Extrem. Manuf.* **2023**, *5*, 032001. [CrossRef]
71. Ostovari Moghaddam, A.; Shaburova, N.A.; Samodurova, M.N.; Abdollahzadeh, A.; Trofimov, E.A. Additive Manufacturing of High Entropy Alloys: A Practical Review. *J. Mater. Sci. Technol.* **2021**, *77*, 131–162. [CrossRef]
72. Ron, T.; Shirizly, A.; Aghion, E. Additive Manufacturing Technologies of High Entropy Alloys (HEA): Review and Prospects. *Materials* **2023**, *16*, 2454. [CrossRef]
73. Cao, L.; Li, J.; Hu, J.; Liu, H.; Wu, Y.; Zhou, Q. Optimization of Surface Roughness and Dimensional Accuracy in LPBF Additive Manufacturing. *Opt. Laser Technol.* **2021**, *142*, 107246. [CrossRef]
74. Nandhakumar, R.; Venkatesan, K. *A Process Parameters Review on Selective Laser Melting-Based Additive Manufacturing of Single and Multi-Material: Microstructure, Physical Properties, Tribological, and Surface Roughness*; Elsevier Ltd.: Amsterdam, The Netherlands, 2023; Volume 35, ISBN 5465151392629.
75. Xia, C.; Pan, Z.; Polden, J.; Li, H.; Xu, Y.; Chen, S. Modelling and Prediction of Surface Roughness in Wire Arc Additive Manufacturing Using Machine Learning. *J. Intell. Manuf.* **2022**, *33*, 1467–1482. [CrossRef]
76. Obilanade, D.; Dordlofva, C.; Törlind, P. Surface Roughness Considerations in Design for Additive Manufacturing—A Literature Review. *Proc. Des. Soc.* **2021**, *1*, 2841–2850. [CrossRef]
77. Zai, L.; Zhang, C.; Wang, Y.; Guo, W.; Wellmann, D.; Tong, X.; Tian, Y. Laser Powder Bed Fusion of Precipitation-Hardened Martensitic Stainless Steels: A Review. *Metals* **2020**, *10*, 255. [CrossRef]
78. Sefene, E.M. State-of-the-Art of Selective Laser Melting Process: A Comprehensive Review. *J. Manuf. Syst.* **2022**, *63*, 250–274. [CrossRef]
79. Gardner, L.; Kyvelou, P.; Herbert, G.; Buchanan, C. Testing and Initial Verification of the World's First Metal 3D Printed Bridge. *J. Constr. Steel Res.* **2020**, *172*, 106233. [CrossRef]
80. Anant Pidge, P.; Kumar, H. Additive Manufacturing: A Review on 3 D Printing of Metals and Study of Residual Stress, Buckling Load Capacity of Strut Members. *Mater. Today Proc.* **2020**, *21*, 1689–1694. [CrossRef]
81. Tan, J.H.; Wong, W.L.E.; Dalgarno, K.W. An Overview of Powder Granulometry on Feedstock and Part Performance in the Selective Laser Melting Process. *Addit. Manuf.* **2017**, *18*, 228–255. [CrossRef]
82. Larimian, T.; Kannan, M.; Grzesiak, D.; AlMangour, B.; Borkar, T. Effect of Energy Density and Scanning Strategy on Densification, Microstructure and Mechanical Properties of 316L Stainless Steel Processed via Selective Laser Melting. *Mater. Sci. Eng. A* **2020**, *770*, 138455. [CrossRef]
83. Chen, J.; Wang, X.; Pan, Y. Influence of Laser Power and Scan Speed on the Microstructure and Properties of GH4169 Alloy Prepared by Selective Laser Melting. *IOP Conf. Series Mater. Sci. Eng.* **2019**, *688*, 033064. [CrossRef]
84. Bremen, S.; Meiners, W.; Diatlov, A. Selective Laser Melting: A Manufacturing Technology for the Future? *Laser Tech J* **2012**, *9*, 33–38. [CrossRef]
85. Mercelis, P.; Kruth, J.P. Residual Stresses in Selective Laser Sintering and Selective Laser Melting. *Rapid Prototyp. J* **2006**, *12*, 254–265. [CrossRef]
86. Zhang, W.; Tong, M.; Harrison, N.M. Scanning Strategies Effect on Temperature, Residual Stress and Deformation by Multi-Laser Beam Powder Bed Fusion Manufacturing. *Addit. Manuf.* **2020**, *36*, 101507. [CrossRef]
87. Miao, X.; Liu, X.; Lu, P.; Han, J.; Duan, W.; Wu, M. Influence of Scanning Strategy on the Performances of GO-Reinforced Ti_6Al_4V Nanocomposites Manufactured by SLM. *Metals* **2020**, *10*, 1379. [CrossRef]
88. Haferkamp, L.; Haudenschild, L.; Spierings, A.; Wegener, K.; Riener, K.; Ziegelmeier, S.; Leichtfried, G.J. The Influence of Particle Shape, Powder Flowability, and Powder Layer Density on Part Density in Laser Powder Bed Fusion. *Metals* **2021**, *11*, 418. [CrossRef]
89. Spierings, A.B.; Herres, N.; Levy, G. Influence of the Particle Size Distribution on Surface Quality and Mechanical Properties in Additive Manufactured Stainless Steel Parts. *Rapid Prototyp. J.* **2010**, *17*, 195–202. [CrossRef]
90. Huck-Jones, D.; Langley, C. Beyond Particle Size: Exploring the Influence of Particle Shape on Metal Powder Performance. *Met. Addit. Manuf.* **2017**, *3*, 99–103.
91. Eddine, S.; Letenneur, M.; Alex, C.; Brailovski, V. Influence of Particle Morphology and Size Distribution on the Powder Flowability and Laser Powder Bed Fusion Manufacturability of Ti-6Al-4V Alloy. *Addit. Manuf.* **2020**, *31*, 100929. [CrossRef]
92. Irrinki, H.; Dexter, M.; Barmore, B.; Enneti, R.; Pasebani, S.; Badwe, S.; Stitzel, J.; Malhotra, R.; Atre, S.V. Effects of Powder Attributes and Laser Powder Bed Fusion (L-PBF) Process Conditions on the Densification and Mechanical Properties of 17-4 PH Stainless Steel. *JOM J. Miner. Met. Mater. Soc.* **2016**, *68*, 860–868. [CrossRef]
93. Rashid, R.; Masood, S.H.; Ruan, D.; Palanisamy, S.; Rahman Rashid, R.A.; Brandt, M. Effect of Scan Strategy on Density and Metallurgical Properties of 17-4PH Parts Printed by Selective Laser Melting (SLM). *J. Mater. Process. Technol.* **2017**, *249*, 502–511. [CrossRef]

94. Nguyen, Q.B.; Luu, D.N.; Nai, S.M.L.; Zhu, Z.; Chen, Z.; Wei, J. The Role of Powder Layer Thickness on the Quality of SLM Printed Parts. *Arch. Civ. Mech. Eng.* **2018**, *18*, 948–955. [CrossRef]
95. Haghdadi, N.; Laleh, M.; Moyle, M.; Primig, S. Additive Manufacturing of Steels: A Review of Achievements and Challenges. *J. Mater. Sci.* **2021**, *56*, 64–107. [CrossRef]
96. Simonelli, M.; Tse, Y.Y.; Tuck, C. Effect of the Build Orientation on the Mechanical Properties and Fracture Modes of SLM Ti–6Al–4V. *Mater. Sci. Eng. A* **2014**, *616*, 1–11. [CrossRef]
97. Guan, K.; Wang, Z.; Gao, M.; Li, X.; Zeng, X. Effects of Processing Parameters on Tensile Properties of Selective Laser Melted 304 Stainless Steel. *Mater. Des.* **2013**, *50*, 581–586. [CrossRef]
98. How Does Part Orientation Affect a 3D Print? Practical Design Tips for Additive Manufacturing. 2024. Available online: https://www.hubs.com/knowledge-base/how-does-part-orientation-affect-3d-print/ (accessed on 7 June 2024).
99. Yadollahi, A.; Shamsaei, N.; Thompson, S.M.; Elwany, A.; Bian, L. Effects of Building Orientation and Heat Treatment on Fatigue Behavior of Selective Laser Melted 17-4 PH Stainless Steel. *Int. J. Fatigue* **2017**, *94*, 218–235. [CrossRef]
100. Hitzler, L.; Janousch, C.; Schanz, J.; Merkel, M.; Heine, B.; Mack, F.; Hall, W.; Öchsner, A. Direction and Location Dependency of Selective Laser Melted AlSi10Mg Specimens. *J. Mater. Process. Technol.* **2017**, *243*, 48–61. [CrossRef]
101. Wang, C.G.; Zhu, J.X.; Wang, G.W.; Qin, Y.; Sun, M.Y.; Yang, J.L.; Shen, X.F.; Huang, S.K. Effect of Building Orientation and Heat Treatment on the Anisotropic Tensile Properties of $AlSi_{10}Mg$ Fabricated by Selective Laser Melting. *J. Alloys Compd.* **2022**, *895*, 162665. [CrossRef]
102. Sufiiarov, V.S.; Popovich, A.A.; Borisov, E.V.; Polozov, I.A.; Masaylo, D.V.; Orlov, A.V. The Effect of Layer Thickness at Selective Laser Melting. *Procedia Eng.* **2017**, *174*, 126–134. [CrossRef]
103. Wan, H.Y.; Zhou, Z.J.; Li, C.P.; Chen, G.F.; Zhang, G.P. Effect of Scanning Strategy on Mechanical Properties of Selective Laser Melted Inconel 718. *Mater. Sci. Eng. A* **2019**, *753*, 42–48. [CrossRef]
104. Giganto, S.; Zapico, P.; Castro-Sastre, M.Á.; Martínez-Pellitero, S.; Leo, P.; Perulli, P. Influence of the Scanning Strategy Parameters upon the Quality of the SLM Parts. *Procedia Manuf.* **2019**, *41*, 698–705. [CrossRef]
105. Forni, D.; Mazzucato, F.; Valente, A.; Cadoni, E. High Strain-Rate Behaviour of as-Cast and as-Build Inconel 718 Alloys at Elevated Temperatures. *Mech. Mater.* **2021**, *159*, 103859. [CrossRef]
106. Brando, G.; Andreacola, F.R.; Capasso, I.; Forni, D.; Cadoni, E. Strain-Rate Response of 3D Printed 17-4PH Stainless Steel Manufactured via Selective Laser Melting. *Constr. Build. Mater.* **2023**, *409*, 133971. [CrossRef]
107. Michla, J.R.J.; Nagarajan, R.; Krishnasamy, S.; Siengchin, S.; Ismail, S.O.; Prabhu, T.R. Conventional and Additively Manufactured Stainless Steels: A Review. *Trans. Indian Inst. Met.* **2021**, *74*, 1261–1278. [CrossRef]
108. Zhu, H.H.; Lu, L.; Fuh, J.Y.H. Study on Shrinkage Behaviour of Direct Laser Sintering Metallic Powder. *Proc. Inst. Mech. Eng. B J. Eng. Manuf.* **2006**, *220*, 183–190. [CrossRef]
109. Ramos, D.; Belblidia, F.; Sienz, J. New Scanning Strategy to Reduce Warpage in Additive Manufacturing. *Addit. Manuf.* **2019**, *28*, 554–564. [CrossRef]
110. Enneti, R.K.; Morgan, R.; Atre, S.V. Effect of Process Parameters on the Selective Laser Melting (SLM) of Tungsten. *Int. J. Refract. Met. Hard. Mater.* **2018**, *71*, 315–319. [CrossRef]
111. Klocke, F.; Wagner, C. Coalescence Behaviour of Two Metallic Particles as Base Mechanism of Selective Laser Sintering. *CIRP Ann.* **2003**, *52*, 117–180. [CrossRef]
112. Simchi, A.; Pohl, H. Effects of Laser Sintering Processing Parameters on the Microstructure and Densification of Iron Powder. *Mater. Sci. Eng. A* **2003**, *359*, 119–128. [CrossRef]
113. Attaran, M. The Rise of 3-D Printing: The Advantages of Additive Manufacturing over Traditional Manufacturing. *Bus. Horiz.* **2017**, *60*, 677–688. [CrossRef]
114. Ali, M.H.; Issayev, G.; Shehab, E.; Sarfraz, S. A Critical Review of 3D Printing and Digital Manufacturing in Construction Engineering. *Rapid Prototyp. J.* **2022**, *28*, 1312–1324. [CrossRef]
115. Hossain, M.A.; Zhumabekova, A.; Paul, S.C.; Kim, J.R. A Review of 3D Printing in Construction and Its Impact on the Labor Market. *Sustainability* **2020**, *12*, 8492. [CrossRef]
116. Buchanan, C.; Gardner, L. Metal 3D Printing in Construction: A Review of Methods, Research, Applications, Opportunities and Challenges. *Eng. Struct.* **2019**, *180*, 332–348. [CrossRef]
117. Riegger, F.; Wenzler, D.L.; Zaeh, M.F. Stud and Wire Arc Additive Manufacturing—Development of a Combined Process for the High-Productivity Additive Manufacturing of Large-Scale Lattice Structures. *J. Adv. Join. Process.* **2024**, *9*, 100189. [CrossRef]
118. Ren, S.; Galjaard, S. ; *Arup Topology Optimisation for Steel Structural Design with Additive Manufacturing. Modelling Behaviour*; Springer: Cham, Switzerland, 2015. [CrossRef]
119. Galjaard, S.; Hofman, S.; Ren, S. Optimizing Structural Building Elements in Metal by Using Additive Manufacturing. *Proc. Int. Assoc. Shell Spat. Struct.* **2015**, *2*, 1–12.
120. Galjaard, S.; Hofman, S.; Ren, S. New Opportunities to Optimize Structural Designs in Metal by Using Additive Manufacturing. In *Advances in Architectural Geometry 2014*; Springer: Cham, Switzerland, 2015. [CrossRef]
121. Gardner, L. Metal Additive Manufacturing in Structural Engineering—Review, Advances, Opportunities and Outlook. *Structures* **2023**, *47*, 2178–2193. [CrossRef]
122. MX3D Bridge. 2024. Available online: https://mx3d.com/industries/mx3d-bridge/ (accessed on 5 July 2024).
123. Connector for Takenaka. Available online: https://mx3d.com/projects/takenaka-connector/ (accessed on 5 July 2024).

124. Mechtcherine, V.; Grafe, J.; Nerella, V.N.; Spaniol, E.; Hertel, M.; Füssel, U. 3D-Printed Steel Reinforcement for Digital Concrete Construction—Manufacture, Mechanical Properties and Bond Behaviour. *Constr. Build. Mater.* **2018**, *179*, 125–137. [CrossRef]
125. Müller, J.; Grabowski, M.; Müller, C.; Hensel, J.; Unglaub, J.; Thiele, K.; Kloft, H.; Dilger, K. Design and Parameter Identification of Wire and Arc Additively Manufactured (WAAM) Steel Bars for Use in Construction. *Metals* **2019**, *9*, 725. [CrossRef]
126. Silvestru, V.A.; Ariza, I.; Taras, A. Structural Behaviour of Point-by-Point Wire Arc Additively Manufactured Steel Bars under Compressive Loading. *J. Constr. Steel Res.* **2023**, *207*, 107982. [CrossRef]
127. Silvestru, V.A.; Ariza, I.; Vienne, J.; Michel, L.; Aguilar Sanchez, A.M.; Angst, U.; Rust, R.; Gramazio, F.; Kohler, M.; Taras, A. Performance under Tensile Loading of Point-by-Point Wire and Arc Additively Manufactured Steel Bars for Structural Components. *Mater. Des.* **2021**, *205*, 109740. [CrossRef]
128. Aboulkhair, N.T.; Simonelli, M.; Parry, L.; Ashcroft, I.; Tuck, C.; Hague, R. 3D Printing of Aluminium Alloys: Additive Manufacturing of Aluminium Alloys Using Selective Laser Melting. *Prog. Mater. Sci.* **2019**, *106*, 100578. [CrossRef]
129. Seabra, M.; Azevedo, J.; Araújo, A.; Reis, L.; Pinto, E.; Alves, N.; Santos, R.; Pedro Mortágua, J. Selective Laser Melting (SLM) and Topology Optimization for Lighter Aerospace Componentes. *Procedia Struct. Integr.* **2016**, *1*, 289–296. [CrossRef]
130. Dimitrov, D.; Uheida, E.; Oosthuizen, G.; Blaine, D.; Laubscher, R.; Sterzing, A.; Blau, P.; Gerber, W.; Damm, O.F.R.A. Manufacturing of High Added Value Titanium Components. A South African Perspective. *IOP Conf. Ser. Mater. Sci. Eng.* **2018**, *430*, 012009. [CrossRef]
131. Saadlaoui, Y.; Milan, J.L.; Rossi, J.M.; Chabrand, P. Topology Optimization and Additive Manufacturing: Comparison of Conception Methods Using Industrial Codes. *J. Manuf. Syst.* **2017**, *43*, 178–186. [CrossRef]
132. Baptista, R.J.S.; Pragana, J.P.M.; Bragança, I.M.F.; Silva, C.M.A.; Alves, L.M.; Martins, P.A.F. Joining Aluminium Profiles to Composite Sheets by Additive Manufacturing and Forming. *J. Mater. Process. Technol.* **2020**, *279*, 116587. [CrossRef]
133. Guo, X.; Kyvelou, P.; Ye, J.; Teh, L.H.; Gardner, L. Experimental Study of DED-Arc Additively Manufactured Steel Double-Lap Shear Bolted Connections. *Eng. Struct.* **2023**, *281*, 115736. [CrossRef]
134. Meng, X.; Zhi, J.; Xu, F.; Gardner, L. Novel Hybrid Sleeve Connections between 3D Printed and Conventional Tubular Steel Elements. *Eng. Struct.* **2024**, *302*, 117269. [CrossRef]
135. Guo, X.; Kyvelou, P.; Ye, J.; Gardner, L. Experimental Investigation of Wire Arc Additively Manufactured Steel T-Stub Connections. *J. Constr. Steel Res.* **2023**, *211*, 108106. [CrossRef]
136. Feucht, T.; Lange, J. 3-D-Printing with Steel: Additive Manufacturing of Connection Elements and Beam Reinforcements. *ce/papers* **2019**, *3*, 343–348. [CrossRef]
137. Lange, J.; Feucht, T.; Erven, M. 3D Printing with Steel: Additive Manufacturing for Connections and Structures. *Steel Constr.* **2020**, *13*, 144–153. [CrossRef]
138. Walton, D.; Moztarzadeh, H. Design and Development of an Additive Manufactured Component by Topology Optimisation. *Procedia CIRP* **2017**, *60*, 205–210. [CrossRef]
139. Hällgren, S.; Pejryd, L.; Ekengren, J. (Re)Design for Additive Manufacturing. *Procedia CIRP* **2016**, *50*, 246–251. [CrossRef]
140. Plocher, J.; Panesar, A. Review on Design and Structural Optimisation in Additive Manufacturing: Towards next-Generation Lightweight Structures. *Mater. Des.* **2019**, *183*, 108164. [CrossRef]
141. Gebisa, A.W.; Lemu, H.G. A Case Study on Topology Optimized Design for Additive Manufacturing. *IOP Conf. Ser. Mater. Sci. Eng.* **2017**, *276*, 012026. [CrossRef]
142. Tyflopoulos, E.; Flem, D.T.; Steinert, M.; Olsen, A. State of the Art of Generative Design and Topology Optimization and Potential Research Needs. In Proceedings of the NordDesign 2018, Linköping, Sweden, 14–17 August 2018.
143. Christensen, P.; Klarbring, A. *An Introduction to Structural Optimization*; Springer: Dordrecht, The Netherlands, 2008; Volume 153.
144. Rozvany, G.I.N. A Critical Review of Established Methods of Structural Topology Optimization. *Struct. Multidiscip. Optim.* **2009**, *37*, 217–237. [CrossRef]
145. Bendsøe, M.P.; Sigmund, O. Material Interpolation Schemes in Topology Optimization. *Arch. Appl. Mech.* **1999**, *69*, 635–654. [CrossRef]
146. Allaire, G.; Jouve, F.; Toader, A.M. Structural Optimization Using Sensitivity Analysis and a Level-Set Method. *J. Comput. Phys.* **2004**, *194*, 363–393. [CrossRef]
147. Wang, M.Y.; Wang, X.; Guo, D. A Level Set Method for Structural Topology Optimization. *Comput. Methods Appl. Mech. Eng.* **2003**, *192*, 227–246. [CrossRef]
148. Maconachie, T.; Leary, M.; Lozanovski, B.; Zhang, X.; Qian, M.; Faruque, O.; Brandt, M. SLM Lattice Structures: Properties, Performance, Applications and Challenges. *Mater. Des.* **2019**, *183*, 108137. [CrossRef]
149. Leary, M.; Mazur, M.; Williams, H.; Yang, E.; Alghamdi, A.; Lozanovski, B.; Zhang, X.; Shidid, D.; Farahbod-Sternahl, L.; Witt, G.; et al. Inconel 625 Lattice Structures Manufactured by Selective Laser Melting (SLM): Mechanical Properties, Deformation and Failure Modes. *Mater. Des.* **2018**, *157*, 179–199. [CrossRef]
150. Panesar, A.; Abdi, M.; Hickman, D.; Ashcroft, I. Strategies for Functionally Graded Lattice Structures Derived Using Topology Optimisation for Additive Manufacturing. *Addit. Manuf.* **2018**, *19*, 81–94. [CrossRef]
151. Liu, R.; Wang, Z.; Sparks, T.; Liou, F.; Newkirk, J. Aerospace Applications of Laser Additive Manufacturing. In *Laser Additive Manufacturing: Materials, Design, Technologies, and Applications*; Woodhead Publishing: Sawston, UK, 2017; pp. 351–371. ISBN 9780081004333.

152. Gao, J.; Folkes, J.; Yilmaz, O.; Gindy, N. Investigation of a 3D Non-Contact Measurement Based Blade Repair Integration System. *Aircr. Eng. Aerosp. Technol.* **2005**, *77*, 34–41. [CrossRef]
153. Leino, M.; Pekkarinen, J.; Soukka, R. The Role of Laser Additive Manufacturing Methods of Metals in Repair, Refurbishment and Remanufacturing—Enabling Circular Economy. *Phys. Procedia* **2016**, *83*, 752–760. [CrossRef]

metals

MDPI

Article

Modelling and Prediction of Process Parameters with Low Energy Consumption in Wire Arc Additive Manufacturing Based on Machine Learning

Haitao Zhang [1], Xingwang Bai [1,*], Honghui Dong [1] and Haiou Zhang [2]

1 School of Mechanical Engineering, University of South China, Hengyang 421001, China; htzhang07@163.com (H.Z.)
2 School of Mechanical Science and Engineering, Huazhong University of Science and Technology, Wuhan 430074, China
* Correspondence: pancard@126.com

Abstract: Wire arc additive manufacturing (WAAM) has attracted increasing interest in industry and academia due to its capability to produce large and complex metallic components at a high deposition rate. One of the basic tasks in WAAM is to determine appropriate process parameters, which will directly affect the morphology and quality of the weld bead. However, the selection of process parameters relies heavily on empirical data from trial-and-error experiments, which results in significant time and cost expenditures. This paper employed different machine learning models, including SVR, BPNN, and XGBoost, to predict process parameters for WAAM. Furthermore, the SVR model was optimized by the Genetic Algorithm (GA) and Particle Swarm Optimization (PSO) algorithms. A 3D laser scanner was employed to obtain the weld bead's point cloud, and the weld bead size was extracted using the point cloud processing algorithm as the training data. The K-fold cross-validation strategy was applied to train and validate machine learning models. The comparison results showed that PSO–SVR predicted process parameters with the highest precision, with the RMSE, R^2, and MAE being 1.1670, 0.9879, and 0.8310, respectively. Based on the process parameters produced by PSO–SVR, an optimal process parameter combination was chosen by taking into comprehensive consideration the impacts of power consumption and efficiency. The effectiveness of the process parameter optimization method was proved through three groups of validation experiments, with the energy consumption of the first two groups decreasing by 10.68% and 11.47%, respectively.

Keywords: WAAM; process parameters; machine learning; point cloud; SVR

1. Introduction

Additive manufacturing (AM) technology has gained significant attention in recent years owing to its notable advantages, including a short lead time, reduced material waste, and the capacity to produce intricate structures. Metal additive manufacturing technology can be categorized into laser additive manufacturing (LAM), electron beam additive manufacturing (EBAM), and WAAM based on the heat sources used. Compared with LAM and EBAM, WAAM can manufacture large, complex parts at higher deposition rates and lower costs [1–3].

WAAM employs an arc as a heat source to systematically construct three-dimensional components by depositing metal material layer by layer. The entire component is generally comprised of numerous weld beads. The weld bead morphology of each layer and each pass affects the subsequent deposition and the final shape and quality of the component. In general, the weld bead morphology of each layer is mainly determined by process parameters [4]. Dinovitzer et al. [5] used the Taguchi method and analysis of variance (ANOVA) to determine the effects of welding speed (WS), wire feed speed (WFS), welding current, and argon flow rate on weld bead shape and size, and they discovered that WFS is the most significant factor. The quality of the weld bead is mainly determined by process parameters, including the welding current, welding voltage, and welding time, while the welding time is mainly related to the welding speed [6]. If the process parameters are not appropriately chosen, excessive heat input can result in deformation, high residual stresses [7], poor surface quality, and splatter phenomena [8]. Karlina et al. [9] emphasized the potential for optimizing process parameters to improve material characteristics. Thus, choosing the appropriate process parameter based on the deposition trajectory yielded by the slicing process is a critical step in the WAAM process.

Many researchers have been devoted to quantifying the association between process parameters and the response variables that characterize the morphology of the weld bead and optimizing process parameters using different mathematical methodologies, such as multiple regression analysis (MRA), finite element modelling (FEM), and machine learning. Sarathchandra et al. [10] evaluated the effects of the WS, welding current, and standoff distance on weld bead characteristics by response surface method in conjunction with ANOVA; they used MRA to establish a model between process parameters and weld bead quality. Le et al. [11] used Grey-Relational Analysis (GRA) and Techniques for Order-Preferences by Similarity-to Ideal Solution (TOPSIS) methods to determine the optimal process parameters. FEM-based modelling can investigate how process parameters affect the dimensional accuracy of the component and optimize process parameters [12,13]. Hanif et al. [14] employed FEM to study the temperature and thermal stress field in the TIG welding process and applied the GRA method to obtain the optimal weld bead geometry by comprehensively considering various factors such as welding current, shielding gas flow rate, and standoff distance on the weld bead.

Machine learning does not require any physics-based equations and only needs to use past experimental data to establish the relationship between input variables and output targets, which can quickly predict output targets. Sharma et al. [15] used three machine-learning algorithms to investigate the influence of WS, welding current, and the number of layers on weld bead morphology. Among them, random forest had the highest prediction accuracy for bead height and width, with 94% and 99% accuracy rates, respectively. Barrionuevo et al. [16] compared the performance of GPR, XGBR, and MLP algorithms in predicting melting efficiency by inputting the wire diameter, nominal power, WFS, and WS variables, finding that GPR had the highest prediction accuracy with an R^2 of 0.9190. Yaseer et al. [17] used random forest and multilayer perceptron algorithms to predict the layer surface roughness in WAAM. Both algorithms could effectively model and predict layer roughness for the same data sets. Still, the random forest was superior to the multilayer perceptron algorithm in terms of accuracy and computational efficiency. Xia et al. [18] compared the performance of different machine learning algorithms in predicting the surface roughness of weld beads. The results showed that GA–ANFIS had the optimal prediction performance, with RMSE, R^2, MAE, and MAPE values of 0.0694, 0.93516, 0.0574, and 14.15%, respectively. Wang et al. [19] established an artificial neural network (ANN) with interlayer temperature, WFS, and WS as input variables to predict the bead width, height, and contact angle in cold metal transfer welding. The average error rate of the model was less than 5.1%. Lee [20] used the Gaussian process regression method to model the process parameters, which improved the productivity of WAAM and the shape and quality of the deposits. Yadav et al. [21] used MRA, FEM, and Back Propagation Neural Network (BPNN) to establish models with WFS, WS, welding voltage, and contact tip-to-substrate distance as input variables. The response surface method guided the experimental design and generates three geometric response variables. The results showed that BPNN had higher accuracy than the MAR and FEM methods. Evidently, the data-driven machine learning model predicted the results much closer to the experimental values than the physics-driven modelling method.

From the above literature review, it can be seen that the study of weld bead geometry has generated a great deal of interest in the WAAM field. However, only some researchers have undertaken the inverse forecast of process parameters from the desired weld bead geometry. In the past, the operator had to choose the proper process parameter settings based on their experience and repeated tests. If the desired weld bead geometry can directly predict the process parameters in advance, not only can the energy consumption of the welder be reduced, but the weld bead quality can also be improved [22,23]. Venkata et al. [24] developed a method with ANN, FEM, and Taguchi-based graph theory to optimize the process parameters for ensuring dimensional accuracy in AM. Karmuhilan et al. [25] established an ANN with bead height and width as inputs and welding voltage, WFS, and WS as outputs. However, the number of response variables exceeds the number of input variables, which can easily lead to unstable prediction results, and the bead width and height are not enough to represent the geometric properties of the entire weld bead. This method of predicting process parameters by reverse modelling is ambiguous because different process parameters may result in the same weld bead geometry [26].

In order to address the issue of ambiguity in predicting process parameters through reverse modelling, this paper employed different machine learning algorithms to predict process parameters for WAAM and determined the optimal process parameters. Initially, a comparative analysis was conducted to evaluate the performance of different machine learning models in predicting process parameters. Then, the model with the highest level of prediction accuracy was utilized to optimize the process parameter. Considering the influence of power consumption and efficiency, the optimal process parameters were selected from the predicted values generated by the optimal machine learning model. Several groups of experiments were conducted utilizing the predicted process parameters to validate the accuracy of the predicted results and the effectiveness of the optimal parameters. This method avoids trial-and-error experiments, greatly shortens the time and cost expenditure, and has a certain guiding significance for selecting process parameters.

2. Methodology

2.1. Experimental Setup

The experiments were conducted on a WAAM system, as illustrated in Figure 1. It consists of a six-axis welding robot (ABB IRB1600, ABB, Zurich, Switzerland), a controller cabinet (ABB IRC5, ABB, Zurich, Switzerland), a MIG welder (SAF-FRODIGIPULS III 420, SHAF Electric, Shanghai, China), a 3D laser scanner (SR7400, SSZN, Shenzhen, China), and a control computer. The specifications for the 3D laser scanner are detailed in Table 1. The depositions were carried out on Q235 substrates with a thickness of 10 mm. Low-carbon steel materials are commonly used in WAAM due to their affordability and consistent mechanical properties. For this reason, the steel CHW-50C6 (Hantai Welding Technology, Changsha, China) with a diameter of 1.2 mm was selected as the filling material, and Q235 steel was used as the substrate. The main components of the welding wire and substrate are shown in Table 2. The shielding gas was a mixture of 20% CO_2 and 80% Ar with a flow rate of 18 L/min. The elongation of the welding wire was 12–15 mm. The welding voltage was set to be 22 V, while the welding current changed with the preset WFS.

Table 1. 3D laser scanner parameters.

Model	Scan Height	Scan X Length	Z-Axis Accuracy	X-Axis Accuracy	Single Line Points
SR7400	200 mm	240 mm	5 μm	90 μm	3200

Table 2. Composition of the used welding wire and substrate (wt. %).

Material	Composition (wt. %)					
	C	Mn	Si	S	P	Cu
CHW-50C6	0.08	1.52	0.92	0.015	0.020	0.20
Q235	≤0.17	≤1.4	≤0.35	≤0.035	≤0.035	-

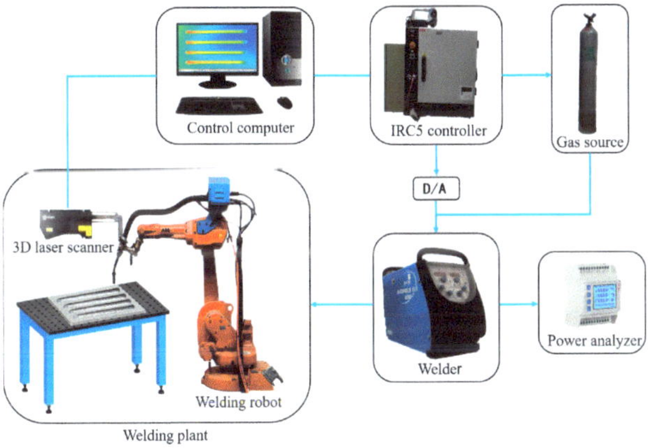

Figure 1. Schematic diagram of the experimental setup.

2.2. Experiment Design

Suitable process parameters ensure the geometric accuracy and performance of the weld. As shown in Figure 2, when the WFS/WS ratio is high, excessive metal materials are filled into the weld in a short period of time, making the molten pool larger and more susceptible to collapse. When the WFS/WS ratio is low, the amount of metal materials filled per unit time is insufficient, resulting in a spheroidization effect that makes the weld bead discontinuous, and a relatively high welding speed may result in a humped weld bead [27]. Only when the WFS/WS is in the proper range can the normal morphology of the weld bead be obtained. The main factor controlling the morphology of the weld bead is WFS/WS [28]. Hence, all possible combinations of 14 different WFS and 6 different WS were used in the experiments, as shown in Table 3. In other words, 84 weld beads were obtained by single-layer single-pass depositions if the process parameter combinations were reasonable. These weld beads were scanned to provide data sets for machine learning and subsequent process parameter optimization.

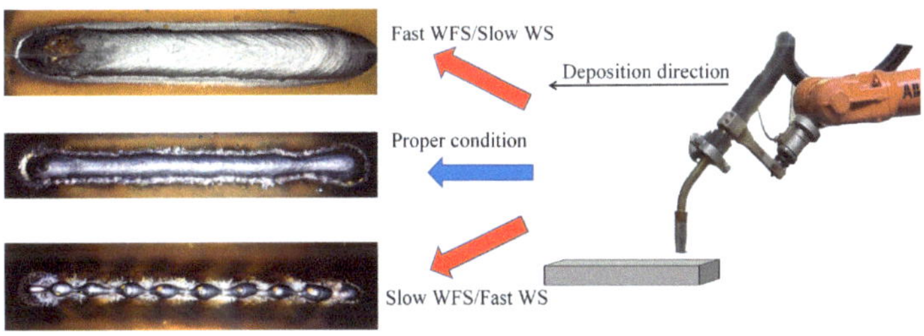

Figure 2. The effect of different WFS/WS on the weld morphology.

Table 3. Experiment design.

Parameters	Value
WFS (m/min)	3, 4, 5, 6, 7, 8
WS (mm/s)	3, 4, 5, 6, 7, 8, 9, 10, 11, 12, 13, 14, 15, 16

2.3. Data Sets Collection for Machine Learning Models

To provide data sets for predicting WFS/WS, the 3D laser scanner was used to extract the cross-section profiles of the weld bead, and then the cross-section profiles were modeled based on the mathematical function curve fitting method, which can obtain the dimensions of the weld bead. Bead width (BW) and bead height (BH) are the geometrical variables used most frequently. However, measuring the BW and BH alone is insufficient to fully characterize its morphological properties. It is necessary to introduce bead cross-section area (BCSA) as the third geometric response variable. As a result, all machine learning models used the BH, BW, and BCSA data as their input set and the corresponding process parameter WFS/WS value as their output set. Additionally, K-fold cross-validation is used to reduce the overfitting of all machine learning models.

A total of 84 welds were deposited in accordance with the experimental design. Figure 3 shows the process of acquiring data sets for machine learning. After each deposition, the point cloud of the weld bead was obtained with the 3D laser scanner. Then, point cloud processing and curve fitting were used to obtain the cross-section profiles and the values of BH, BW, and BCSA. However, certain parameter combinations resulted in unsuccessful deposition with defects such as humps and pits. Those inapplicable combinations were ruled out, and only 64 weld beads were scanned to provide data sets. The detailed procedure for extracting the cross-section profiles and curve fitting will be discussed below.

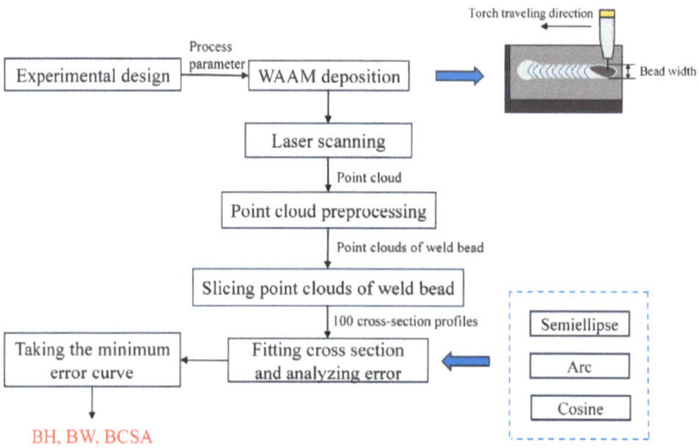

Figure 3. Schematic diagram to obtain machine learning data sets.

2.3.1. Extracting the Cross-Section Profiles of Weld Bead

As shown in Figure 4, the 3D laser scanner emitted laser stripes oriented perpendicularly to the deposition direction, and it employed the triangulation principle to obtain the coordinate information of the surface points, enabling the reconstruction of the three-dimensional morphology by the point cloud. The obtained point cloud data were processed using functions in the open-source Point Cloud Library (PCL) to extract the cross-section profiles of the weld bead. Firstly, the pass-through filter was used to segment the part of the weld to be detected. Secondly, the RANSAC algorithm was used to segment the weld bead from the substrate. Thirdly, sparse points near the weld bead were removed by applying the statistical filter to get a clean point cloud of the weld bead. Finally, 100 cross-section profiles were obtained by slicing the weld bead at 100 locations along the length direction. Generally, if the welding parameter stays constant, the cross-section profiles at different locations are similar. However, the slicing location should not be in the start and end areas of the weld bead because the ignition and quenching of the arc lead to unstable weld bead morphology. Even if not in the start and end areas, the profiles are still likely to vary slightly due to a variety of factors. Obtaining 100 cross-section profiles can significantly decrease accidental errors.

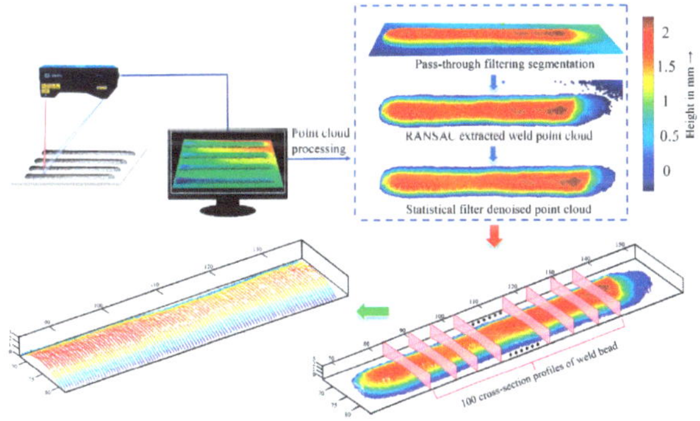

Figure 4. Schematic diagram to extract the cross-section profiles of weld bead.

2.3.2. Curve Fitting of Weld Bead Profiles

The cross-section profiles are generally arc-like curves. In many previous studies, the profiles were modeled using mathematical functions or geometrical parameters [29–31]. In this paper, the cross-section profiles of weld beads were modeled based on the methods of semi-ellipse, arc, and cosine function curve fitting, as shown in Table 4. Based on the acquired cross-section profiles, the process of curve fitting was carried out using three different mathematical models. Figure 5a shows that the semi-ellipse model accurately matched the actual profile on both sides but exhibited a downward tilt at the top, and the cosine model overshot the actual profile at the top, while the arc model best matched the actual profile at both the sides and the top.

Table 4. Mathematical model of weld bead profile.

Profile Model	Functional Model	Bead Cross-Section Area (BCSA)
Cosine	$y = a\cos(bx)$	$\int_{\frac{-w}{2}}^{\frac{w}{2}} a\cos(bx)dx$
Arc	$y = \sqrt{b^2 - x^2} - a$	$\int_{\frac{-w}{2}}^{\frac{w}{2}} (\sqrt{b^2 - x^2} - a)dx$
Semi-ellipse	$y = \sqrt{b^2 - \frac{b^2x^2}{a^2}}$	$\int_{\frac{-w}{2}}^{\frac{w}{2}} \sqrt{b^2 - \frac{b^2x^2}{a^2}}dx$

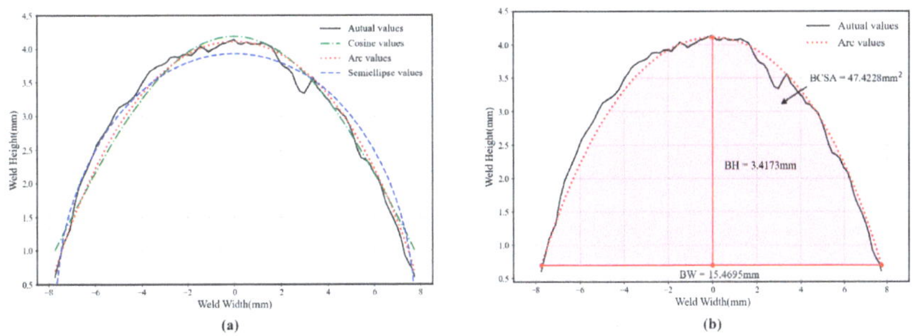

Figure 5. Weld bead modeling process. (**a**) One of the cross-section profiles and its fitted curves using three different functions; (**b**) calculating BH, BW, and BSCA after curve fitting of the arc function.

As mentioned in Section 2.3.1, 100 cross-sectional profiles with a length of 200 mm were collected and used for curve fitting to avoid accidental errors. Figure 6 shows the Root Mean Square Errors (RMSE) of the fitting curves for each profile. The arc model had the RMSE closest to 0, indicating the highest accuracy in the fitting. The semi-ellipse model showed the most significant error, particularly at the 22nd profile, where the RMSE exceeded 0.7. The cosine model's error was between the arc and semi-elliptical models.

Thus, the arc function model was chosen as the final curve fitting model to determine the BH, BW, and BCSA, as shown in Figure 5b. Since 100 cross-section profiles were extracted for each weld bead, 100 groups of values for BH, BW, and BCSA could be obtained. The average values of them were used for data sets. As mentioned above, only 64 weld beads were scanned, and the collected whole data sets are shown in Table 5. When the WS value remained constant, the BH, BW, and BCSA increased as the WFS increased. Similarly, when the WFS value remained unchanged, the BH, BW, and BCSA decreased with the increase in WS, consistent with the experimental findings reported in [6,32,33].

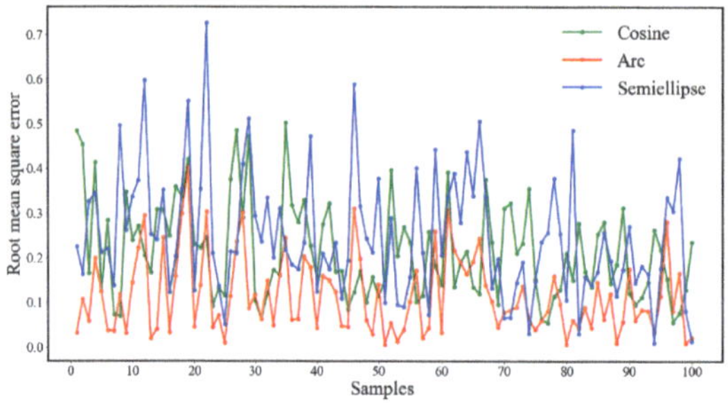

Figure 6. Root Mean Square Error of different mathematical functions.

Table 5. Data sets from experiments for each design.

No.	WFS (m/min)	WS (mm/s)	WFS/WS	BW (mm)	BH (mm)	BCSA (mm^2)	No.	WFS (m/min)	WS (mm/s)	WFS/WS	BW (mm)	BH (mm)	BCSA (mm^2)
1	3	3	16.67	7.46	2.63	14.33	33	6	8	12.50	5.82	2.45	10.76
2	4	3	22.22	9.33	2.76	18.35	34	7	8	14.58	6.54	2.78	13.71
3	5	3	27.78	10.08	3.26	23.64	35	8	8	16.67	6.89	2.95	15.36
4	6	3	33.33	11.08	3.43	27.18	36	4	9	7.41	4.04	1.91	5.99
5	7	3	38.89	12.49	3.81	34.01	37	5	9	9.26	4.65	2.21	7.96
6	8	3	44.44	13.56	3.67	35.09	38	6	9	11.11	4.85	2.44	9.30
7	3	4	12.50	6.07	2.42	10.92	39	7	9	12.96	6.09	2.65	12.23
8	4	4	16.67	8.34	2.63	15.76	40	8	9	14.81	6.38	2.81	13.63
9	5	4	20.83	8.84	2.86	18.16	41	4	10	6.67	3.77	1.94	5.77
10	6	4	25.00	9.38	3.13	21.24	42	5	10	8.33	4.68	2.12	7.59
11	7	4	29.17	10.87	3.57	27.93	43	6	10	10.00	4.73	2.19	7.96
12	8	4	33.33	11.40	3.62	29.62	44	7	10	11.67	5.80	2.59	11.46
13	3	5	10.00	4.87	2.30	8.68	45	8	10	13.33	5.98	2.70	12.38
14	4	5	13.33	6.37	2.44	11.49	46	4	11	6.06	3.65	1.93	5.60
15	5	5	16.67	7.45	2.86	15.78	47	8	11	12.12	5.55	2.76	11.99
16	6	5	20.00	7.90	2.77	15.94	48	4	12	5.56	3.34	1.87	5.06
17	7	5	23.33	9.22	3.25	21.87	49	5	12	6.94	3.78	1.98	5.94
18	8	5	26.67	10.09	3.35	24.41	50	6	12	8.33	4.00	1.99	6.25
19	3	6	8.33	4.24	2.17	7.27	51	7	12	9.72	4.95	2.15	8.06
20	4	6	11.11	5.24	2.14	8.38	52	8	12	11.11	5.21	2.49	10.09
21	5	6	13.89	6.48	2.72	13.24	53	5	13	6.41	3.66	1.93	5.63
22	6	6	16.67	7.10	2.75	14.47	54	6	13	7.69	5.42	2.04	8.13
23	7	6	19.44	7.93	3.06	17.99	55	7	13	8.97	4.54	2.14	7.50
24	8	6	22.22	8.97	3.22	21.11	56	8	13	10.26	5.11	2.38	9.37
25	3	7	7.14	3.95	2.05	6.40	57	5	14	5.95	3.37	1.93	5.32
26	4	7	9.52	4.87	2.21	8.26	58	6	14	7.14	4.68	1.89	6.59
27	5	7	11.90	5.69	2.48	10.74	59	7	14	8.33	4.27	2.05	6.80
28	6	7	14.29	6.31	2.57	12.12	60	8	14	9.52	4.76	2.34	8.69
29	7	7	16.67	7.22	2.95	15.92	61	7	15	7.78	4.25	2.07	6.87
30	3	8	6.25	3.57	1.92	5.50	62	8	15	8.89	4.71	2.33	8.60
31	4	8	8.33	4.50	2.10	7.27	63	7	16	7.29	3.86	1.97	6.01
32	5	8	10.42	5.37	2.34	9.52	64	8	16	8.33	4.38	2.25	10.76

2.4. Process Parameters Optimization Procedure

The procedure for process parameter optimization is illustrated in Figure 7. Three steps are included as follows:

(1) Prediction of WFS/WS: Firstly, the BH and BW were set up to create the desired surface morphology, and the BSCA value was then calculated according to the arc mathematical function. Subsequently, the trained machine learning model was used to predict WFS/WS. The primary reason for choosing WFS/WS was to ensure the deposition quality. As discussed in Section 2.2, the morphology and success of the deposition were highly related to WFS/WS.

(2) Calculation of candidate process parameters: Referring to the welding-feasible region diagram obtained from the previous experiment (Figure 8), multiple sets of process parameter combinations were generated by cyclic iteration and taken as candidate process parameters. Figure 8 demonstrates that even when the WFS/WS value is identical, the formation quality of the weld bead will have significant variations if the WFS or WS value is not chosen correctly. Hence, it was crucial to compute the WFS and WS values based on the summarized range of the welding-feasible area.

(3) Choosing optimal process parameters among candidates: A machine learning model was established with process parameters as input to forward predict the BW and BH. The difference between the predicted and preset values was analyzed. If the error exceeded 5%, the corresponding combination would be removed. Finally, the optimal parameter combination was selected by maximizing the effective deposition volume per power (EDVP).

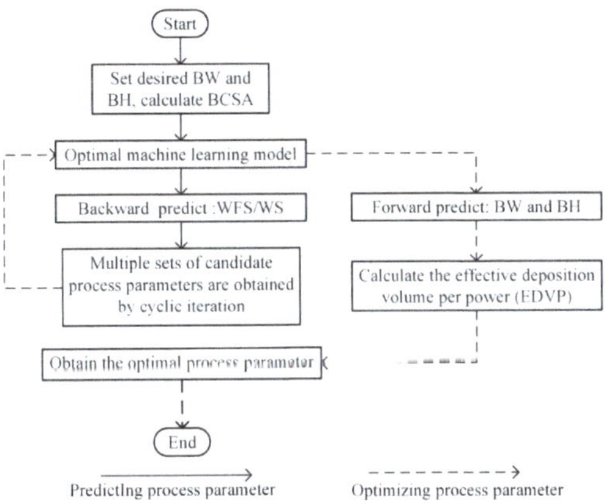

Figure 7. Flow chart of process parameter optimization.

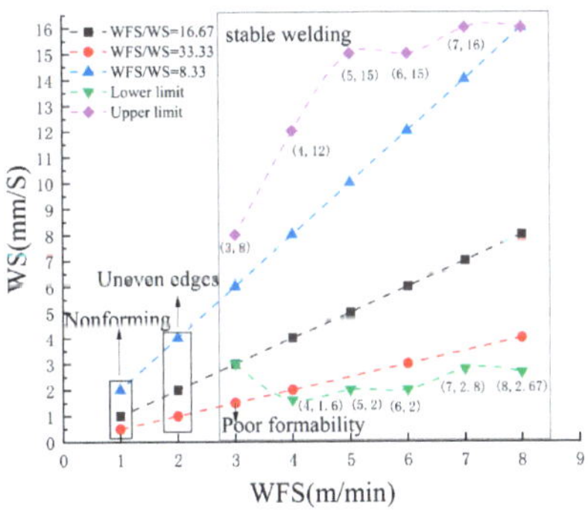

Figure 8. Diagram of a welding-feasible region.

It is worth mentioning that multiple ML algorithms were used for process parameter planning. The performances of those algorithms are compared, as presented in Section 4.1. The optimal algorithm was then chosen to perform the process parameter planning for the experiments, as shown in Section 4.3.

EDVP is an index regarding the influence of power consumption and efficiency. The value of EDVP can be calculated based on the following Equations (1)–(5).

$$V_{\text{effective}} = S \times L = BCSA \times L \tag{1}$$

where $V_{\text{effective}}$, S, and L represent the effective deposition volume of the deposited sample, the effective area, and the sample effective length, respectively. The effective area is $BSCA$, which the arc mathematical function calculates.

$$V_{\text{total}} = WFS \times T \times \pi \times r^2 = WFS \times \frac{L}{WS} \times \pi \times r^2 \tag{2}$$

where V_{total}, WFS, and T represent the total deposited volume, the wire feeding speed, and the duration of each welding cycle, determined by the total length and welding speed. And r indicates the wire radius, which is 0.6 mm in the experiments. The formula for computing the effective deposition rate (EDR) is as follows:

$$EDR(\%) = \frac{V_{\text{effective}}}{V_{\text{total}}} \times 100\% \tag{3}$$

The calculation of the effective deposition rate per power ($EDRP$) is performed using the following formula:

$$EDRP(\%/W) = EDR/P \tag{4}$$

where P represents the power of the welding supply. The formula for computing the effective deposition volume per power ($EDVP$) is as follows:

$$EDVP(\text{mm}^3/\text{W}) = EDRP \times V_{\text{effective}} \tag{5}$$

3. Machine Learning Algorithms

3.1. Support Vector Regression

Support vector regression (SVR) is a powerful machine learning algorithm widely used in regression analysis. Its principle is to use the kernel function to map input data to high-dimensional space to solve the optimal hyperplane and transform the nonlinear relationship between input and output variables into a linear relationship for prediction [34]. The expression of the SVR model is as follows:

$$f(x) = \sum_{i=1}^{m} (\alpha_i - \alpha_i^*) K(x_i, x_j) + b \tag{6}$$

where α_i and α_i^* are Lagrange multipliers; b denotes intercept; $K(x_i, x_j)$ is the kernel function. In order to improve the nonlinear processing ability of the SVR model, the Gaussian radial basis function (RBF) is usually used as the kernel function. The RBF function is as follows:

$$K(x_i, x_j) = \exp\left(-\frac{\| x_i - x_j \|^2}{g^2}\right) \tag{7}$$

In the SVR model, the penalty factor C and RBF function parameter g directly determine the prediction performance of the model. The generalization of the model is influenced by the parameter C, while the training speed of the model is influenced by the parameter g. Therefore, it is important to reasonably select appropriate values for these two parameters. This study employs search ranges of (0.1, 200) for the C and (0.001, 1) for the g.

3.2. XGBoost

XGBoost reduces the error of the previous prediction by continuously generating new regression trees, gradually narrows the gap between the true value and the predicted value, and ultimately improves the model's prediction accuracy [35]. The prediction function of XGBoost is as follows:

$$\hat{y}_i = \sum_{k=1}^{K} f_k(x_i), f_k \in \Gamma \tag{8}$$

where $\hat{y}_i$ the predicted output value of the ith sample; K is the total number of regression trees; f_k is the predicted value of the kth model in the ith sample; Γ is the space of the regression tree. The objective function is as follows:

$$Obj = \sum_{i=1}^{n} l(y_i, \hat{y}_i) + \sum_{k=1}^{K} \Omega\left(f_k\right) \tag{9}$$

where Obj is the objective function; $\sum_{i=1}^{n} l(y_i, \hat{y}_i)$ is the loss function, indicating the fitting degree of the model; $\Omega\left(f_k\right)$ is a penalty function to reduce the risk of overfitting. The number of decision trees, the maximum depth of the tree, and the learning rate are the three key hyperparameters of XGBoost. In this paper, the search range of the number of decision trees, the maximum depth of the tree, and the learning rate are [50, 100, 200, 400, 500], [3, 4, 5, 6, 7, 8, 9, 10], and [0.01, 0.05, 0.1, 0.15, 0.2], respectively.

3.3. Back Propagation Neural Network

BPNN is a multi-layer feedforward neural network with error back-propagation. Through the back-propagation of errors, the weights and thresholds of the network are constantly adjusted to minimize the mean square error of the network [36]. As shown in Figure 9, the structure of the BP neural network model consists of an input layer, a hidden layer, and an output layer. In this experiment, the input layer nodes are BW, BH, and BCSA, respectively, and the output layer nodes are WFS and WS. When there is information input, the input information is sent to the input node, and after the hidden layer is processed by the function, it is sent to the output node. When the neural network has no activation function, the final output result is linear, and nonlinear data prediction cannot be performed [37]. Thus, it is necessary to add activation functions to the neural network model. This paper used the sigmoid function, the tanh function, and the ReLU function as the activation functions of the hidden layer. The neuron nodes of the hidden layer were in the range of 10–40, and different optimization algorithm solvers (SGD, LBFGS, Adam) were used to find the optimal weight to minimize the loss function. The number of iterations was 500.

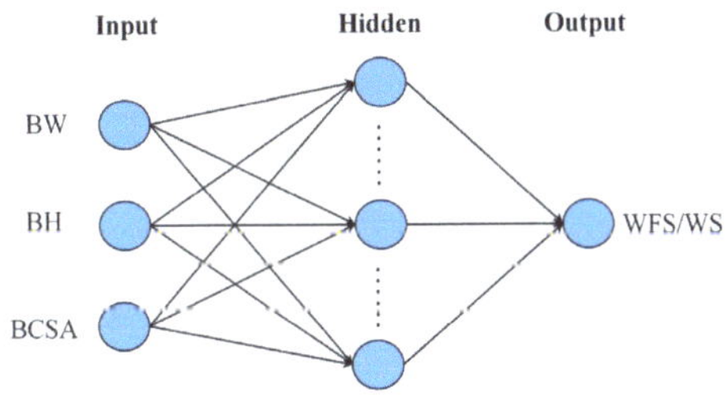

Figure 9. Back Propagation Neural Network architecture.

3.4. Machine Learning Tools

This paper mainly uses PyCharm software (Community Edition 2022.2.2) to write machine learning code based on Python. Python is an open-source programming language widely used to develop machine learning algorithms. PyCharm software provides powerful code analysis tools to help developers improve development efficiency and code quality. The essential libraries, including NumPy, Pandas, Scikit-learn, XGboost, and Matplotlib, were imported into PyCharm before the machine learning codes and data visualization were executed.

3.5. Evaluation of Machine Learning Algorithms

To better evaluate the established machine learning model, this paper employs metrics such as Mean Absolute Error (MAE), Root Mean Square Error (RMSE), and the Coefficient of Determination (R^2) for evaluation and comparative analysis. The MAE is usually used to weigh the absolute error between the actual and predicted values of the model. The disadvantage of MAE is that it only considers the average of absolute errors and is not sensitive to outliers. The RMSE is used to weigh the deviation between predicted and actual values. The R^2 evaluates the prediction accuracy of different models. Among them, the smaller the MAE and RMSE values, the closer the model predicted value is to the actual value, the closer the R^2 value is to 1, and the better the model fitting performance. The expression of the three metrics is as follows:

$$MAE = \frac{1}{n}\sum_{i=1}^{n} \mid y_i - \hat{y}_i \mid \tag{10}$$

$$RMSE = \sqrt{\frac{1}{n}\sum_{m}^{i=1} (y_i - \overline{y}_i)^2} \tag{11}$$

$$R^2 = 1 - \frac{\sum_{i=1}^{n} (y_i - \hat{y}_i)^2}{\sum_{i=1}^{n} (y_i - \overline{y})^2} \tag{12}$$

where y_i, $\hat{y}_i$, $\overline{y}$, and n represent the actual value, the predicted value, the average of actual values, and the total of input data, respectively.

4. Results and Discussion

4.1. Comparing Different Machine Learning Models Prediction Results

Multiple algorithms, including SVR, XGBoost, and BPNN, were employed to predict the process parameters of WAAM. The K-fold cross-validation method was used to train and validate the model using the data sets obtained from the experiments. K-fold cross-validation is a resampling procedure employed to evaluate the precision of models when dealing with limited data sets [38]. This method randomly divides the original data sets into K discrete subsets. One of the subsets is designated as the validation data set, while the remaining K−1 subsets are dedicated to training. This process is repeated K times, and each subset takes turns as the validation set. In this study, the value of K was selected as 5, and 80% of the data sets were used for training, while the remaining 20% were reserved for model regression testing. The prediction results of different machine learning models were compared, and the model with the highest prediction accuracy was selected to plan process parameters.

The grid search is used to find the hyperparameters of the above machine learning, and the results are shown in Table 6. As shown in Figure 10, the predicted value of SVR was in good agreement with the actual value, while XGBoost had the lowest prediction accuracy. BPNN displayed signs of overfitting during the training process, resulting in unstable prediction outcomes. Although most points predicted by BPNN were consistent, the individual point deviation was relatively substantial, making the overall prediction accuracy not ideal. As shown in Table 7, the R^2 value of the BPNN test set was 0.9170, while the R^2 value of the training set was 0.9961. The R^2 value of the test set was lower than the R^2 value of the training set, and there was a potential overfitting phenomenon. The RMSE values of SVR, XGBoost, and BPNN were 1.8087, 2.0739, and 3.0545, respectively. The SVR achieved a higher prediction accuracy, but it still fell short of meeting the accuracy requirements for welding process parameter planning. Consequently, there was a decision to optimize SVR through hyper-parameter tuning to enhance prediction accuracy and stability.

Table 6. Optimal hyper-parameters in different machine learning algorithms.

Machine Learning Model	Parameter	Value	Parameter	Value	Parameter	Value
SVR	kernel	RBF	C	100	Gamma	0.0001
XGBoost	max_depth	6	learning_rate	0.05	n_estimators	400
BPNN	activation	tanh	Hidden layer sizes	10	solver	lbfgs

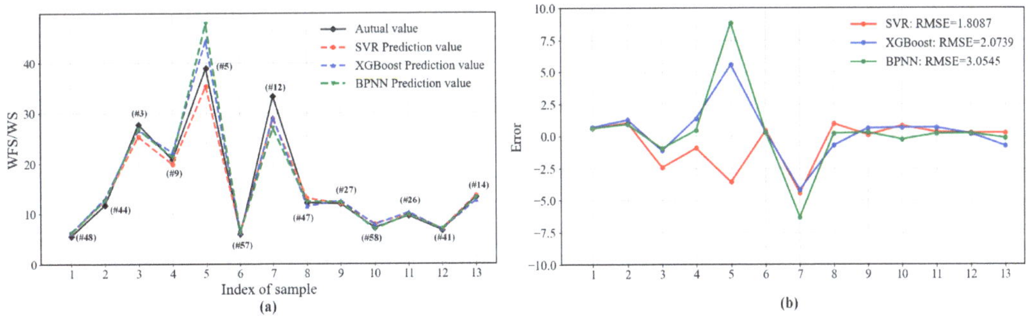

Figure 10. Predictions of different machine learning. (**a**) Comparison of the predicted and actual values of WFS/WS (# represents the corresponding serial number in Table 5); (**b**) error analysis of WFS/WS prediction.

Table 7. Performance comparison of training sets and test sets in different machine learning algorithms.

	Training			Testing		
	RMSE	MAE	R^2	RMSE	MAE	R^2
SVR	1.5165	0.8077	0.9663	1.8087	1.2413	0.9709
XGBoost	2.6030	1.2569	0.9196	2.0739	1.3916	0.9617
BPNN	0.6478	0.4667	0.9961	3.0545	1.5200	0.9170

4.2. Comparing the Effect of GA and PSO on SVR

The traditional hyper-parameter tuning methods mainly use manual tuning, grid searching, and random searching [39]. For example, the above three machine learning models derive their hyperparameters through grid search. However, this method has corresponding defects, such as a long calculation time, an inability to deal with continuous parameters, ignoring the correlation between parameters, etc. Grid search is unsuitable for non-convex optimization problems and may also produce overfitting. Random search offers higher computational efficiency and broader coverage of the search space than grid search. However, it still exhibits instability with fewer data samples and cannot ensure a globally optimal solution exists. As a result, this study employs PSO and GA to optimize SVR parameters in the training scheme for complicated issues to improve the global search capabilities of the model and circumvent the local optimum, as shown in Figure 11.

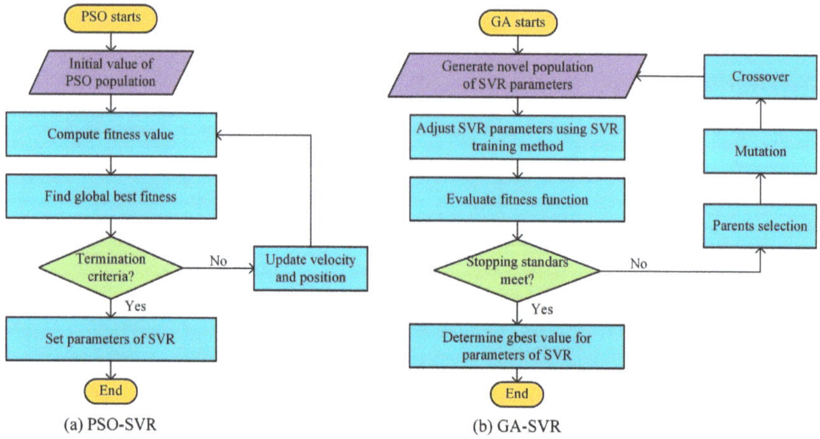

Figure 11. Scheme of GA–SVR and PSO–SVR algorithm. (**a**) PSO–SVR; (**b**) GA–SVR.

The primary parameters of the PSO algorithm include the number of particle swarms, the maximum number of iterations, the acceleration constants C1 and C2, and the weight coefficient. The main parameters of the GA algorithm include population size, the maximum number of iterations, crossover rate, mutation rate, and DNA size. Population size significantly impacts the performance and effectiveness of GA and PSO. Therefore, this study observed the performance of PSO–SVR and GA–SVR under different population sizes using the trial-and-error method, as shown in Table 8.

Table 8. Performance of the model for various population sizes.

Population Size	PSO–SVR Result		GA–SVR Results	
	RMSE	R^2	RMSE	R^2
20	2.0522	0.9614	3.3094	0.9025
30	2.0830	0.9690	3.1986	0.9090
40	1.8269	0.9703	3.0348	0.9180
50	1.8464	0.9697	2.7042	0.9349
60	1.1670	0.9879	2.6308	0.9384
70	1.2653	0.9858	2.5409	0.9425
80	1.5696	0.9781	2.1823	0.9576
90	1.5841	0.9777	2.6461	0.9377

In the PSO–SVR algorithm, it can be observed that the increase in population size from 20 to 60 corresponded to the decrease in RMSE value. However, when the population increased from 60 to 90, the RMSE value increased. Therefore, PSO–SVR performed the best prediction when the population size was 60. In the GA–SVR algorithm, the RMSE gradually reduced as the population size rose from 20 to 80. When the population was 80, the RMSE value reached its lowest value. As a result, the ultimate population sizes for PSO and GA were determined to be 60 and 50, respectively. In addition, other parameters were also determined by the trial-and-error method, as shown in Table 9. The performance of PSO–SVR and GA–SVR is shown in Figures 12 and 13, respectively. Among them, PSO–SVR achieved the highest predictive accuracy, with RMSE values of 1.1670. The RMSE value of GA–SVR was 2.1823. Compared with the normal SVR in Figure 10, PSO–SVR can obtain better prediction performance. On the contrary, GA–SVR's prediction accuracy was lower than the normal SVR's, which did not play an optimization role.

Table 9. Model parameters of GA–SVR and PSO–SVR.

PSO		GA	
Number of particle swarm	60	Population size	80
Maximum number of iterations	220	Maximum number of iterations	700
Cognitive acceleration C1	1.5	Crossover rate	0.85
Social acceleration C2	3	Mutation rate	0.097
Initial inertia weight W1	0.85	DNA size	25

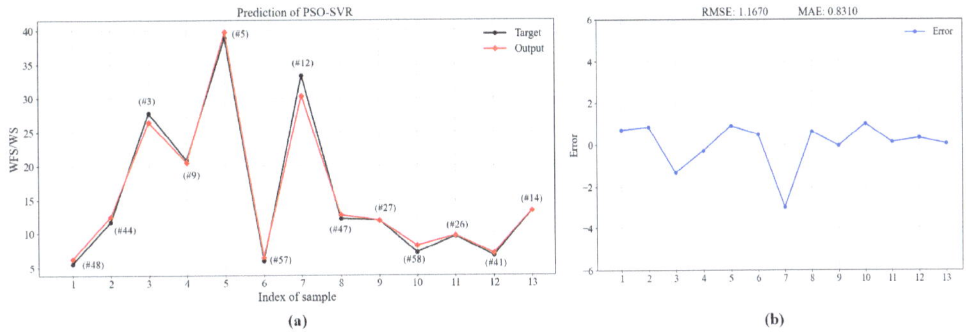

Figure 12. Prediction of PSO–SVR. (**a**) Comparison of the predicted and actual values of WFS/WS (# represents the corresponding serial number in Table 5); (**b**) error analysis of WFS/WS prediction.

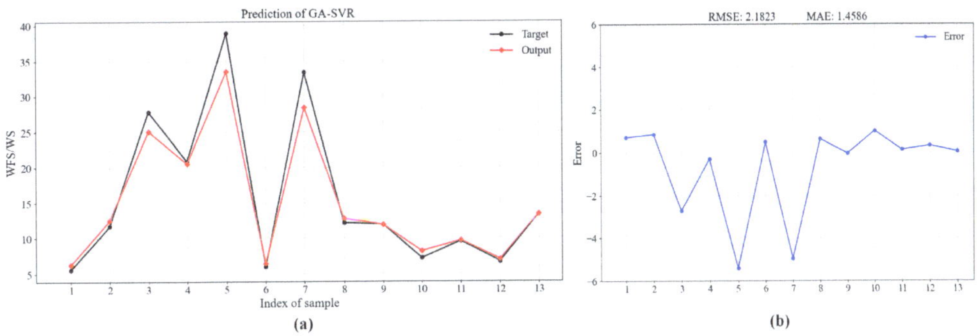

Figure 13. Prediction of GA–SVR. (**a**) Comparison of the predicted and actual values of WFS/WS; (**b**) error analysis of WFS/WS prediction.

In order to facilitate a more comprehensive comparison of the results, it was necessary to calculate the R^2 of the prediction model using Equation (12). As shown in Figure 14, the correlation between the actual WFS/WS and the predicted WFS/WS values generated using different machine learning models is intuitively depicted. The distance between the surrounding scattered data points and the fitted line (representing the actual WFS/WS) shows the degree of correlation between the model-predicted value and the observed value. During the training and testing of the predictive models using K-fold cross-validation, the R^2 values of SVR, XGBoost, BPNN, PSO–SVR, and GA–SVR were 0.9709, 0.9617, 0.9170, 0.9879, and 0.9576, respectively. The results show that the correlation coefficient R^2 value of PSO–SVR was closer to 1 than that of GA–SVR, SVR, XGBoost, and BPNN. It can be inferred that among all the applied machine learning models, the PSO–SVR performed better in predicting the welding process parameters (WFS/WS) in WAAM. In addition, the PSO–SVR was also used to forward predict the bead height and width.

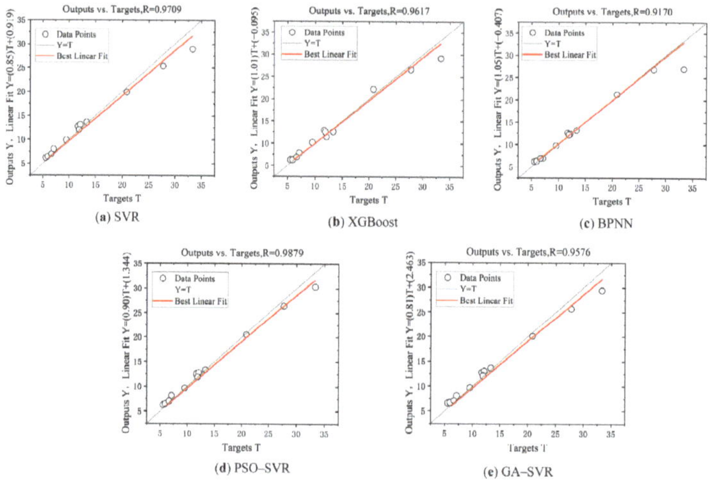

Figure 14. Regression plot for different models 'predictions. (**a**) SVR; (**b**) XGBoost; (**c**) BPNN; (**d**) PSO–SVR; (**e**) GA–SVR.

4.3. Validation Experiment

To assess the efficacy of the process parameter optimization method, this study conducted three sets of verification experiments. The first group consisted of single-layer single-pass experiments, while the second group involved multi-layer single-pass curve experiments, and the third group is multi-layer multi-pass experiments. The process parameter optimization was performed based on the predicted results of the PSO–SVR algorithm, which demonstrated the highest prediction performance, as discussed in Section 4.2. In the experimental verification section, we used the power analyzer to measure the energy consumption of all candidate processes in the actual deposition process. The percentage of maximum energy savings can be calculated by comparing the energy consumption of the optimized process parameters with other parameters. The formula is as follows:

$$\eta = \frac{E_{\max} - E_o}{E_{\max}} \times 100\% \quad (13)$$

where E_o is the actual energy consumption of the optimized process parameters, and $E_{\max}$ is the largest energy consumption value among other process parameters.

4.3.1. Single-Layer Single-Pass Deposition

Single-layer single-pass experiments used a sample weld with bead dimensions of 11.4 mm in width, 3.6 mm in height, and 180 mm in length to verify the effectiveness of the process parameter optimization method. The BCSA can be determined as 29.43 mm^2 using the arc mathematical function. Firstly, the trained PSO–SVR machine learning model was utilized to input the BH (3.6 mm), BW (11.4 mm), and BCSA (29.43 mm^2) to obtain a WFS/WS value of 33.008. Secondly, 12 sets of candidate process parameters can be determined by cyclic iteration based on the welding-feasible region diagram in Figure 8. Thirdly, these candidate process parameters were input to forward predict the bead width and height based on the PSO–SVR model and compare the error between the predicted size and the preset value. If the comprehensive error of the height and width of the weld bead exceeded 5%, the parameter was removed from the candidate process parameters. As shown in Figure 15c, the total error of the first set of process parameters was 7.03%, which did not meet the requirements of the error range. Lastly, considering the relationship between EDVP and welder energy consumption, the optimal process parameters were selected from the candidate process parameters by maximizing EDVP. The calculation of EDVP can be determined by Equations (1)–(5). As shown in Table 10, it can be observed that the group labelled No. 7 exhibited the highest value for the EDVP, indicating that it represents the optimal combination of process parameters. Conversely, the group identified as No. 2 displayed the lowest value for the EDVP, indicating that it represents the worst combination of process parameters. Subsequently, experiments were conducted on individual candidate process parameters, and the welding energy consumption during the deposition process was measured with a power analyzer to verify the accuracy of the optimal process parameters.

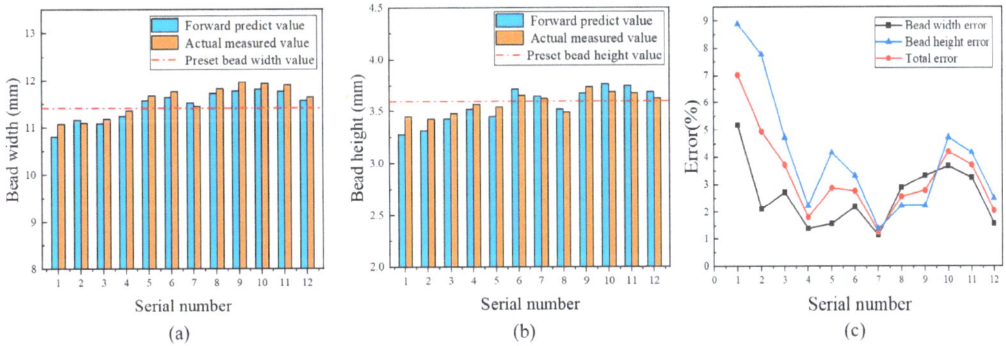

Figure 15. The comparison of preset the size of weld bead, forward predicted bead size, and actual measured size. (**a**) Comparison of weld bead width; (**b**) comparison of weld bead height; (**c**) error analysis between preset value and forward predictive value.

Table 10. Single-layer single-pass experimental design and results.

No.	WFS (m/min)	WS (mm/s)	BW (mm)	BH (mm)	BCSA (mm^2)	EDRP (%W)	EDVP (mm^3/W)	E (Wh)
1	3	1.5	10.81	3.28	-	-	-	-
2	3.5	1.75	11.16	3.32	26.17	0.005184	24.4178	76.2173
3	4	2	11.09	3.43	27.21	0.005237	25.6443	74.4883
4	4.5	2.25	11.21	3.52	28.31	0.005270	26.8870	72.2443
5	5	2.5	11.58	3.45	28.44	0.005120	26.2076	72.4333
6	5.5	2.75	11.65	3.72	31.13	0.005264	29.4933	70.9679
7	6	3	11.53	3.65	30.19	0.005564	30.2357	68.1638
8	6.5	3.25	11.73	3.52	29.42	0.004829	25.5683	75.4061
9	7	3.5	11.78	3.68	31.04	0.004950	27.6606	71.8503
10	7.5	3.75	11.82	3.77	32.00	0.004961	28.5747	71.3896
11	8	4	11.77	3.75	31.69	0.004654	26.5502	72.3520
12	8.5	4.25	11.58	3.69	30.68	0.004667	25.7723	73.1400

The results of the weld bead deposition utilizing the candidate process parameters are shown in Figure 16. It can be seen that the actual weld morphology and quality under different process parameter combinations were similar, aligning with the expected setting. As shown in Figure 15a,b, the actual measured bead width and height value fluctuated roughly around the preset value, and the predicted value was mostly close to the actual value. The power analyzer record indicates that the minimum energy consumption of the No. 7 group was 68.1638 Wh, while the maximum energy consumption of the No. 2 group was recorded at 76.2173 Wh. The process parameter combination of the minimum energy consumption index was consistent with the previously planned process parameter combination. Under the equivalent welding quality and morphology conditions, the optimized process parameters can save up to 10.68% of energy (Equation (13)).

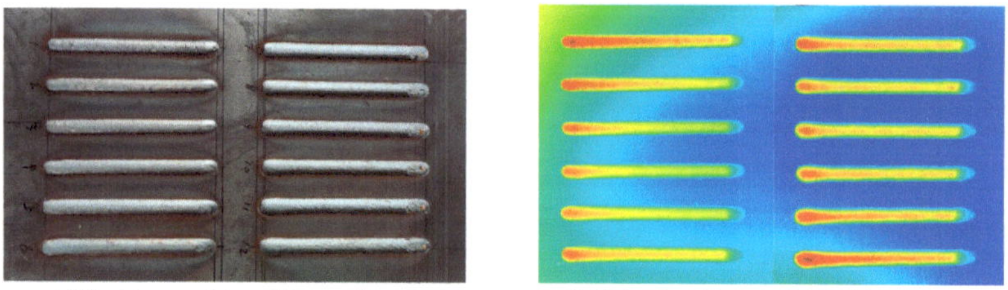

(a) Real image of deposition results (b) Point cloud image of deposition results

Figure 16. Single-layer single-pass deposition diagram (The serial number in subfigure (**a**) corresponds to Table 10).

4.3.2. Multi-Layer Single-Pass Curve Deposition

Multi-layer single-pass curve deposition was conducted to assess the effectiveness of the process parameter planning method for complex-shaped components. Firstly, the preset width of the weld bead was 7 mm, the height of the weld bead was 3 mm, the length of the curve weld bead was 800 mm, and the number of bead layers was 10. The BCSA can be calculated to be 15.88 mm^2 by the arc mathematical function. The WFS/WS predicted by the trained PSO-SVR model was 16.42. Secondly, 10 sets of candidate process parameters can be obtained by cyclic iteration, as shown in Table 11. Thirdly, the optimal process parameters obtained by the process parameter optimization method were in the No. 4 group. Then, experimental validation was conducted, whereby distinct candidate process parameter combinations were utilized for each layer of multi-layer single-pass curve deposition. The outcomes of the deposition are shown in Figure 17. The measurement results of the power analyzer show that the lowest energy consumption of No. 4 was 145.73 Wh, which is consistent with the planned combination of process parameters. In this multi-layer single-pass verification experiment, the process parameter planning method can save up to 11.95% of energy by Equation (13), which once again verifies the effectiveness of the process parameter optimization method.

Table 11. Multi-layer single-pass experimental design and results.

No.	WFS (m/min)	WS (mm/s)	BW (mm)	BH (mm)	BCSA (mm^2)	EDRP (%W)	EDVP (mm^3/W)	E (Wh)
1	3	3	6.64	2.80	13.98	0.005952	66.5557	149.17
2	4	4	7.59	2.57	14.13	0.005833	65.9297	154.56
3	4.5	4.5	7.62	2.60	14.37	0.005871	67.4775	147.13
4	5	5	7.58	2.73	15.14	0.005674	68.7044	145.73
5	5.5	5.5	7.40	2.78	15.16	0.005383	65.2717	155.69
6	6	6	7.36	2.84	15.48	0.004871	60.3053	158.91
7	6.5	6.5	7.25	2.92	15.80	0.004399	55.6153	161.51
8	7	7	7.12	2.90	15.45	0.003946	48.7675	162.53
9	7.5	7.5	7.06	2.97	15.79	0.003848	48.6187	164.14
10	8	8	7.02	3.04	16.18	0.003635	47.0386	165.50

(a) Real image of deposition results (b) Point cloud image of deposition results

Figure 17. Multi-layer single pass curve deposition diagram.

4.3.3. Multi-Layer Multi-Pass Deposition

To further verify the actual effect of the process parameter optimization method in this paper, multi-layer multi-pass verification experiments were carried out. Three $100 \times 60 \times 8$ mm^3 cuboid components were deposited in the multi-layer multi-pass experiment, and the process parameters #4, #7, and #12 in Table 10 from the single-layer single-pass experiment were respectively used for the experiment. The process parameters in #7 are the optimal combination of process parameters after optimization. The process parameters of the three groups of experiments were predicted by the same size of weld bead shape. It can be seen that the bead width and height of the actual deposited single bead were basically the same as the preset value in Section 4.3.1, and the bead height and width were 3.6 mm and 11.40 mm, respectively. According to the size of the single bead, to realize the formation of $100 \times 60 \times 8$ mm^3 cuboid components, each layer needed to deposit five welds, the overlap rate between adjacent welds was 60%, and the Z-shaped reciprocating deposition had three layers, as shown in Figure 18a. During the deposition process, the power analyzer was used to record the energy consumption and time of each group of experiments, and each layer was cooled for 3–5 min after the end of the deposition to avoid excessive heat accumulation. The 3D laser scanner can also be utilized to measure the geometric size of each layer of weld bead in the cooling period, as shown in Figure 18b. As shown in Figure 19, it can be found that the surface shape and quality of the weld bead after deposition of the process parameters of #7 were better than those of the other two sets of parameters. For example, the surface of the weld bead in Figure 19a was uneven, and even pores appeared, indicating that the selection of reasonable process parameters can even play a role in inhibiting defects. As shown in Table 12, the size of the deposited components of #7 process parameters was closest to the expected size, and the time and energy consumption of the process were significantly lower than those of the other two groups of experiments. In terms of energy consumption, #7 was 8.45% and 7.48% lower than #4 and #12, respectively. It was proved again that the process parameter optimization method in this paper can save energy consumption, reduce production costs, and achieve green manufacturing.

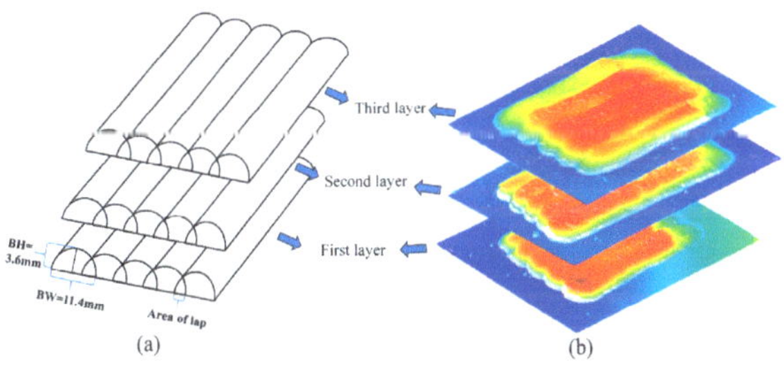

Figure 18. (**a**) Multi-layer multi-pass deposition schematic diagram; (**b**) the point cloud diagram of #7 process parameter deposition results measured by the 3D laser scanner.

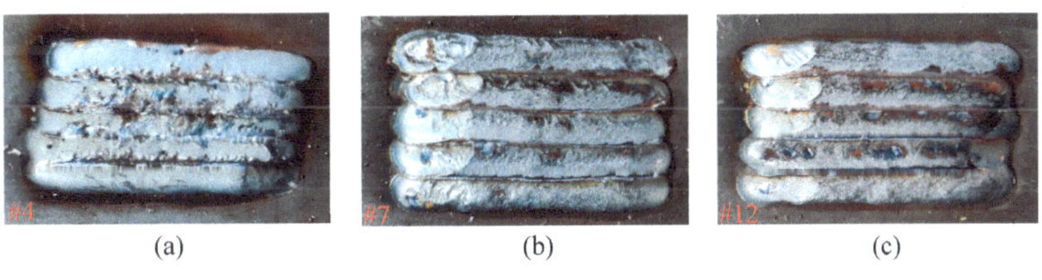

Figure 19. Multi-layer multi-pass deposition results. (**a**) #4 process parameter deposition results; (**b**) #7 process parameter deposition results; (**c**) #12 process parameter deposition results.

Table 12. Multi-layer multi-pass experimental results.

Group	WFS (m/min)	WS (mm/s)	Length (mm)	Width (mm)	Heigh (mm)	Processing Time (min)	E (Wh)
#4	4.50	2.25	96.96	58.83	8.54	10.53	549.38
#7	6.00	3.00	97.85	63.31	7.52	5.75	502.94
#12	8.50	4.25	100.24	61.37	8.16	5.00	543.62

5. Conclusions

This paper presents a method of process parameter optimization for WAAM based on machine learning. Different machine learning algorithms were utilized and compared in terms of performance. With machine learning models, the WFS/WS ratio was first predicted, and then the optimal process parameters were chosen by taking into account the impacts on power consumption and efficiency. The following conclusions can be drawn from this study:

(1) Not only bead width (BW) and bead height (BH) but also Bead Cross-Section Area (BCSA) were used as geometric response variables in machine learning models. To calculate BCAS quickly, three mathematical functions were utilized to describe the profile of weld beads. Among them, the arc mathematical function was the closest to the actual cross-sectional profile, and the fitting accuracy was the highest, followed by the semi-elliptic and cosine functions.
(2) K-fold cross-validation was used to assess the prediction performance of the machine learning models to maximize the use of training data. The results revealed that the SVR model had the highest prediction accuracy, with an RMSE of 1.8087 and an R^2 of 0.9709. Conversely, XGBoost demonstrated the lowest accuracy. Notably, BPNN tends to overfit when working with small sample data sets, resulting in lower prediction accuracy for the test set than the training set.
(3) To enhance the performance of the SVR, GA and PSO were applied to optimize the parameters of the SVR. The results showed that PSO–SVR has the highest prediction performance among the developed models, with an RMSE of 1.1670 and an R^2 of 0.9879. Compared with SVR, the prediction accuracy is greatly improved.
(4) The selection of the optimal process parameter considering the effective deposition volume per power can reduce the welding energy consumption to some extent. The optimized process parameters in the first single-layer single-pass experiment can save up to 10.68% energy. In the multi-layer single-bead validation experiment, the optimized parameters realized energy savings of up to 11.47%. The third set of verification experiments further verified the effectiveness of the process parameter optimization method.

In the future, more machine learning algorithms can be introduced and trained with larger amounts of data to improve the accuracy and robustness of predicting. By combining the slicing program with this parameter planning, a fully automatic WAAM system can be developed to achieve higher efficiency and lower energy consumption.

Author Contributions: Conceptualization, X.B. and H.Z. (Haiou Zhang); methodology, H.Z. (Haitao Zhang), X.B. and H.D.; software, H.Z. (Haitao Zhang) and H.D.; validation, H.Z. (Haitao Zhang); formal analysis, H.Z. (Haitao Zhang), and X.B.; investigation, H.Z. (Haitao Zhang); resources, H.Z. (Haiou Zhang) and X.B.; data curation, H.Z. (Haitao Zhang); writing—original draft preparation, H.Z. (Haitao Zhang); writing—review and editing, X.B.; visualization, H.Z. (Haitao Zhang) and X.B.; supervision, X.B.; project administration, H.Z. (Haitao Zhang) and X.B.; funding acquisition, X.B. All authors have read and agreed to the published version of the manuscript.

Funding: This work was supported by the Natural Science Foundation of Hunan Province (Grant No. 2023JJ30510), National Natural Science Foundation of China (Grant No. 51975270), key project of the Education Department of Hunan Province, China (Grant No. 21A0257).

Data Availability Statement: The original contributions presented in the study are included in the article, further inquiries can be directed to the corresponding author.

Conflicts of Interest: The authors declare no conflicts of interest.

References

1. Chakraborty, D.; Tirumala, T.; Chitral, S.; Sahoo, B.N.; Kiran, D.V.; Kumar, P.A. The State of the Art for Wire Arc Additive Manufacturing Process of Titanium Alloys for Aerospace Applications. *J. Mater. Eng. Perform.* **2022**, *31*, 6149–6182. [CrossRef]
2. Cunningham, C.R.; Flynn, J.M.; Shokrani, A.; Dhokia, V.; Newman, S.T. Invited Review Article: Strategies and Processes for High Quality Wire Arc Additive Manufacturing. *Addit. Manuf.* **2018**, *22*, 672–686. [CrossRef]
3. IvánTabernero; Paskual, A.; Álvarez, P.; Suárez, A. Study on Arc Welding Processes for High Deposition Rate Additive Manufacturing. *Procedia CIRP* **2018**, *68*, 358–362. [CrossRef]
4. Sharma, S.K.; Maheshwari, S. Arc Characterization Study for Submerged Arc Welding of HSLA (API X80) Steel. *J. Mech. Sci. Technol.* **2017**, *31*, 1383–1390. [CrossRef]
5. Dinovitzer, M.; Chen, X.; Laliberte, J.; Huang, X.; Frei, H. Effect of Wire and Arc Additive Manufacturing (WAAM) Process Parameters on Bead Geometry and Microstructure. *Addit. Manuf.* **2019**, *26*, 138–146. [CrossRef]
6. Papacharalampopoulos, A.; Sabatakakis, K.; Stavropoulos, P. Quality Monitoring of RSW Processes. The Impact of Vibrations. *Procedia CIRP* **2023**, *118*, 895–900. [CrossRef]
7. Gurmesa, F.D.; Lemu, H.G. Literature Review on Thermomechanical Modelling and Analysis of Residual Stress Effects in Wire Arc Additive Manufacturing. *Metals* **2023**, *13*, 526. [CrossRef]
8. Jafari, D.; Vaneker, T.H.J.; Gibson, I. Wire and Arc Additive Manufacturing: Opportunities and Challenges to Control the Quality and Accuracy of Manufactured Parts. *Mater. Des.* **2021**, *202*, 109471. [CrossRef]
9. Karlina, A.I.; Karlina, Y.I.; Kondratiev, V.V.; Kononenko, R.V.; Breki, A.D. Study of Wear of an Alloyed Layer with Chromium Carbide Particles after Plasma Melting. *Crystals* **2023**, *13*, 1696. [CrossRef]
10. Sarathchandra, D.T.; Davidson, M.J.; Visvanathan, G. Parameters Effect on SS304 Beads Deposited by Wire Arc Additive Manufacturing. *Mater. Manuf. Process.* **2020**, *35*, 852–858. [CrossRef]
11. Le, V.T.; Doan, Q.T.; Mai, D.S.; Bui, M.C.; Tran, H.S.; Van Tran, X.; Nguyen, V.A. Prediction and Optimization of Processing Parameters in Wire and Arc-Based Additively Manufacturing of 316L Stainless Steel. *J. Braz. Soc. Mech. Sci. Eng.* **2022**, *44*, 394. [CrossRef]
12. Srivastava, S.; Garg, R.K.; Sharma, V.S.; Sachdeva, A. Measurement and Mitigation of Residual Stress in Wire-Arc Additive Manufacturing: A Review of Macro-Scale Continuum Modelling Approach. *Arch. Comput. Methods Eng.* **2021**, *28*, 3491–3515. [CrossRef]
13. Geng, R.; Du, J.; Wei, Z.; Xu, S.; Ma, N. Modelling and Experimental Observation of the Deposition Geometry and Microstructure Evolution of Aluminum Alloy Fabricated by Wire-Arc Additive Manufacturing. *J. Manuf. Process.* **2021**, *64*, 369–378. [CrossRef]
14. Hanif, M.; Shah, A.H.; Shah, I.; Mumtaz, J. Optimization of Bead Geometry during Tungsten Inert Gas Welding Using Grey Relational and Finite Element Analysis. *Materials* **2023**, *16*, 3732. [CrossRef] [PubMed]
15. Sharma, R.; Raj Paul, A.; Mukherjee, M.; Ram Krishna Vadali, S.; Kumar Singh, R.; Kumar Sharma, A. Forecasting of Process Parameters Using Machine Learning Techniques for Wire Arc Additive Manufacturing Process. *Mater. Today Proc.* **2023**, *80*, 248–253. [CrossRef]
16. Barrionuevo, G.O.; Sequeira-Almeida, P.M.; Ríos, S.; Ramos-Grez, J.A.; Williams, S.W. A machine learning approach for the prediction of melting efficiency in wire arc additive manufacturing. *Int. J. Adv. Manuf. Technol.* **2022**, *120*, 3123–3133. [CrossRef]
17. Yaseer, A.; Chen, H. Machine Learning Based Layer Roughness Modeling in Robotic Additive Manufacturing. *J. Manuf. Process.* **2021**, *70*, 543–552. [CrossRef]
18. Xia, C.; Pan, Z.; Polden, J.; Li, H.; Xu, Y.; Chen, S. Modelling and prediction of surface roughness in wire arc additive manufacturing using machine learning. *J. Intell. Manuf.* **2021**, *33*, 1467–1482. [CrossRef]
19. Wang, Z.; Zimmer-Chevret, S.; Léonard, F.; Abba, G. Prediction of Bead Geometry with Consideration of Interlayer Temperature Effect for CMT-Based Wire-Arc Additive Manufacturing. *Weld. World* **2021**, *65*, 2255–2266. [CrossRef]
20. Lee, S.H. Optimization of Cold Metal Transfer-Based Wire Arc Additive Manufacturing Processes Using Gaussian Process Regression. *Metals* **2020**, *10*, 461. [CrossRef]
21. Yadav, A.; Srivastava, M.; Jain, P.K.; Rathee, S. Investigation of Bead Morphology and Mechanical Behaviour for Metal Inert Gas Welding-Based WAAM in Pulsed Mode Metal Transfer on 316LSi Stainless Steel. *J. Adhes. Sci. Technol.* **2024**, *38*, 738–769. [CrossRef]
22. Oh, W.-J.; Lee, C.-M.; Kim, D.-H. Prediction of Deposition Bead Geometry in Wire Arc Additive Manufacturing Using Machine Learning. *J. Mater. Res. Technol.* **2022**, *20*, 4283–4296. [CrossRef]
23. Xia, C.; Pan, Z.; Polden, J.; Li, H.; Xu, Y.; Chen, S.; Zhang, Y. A Review on Wire Arc Additive Manufacturing: Monitoring, Control and a Framework of Automated System. *J. Manuf. Syst.* **2020**, *57*, 31–45. [CrossRef]
24. Venkata Rao, K.; Parimi, S.; Suvarna Raju, L.; Suresh, G. Modelling and Optimization of Weld Bead Geometry in Robotic Gas Metal Arc-Based Additive Manufacturing Using Machine Learning, Finite-Element Modelling and Graph Theory and Matrix Approach. *Soft Comput.* **2022**, *26*, 3385–3399. [CrossRef]

25. Karmuhilan, M.; Sood, A.K. Intelligent Process Model for Bead Geometry Prediction in WAAM. *Mater. Today Proc.* **2018**, *5*, 24005–24013. [CrossRef]
26. Hu, Z.; Qin, X.; Li, Y.; Ni, M. Welding Parameters Prediction for Arbitrary Layer Height in Robotic Wire and Arc Additive Manufacturing. *J. Mech. Sci. Technol.* **2020**, *34*, 1683–1695. [CrossRef]
27. Yuan, L.; Pan, Z.; Ding, D.; He, F.; Van Duin, S.; Li, H.; Li, W. Investigation of Humping Phenomenon for the Multi-Directional Robotic Wire and Arc Additive Manufacturing. *Robot. Comput. -Integr. Manuf.* **2020**, *63*, 101916. [CrossRef]
28. Yildiz, A.S.; Davut, K.; Koc, B.; Yilmaz, O. Wire arc additive manufacturing of high-strength low alloy steels: Study of process parameters and their influence on the bead geometry and mechanical characteristics. *Int. J. Adv. Manuf. Technol.* **2020**, *108*, 3391–3404. [CrossRef]
29. Chen, C.; He, H.; Zhou, J.; Lian, G.; Huang, X.; Feng, M. A Profile Transformation Based Recursive Multi-Bead Overlapping Model for Robotic Wire and Arc Additive Manufacturing (WAAM). *J. Manuf. Process.* **2022**, *84*, 886–901. [CrossRef]
30. Ding, D.; Pan, Z.; Cuiuri, D.; Li, H. A Multi-Bead Overlapping Model for Robotic Wire and Arc Additive Manufacturing (WAAM). *Robot. Comput. -Integr. Manuf.* **2015**, *31*, 101–110. [CrossRef]
31. Chen, C.; He, H.; Zhou, S.; Lian, G.; Huang, X.; Feng, M. Prediction of Multi-Bead Profile of Robotic Wire and Arc Additive Manufactured Components Recursively Using Axisymmetric Drop Shape Analysis. *Virtual Phys. Prototyp.* **2023**, *18*, e2254563. [CrossRef]
32. Lambiase, F.; Scipioni, S.I.; Paoletti, A. Accurate Prediction of the Bead Geometry in Wire Arc Additive Manufacturing Process. *Int. J. Adv. Manuf. Technol.* **2022**, *119*, 7629–7639. [CrossRef]
33. Zhao, Y.; Li, F.; Chen, S.; Lu, Z. Unit Block–Based Process Planning Strategy of WAAM for Complex Shell–Shaped Component. *Int. J. Adv. Manuf. Technol.* **2019**, *104*, 3915–3927. [CrossRef]
34. Castro-Neto, M.; Jeong, Y.-S.; Jeong, M.-K.; Han, L.D. Online-SVR for Short-Term Traffic Flow Prediction under Typical and Atypical Traffic Conditions. *Expert Syst. Appl.* **2009**, *36*, 6164–6173. [CrossRef]
35. Chen, T.; He, T.; Benesty, M.; Khotilovich, V.; Tang, Y.; Cho, H. Xgboost: Extreme gradient boosting. *R Package Version 0.4–2* **2015**, *1*, 1–4.
36. Hecht-Nielsen, R. Theory of the backpropagation neural network. In Proceedings of the International Joint Conference on Neural Networks, Washington, DC, USA, 18–22 June 1989; Volume 1, pp. 593–605.
37. Nawi, N.M.; Khan, A.; Rehman, M.Z. A New Back-Propagation Neural Network Optimized with Cuckoo Search Algorithm. In *Computational Science and Its Applications—ICCSA 2013: 13th International Conference, Ho Chi Minh City, Vietnam, 24–27 June 2013*; Lecture Notes in Computer, Science; Murgante, B., Misra, S., Carlini, M., Torre, C.M., Nguyen, H.-Q., Taniar, D., Apduhan, B.O., Gervasi, O., Eds.; Springer: Berlin/Heidelberg, Germany, 2013; Volume 7971, pp. 413–426. ISBN 978-3-642-39636-6.
38. Anandan, B.; Manikandan, M. Machine Learning Approach with Various Regression Models for Predicting the Ultimate Tensile Strength of the Friction Stir Welded AA 2050-T8 Joints by the K-Fold Cross-Validation Method. *Mater. Today Commun.* **2023**, *34*, 105286. [CrossRef]
39. Yu, T.; Zhu, H. Hyper-parameter optimization: A review of algorithms and applications. *arXiv* **2020**, arXiv:2003.05689.

metals

MDPI

Article

Effect of Powder Reuse on Powder Characteristics and Properties of DED Laser Beam Metal Additive Manufacturing Process with Stellite® 21 and UNS S32750

Juan Carlos Pereira [1,*], Uxue Irastorza [1], Ane Solana [2], Carlos Soriano [2], David García [3], José Exequiel Ruiz [4] and Aitzol Lamikiz [4]

1 LORTEK Technological Centre, Basque Research and Technology Alliance BRTA, Arranomendia Kalea 4A, 20211 Ordizia, Gipuzkoa, Spain; uirastorza@lortek.es
2 TEKNIKER, Basque Research and Technology Alliance BRTA, Calle Iñaki Goenaga 5, 20600 Eibar, Gipuzkoa, Spain; ane.solana@tekniker.es (A.S.); carlos.soriano@tekniker.es (C.S.)
3 AZTERLAN Metallurgy Research Centre, Basque Research and Technology Alliance BRTA, Aliendalde Auzunea 6, 48200 Durango, Bizkaia, Spain; dgarcia@azterlan.es
4 Department of Mechanical Engineering, University of the Basque Country UPV/EHU, 48013 Bilbao, Bizkaia, Spain; joseexequiel.ruiz@ehu.eus (J.E.R.); aitzol.lamikiz@ehu.eus (A.L.)
* Correspondence: jcpereira@lortek.es; Tel.: +34-943-882-303

Abstract: In this work, the influence of powder reuse up to three times on directed energy deposition (DED) with laser processing has been studied. The work was carried out on two different gas atomized powders: a cobalt-based alloy type Stellite® 21, and a super duplex stainless steel type UNS S32750. One of the main findings is the influence of oxygen content of the reused powder particles on the final quality and densification of the deposited material and the powder catch efficiency of the laser deposition process. There is a direct relationship between a higher surface oxidation of the particles and the presence of oxygen content in the particles and in the as-built materials, as well as oxides, balance of phases (in the case of the super duplex alloy), pores and defects at the micro level in the laser-deposited material, as well as a decrease in the amount of material that actually melts, reducing powder catch efficiency (more than 12% in the worst case scenario) and the initial bead geometry (height and width) that was obtained for the same process parameters when the virgin powder was used (without oxidation and with original morphology of the powder particles). This causes some melting faults, oxides and formation of undesired oxide compounds in the microstructure, and un-balance of phases particularly in the super duplex stainless steel material, reducing the amount of ferrite from 50.1% to 37.4%, affecting in turn material soundness and its mechanical properties, particularly the hardness. However, the Stellite® 21 alloy type can be reused up to three times, while the super duplex can be reused only once without any major influence of the particles' surface oxidation on the deposited material quality and hardness.

Keywords: powder reuse; additive manufacturing; directed energy deposition; laser metal deposition; Stellite® 21 alloy; super duplex stainless steel

1. Introduction

One of the most promising metal additive manufacturing (AM) technologies for large and multimaterial components is directed energy deposition and blown metal powder (DED-LB/M powder), whose acronym has recently been established in the ASTM 52900 standard guidelines [1], and which is also known as Laser Metal Deposition (LMD) in the industrial sector. The main barrier to the adoption of DED-LB/M powder is the efficiency of the material deposition, because due to the nature and the complex dynamics of the process, not all of the powder that is dispensed through coaxial and discrete nozzles and then interacts with the laser beam is melted and transformed into metal after deposition, so a percentage of the feedstock material (in this case powder) is wasted. One way to minimize the impact of the loss of the unmelted material is to study how to confine its disposal in the manufacturing cell or machine, collect it, dry it, sieve it, and reuse it for further use in a subsequent manufacturing batch. However, one of the drawbacks in the case where the process is carried out in an uncontrolled atmosphere (e.g., workstations/cells open to the atmosphere), is the surface oxidation of the powder and the change of particle morphology, in which some partially melted particles join other unmelted particles in a non-controlled manner, and therefore do not have the same size, morphology and flowability as the virgin particles that have not been used. There is also a shortage of reports in the literature on how the reuse of the collected post-process powder influences the microstructure, quality and properties of certain non-common materials deposited with reused powder.

Recently, some research papers have been published showing new methodologies for the recycling and reuse of powder collected after additive manufacturing by DED-LB, among which are the studies conducted with austenitic stainless steels by Gutjahr et al. [1] and Terrassa et al. [2]. Gaining new knowledge on how to reuse metal powder in DED-LB/M and how its reuse influences the quality of fabrications in SS 316L has also been studied by Li et al. [3]. Interesting results of the environmental impact assessment and sustainability of the powder feedstock were studied by Verdi et al. [4] and Joju et al. [5]. In other metal AM technologies like powder bed fusion with laser beam (PBF-LB/M), more abundant studies are available in the literature [6–8]. A review paper has recently been published highlighting the main challenges associated with powder recycling, such as maintaining a uniform particle size distribution and shape for reuse, contamination management and mitigation of the main degradation effects of repeated powder use, such as particle deformation, contamination, oxide deposits, fragmentation, wear, sintering, dealloying, and surface oxidation [9].

In this study, two different alloys were evaluated, Stellite® 21 type alloy and super duplex stainless steel. According to the literature review, both alloys are not reported in DED-LB powder reuse studies and scientific reports published. Cobalt-based alloys are commonly used in the DED-LB process for hardfacing coatings and manufacturing. They are recommended for applications involving wear, galling or corrosion and retain these properties at high temperatures. They have a wide range of applications; for example, Stellite® 6 is a common material for the seat surface enhancement of various control valves, while Stellite® 21 is often used for valve trims under high-pressure steam and harsh conditions [10]. Stellite® alloy compositions have also been used to remanufacture components [11,12], for example, rail components [13]. Due to its good sliding wear and impact resistance, these alloys have been widely used in the building up and repairing of forging or hot stamping die components [14]. Stellite® 21 is used in applications that require high wear resistance, as well as retaining these properties at high temperatures. The hardness of the material is directly related to its wear resistance. This is why it is important that the hardness and performance of the material is maintained throughout the different powder reuse cycles.

On the other hand, the super duplex stainless steels have an austeno-ferritic microstructure with an average fraction of each phase of approximately 50 wt.%. This duplex microstructure improves simultaneously the mechanical properties and corrosion resistance. Welding of these steels is often a critical operation [15]. The second generation of super duplex stainless steels (SDSS) such as type UNS S32750 (DIN 1.4410/Sandvik SAF2507) is increasingly being used in oil and gas and petrochemical applications because of its good corrosion resistance and high strength due to its dual phase austenite/ferrite microstructure [16]. Recently, a high interest in the DED-LB processing of SDSS has been observed. Authors like Jiang et al. [17], Iams et al. [18,19] and Brázda et al. [20] have published their research work with type UNS S32750 super duplex stainless steel and DED processes, giving their deep analysis and conclusions regarding the austenitic formation mechanisms, duplex microstructure balance and understanding, mechanical properties obtained, and potential heat treatments to be applied in this additively manufactured material.

Both materials addressed in this study are expensive in terms of their cost per kg or powder and their reuse in the DED-LB process makes good sense, in particular, for metal AM of medium to large parts. The aim of our research work is to study the influence of powder reuse on particle morphology, particle surface oxidation, internal defectology, microstructure and mechanical properties. This was done by collecting unmelted powder particles after processing with DED-LB (LMD) up to three successive times, with no use of virgin powder or mixing, and applying a specific methodology for reuse involving sieving and drying in the same storage canister. In doing so, we aim to provide the scientific community and industry with detailed technical and practical knowledge of the reuse of these two very different families of materials processed by laser using DED process.

2. Materials and Methods

This section summarizes all feedstock materials, microstructural characterization procedures, hardness testing equipment, laser material processing equipment, and methods that were used in the experimental work conducted.

2.1. Materials Characterization—Equipment and Methods Used in This Work

In this research work, two metal alloys were selected for the study, a cobalt-based alloy type Stellite® 21 and a super duplex stainless steel. The macro and microstructural analysis of the Super Duplex samples was carried out using light optical microscopy (LOM) at different magnifications (from 100× to 1000×) with an Olympus GX51 optical microscope (Shinjuku City, Japan) with an image acquisition system via digital camera. The chemical etching in the case of the SDSS was carried out using a manual etching with Beraha's reagent (20 mL HCl + 100 mL H_2O + 1 g $K_2S_2O_5$) for 12 s. For more advanced studies in the microstructure, a field emission scanning electron microscope (FESEM) Zeiss Ultra Plus model (Oberkochen, Germany) equipped with an X-ray detector from Oxford instruments (X-Max) (Abingdon, UK) was also used. The area fraction of the main phases (in percentage) was quantified by making measurements via image analysis from the micrographs. In this case, five LOM images were taken at 200× magnification of different areas of the central part of the cubes manufactured, and then the images were binarized to contrast each phase, and the area represented by each phase was thereafter measured. The method used to quantify the austenite and ferrite phases is based on determining the volume fraction by systematic manual counting of points in the analysed area of the cross-section in the samples studied, following the guidelines of ASTM E562-19 [21]. Microhardness Vickers measurements were taken in an EmcoTest DuraScan durometer (Kuchl, Austria) using a load of 100 g (HV0.1 scale).

The metallurgical characterization of Stellite® 21 powder and manufactured samples was based on light optical microscopy (LOM) using a Leica Microsystems microscope (Wetzlar, Germany). Microhardness measurements were carried out by a Beortek Future-Tech FM700 Vickers hardness tester (Erandio, Spain), with a load of 500 g (HV0.5 scale). To identify the morphology of the powder and microstructure analysis in as-built cube samples, a Zeiss Ultra Plus FESEM (Oberkochen, Germany) was used for the analysis and examination. All manufactured samples were cut transversally, ground, polished and chemically etched with nitro-hydrochloric acid for variable duration (from 5 to 30 min).

For the rigorous chemical composition measurements carried out for each element in both alloys including the oxygen content, different techniques were employed. Carbon and sulfur measurements were determined by an Automatic Combustion Analyzer and infrared detection, using the CS 744 procedure based on ASTM E1019-18 guidelines [22], while oxygen content was determined using the ON 736 procedure based on ASTM E1019-18 [22] for both the Stellite® 21 and for super duplex SS. Remaining elements were measured using the Inductively Coupled Plasma Optical Emission spectroscopy (ICP-OES) technique. This is a multi-elemental analysis technique. The sample must be introduced into the equipment as a liquid, so it is necessary to perform a prior digestion of the sample using a Thermo Scientific Icap 7400 V (Waltham, MA, USA). The procedure used is applicable for nickel-based and cobalt-based materials like the Stellite® 21 type alloy. For the remaining elements in super duplex stainless steel, ICP-OES technique was also used but following the specific procedure indicated in the UNE-EN 10361:2016 standard [23].

2.2. Feedstock—Virgin Powders Characterization

In this work, two commercial gas-atomized powders were used. One was the super duplex stainless steel type SAF 2507 (UNS S32750) manufactured by Sandvik Osprey®. Powder particles were sieved after atomisation to achieve a particle size of +45–90 μm according to the certificate provided by the manufacturer. The chemical compositions reported by the powder manufacturer are shown in Table 1.

Table 1. Chemical composition of powder batch used in the study (manufacturers certificate).

Powder	Chemical Composition (wt.%)													
	Co	Cr	Ni	Mo	Mn	Si	Ti	Al	C	Fe	P	S	N	Others
SAF 2507	---	24.8	7.1	3.92	0.80	0.50	0.009	0.015	0.02	Bal.	0.008	0.006	0.3	0.11
Metco 1221A	Bal.	27.2	3.13	5.65	0.69	0.74	---	0.11	0.23	<0.10	---	---	---	0.017

As a feedstock material quality check procedure, the powder particle batch was characterized and analysed using LOM and FESEM images. The particles' morphology is shown in the FESEM micrographs of Figure 1a, and the calculated particle size distribution (PSD curve) is shown in Figure 1b. The particles showed a spherical shape, a morphology commonly obtained from the gas atomization procedure in their manufacturing. After the image analysis of the particles, the results revealed 10% by volume of the particles presented a size less than 56.51 μm in diameter; 50% of the particles were less than 77.6 μm in diameter; and a cumulative 90% by volume of the powder particles had a diameter less than 94.57 μm. As shown in Figure 1b, the particle size distribution is almost normal, like a gaussian distribution.

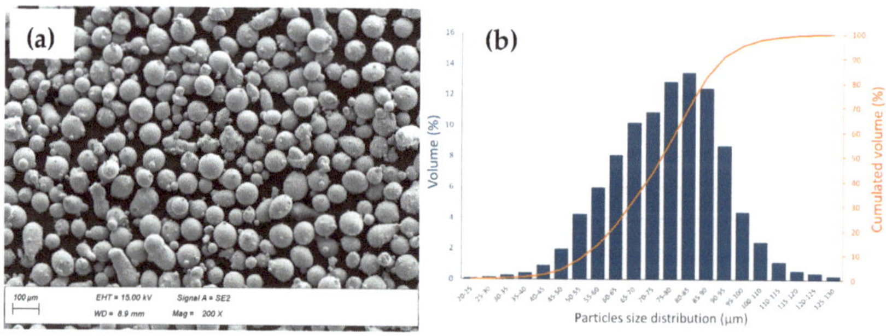

Figure 1. Characterization of SAF 2507 virgin powder particles. (**a**) FESEM micrographs (200× and SE2 mode) and (**b**) histogram of the particle size distribution (PSD).

The second commercial gas-atomized cobalt-based powder used was a Stellite® 21 type alloy powder from Oerlikon (Schwyz, Switzerland) (Metco 1221A), with a particle size of +45–125 μm according to the certificate provided by the manufacturer. It is a little larger in size and coarser than the other powder, and the particle size distribution seems to have a non-gaussian distribution (see Figure 2b). The chemical composition reported for this Stellite® 21 type alloy powder is also shown in Table 1.

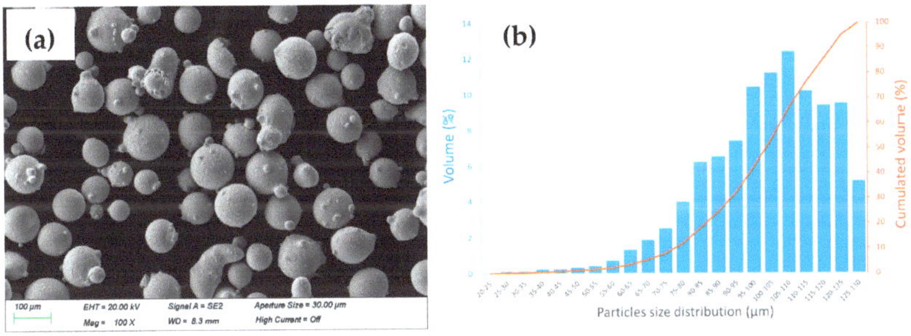

Figure 2. Characterization of Stellite® 21 virgin powder particles. (**a**) FESEM micrographs (100× and SE2 mode) and (**b**) histogram of the particle size distribution (PSD).

2.3. *Laser Material Deposition—Equipment and Experimental Set-Up*

The super duplex stainless steel powder was processed by LORTEK using a robotic LMD cell equipped with a 6-axis Fanuc robot arm and a 2-axis positioner table. A solid-state 5 kW disc laser source (Trumpf TruDisk 6002, Ditzingen, Germany) operated in continuous wave (CW) mode with a wavelength of 1030 nm was also used. The laser beam is guided through an optic fibre of 400 µm. The configuration of the LMD robotic station includes a fixed optic head (Trumpf BEO D70-90°, Ditzingen, Germany) with collimation/focal length of 200/200 mm. For powder delivery, the LMD station has a twin powder feeder with two 5 L heated hoppers (Oerlikon-Metco Twin 150, Wohlen, Switzerland) and a 3-jet discrete nozzle (3-Jet-SO16-F manufactured by FhG ILT, Aachen, Germany). Argon was used as protective (8.5 L/min) and carrier gas (2 L/min flow at 2 bar) for powder particles delivery.

The manufacturing and study of Stellite® 21 samples were performed in TEKNIKER (Eibar, Spain) facilities and equipment with an LMD robotized cell, consisting of an optical head (Precitec YC50, Gaggenau, Germany) with collimation/focal length of 200/200 mm and a 3-jet powder nozzle cladding head (FhG ILT, Aachen, Germany) attached to an industrial 6-axis robot arm (ABB, Zurich, Switzerland) and a CW 2 kW laser source (IPG YLS-2000-CT-Y17, Burbach, Germany). The laser beam is guided through a 600 µm diameter optical fibre. The powder was fed during the deposition process by means of a powder feeder (Sulzer-Metco Twin 10C, Wohlen, Switzerland). Argon was used as shielding (15 L/min) and powder carrier gas (8 L/min flow at 2.8 bar).

2.4. *Manufacturing Process, Powder Collection and Reuse Methodology*

The very first step is to prepare a series of C45 steel base plates and by grinding them on both sides, ensuring parallelism and flatness. In this way, all the manufacturing cycles performed maintain the same distance between the nozzle and the part and the initial surface finish of the build plate substrate. Subsequently, each cycle consists of manufacturing a series of prismatic specimens by DED-LB (LMD) in an uncontrolled atmosphere, under normal conditions of pressure and temperature. The specimens' sizes were 20 × 20 × 10 mm^3 (16 layers) cubes for super duplex builds and 15 × 15 × 10 mm^3 (14 layers) cubes for Stellite® 21 builds, repeating the necessary fabrications to obtain enough residual powder to manufacture samples in the following cycles with powder re-used. In addition, a single track is also deposited for each cycle as a reference track for dilution, height and width measurements.

During the DED-LB process, the non-deposited powder was collected using a collecting tray. This collecting tray is cleaned with isopropyl alcohol before each cycle to avoid cross-contamination with powder from other cycles. Once collected, the powder is manually collected and transferred to a suitable hermetically sealed container for storage (with two silica bags inside) until the next manufacturing and sieving cycle. The sieving of each cycle is carried out after the manufacturing of all the test specimens. The objective is to sieve the collected powder to remove impurities, spatter or clusters of adhered particles, whereby particles that are too fine or too coarse are separated, adjusting the particle size to a typical Laser DED distribution +53–150 μm.

A vibrating gravity sieve column is used for this step, which is schematically represented in Figure 3. The setup consists of two sieves and a container for the collection of finer powder. At the end of each cycle, the powder is poured out of the top of the column and the vibrating sieve is activated. The vibration is maintained for at least 20 min to ensure that the entire powder sample is processed. Once finished, the powder retained on the finest sieve is collected so that the particle size of the sample is between 0.053 mm (53 μm) and 0.150 mm (150 μm).

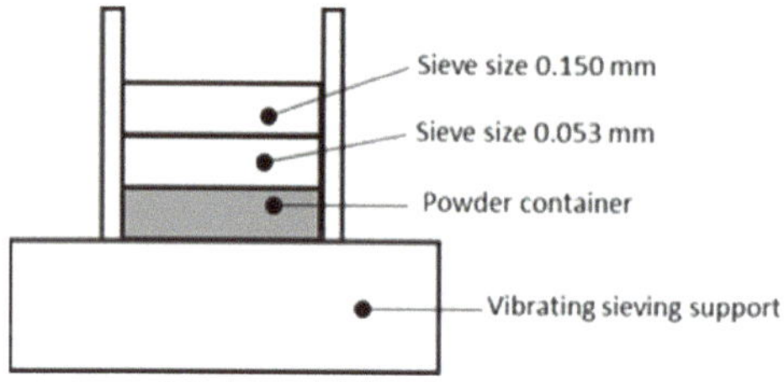

Figure 3. Vibrating sieving column scheme.

Next step is the extraction of a powder sample (25 g) for analysis in the FESEM and for chemical analysis by ICP-OES and oxygen measurement, while the remaining powder is stored in a hermetic and standard container for the next cycle. Finally, to conclude the reuse cycle and avoid cross-contamination, all containers are cleaned with isopropyl alcohol and the sieves are placed into an ultrasonic bath for 20 min. Recycled powders were stored in containers with two silica bags for drying.

The collected and recycled powder is reused and added to a clean hopper for subsequent builds, so that this powder is not mixed with virgin powder (as is often done in the reuse methodology of PBF-LB/M powder). The reuse of all collected (and not melted) powder has been studied, with up to three cycles of use. The full workflow proposed for the powder reuse methodology including collection and stages for characterization and analysis is shown in Figure 4.

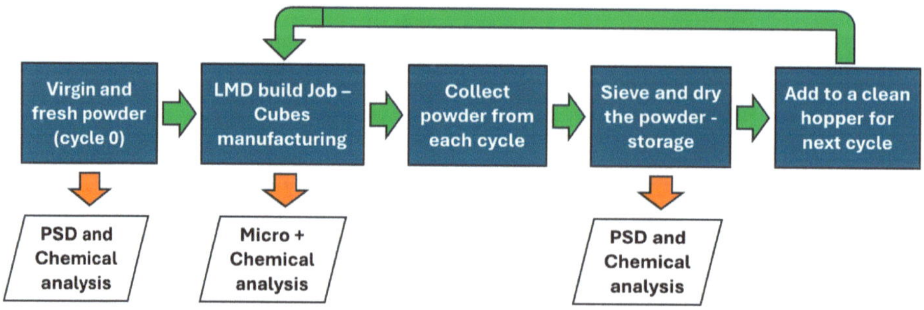

Figure 4. Full workflow methodology proposed for the powder reuse in DED-LB (LMD) process. This methodology was used for the three cycles of use studied. Green arrows denote process steps and orange arrows denote characterization, evaluations and analysis steps.

3. Results and Their Analysis

This section describes the main findings after deep material characterization has been performed on the virgin and reused powders with the methodology previously described. It also details the manufacturing trials conducted, and geometries selected as coupons, and the results and analysis regarding chemical composition variation, microstructure, internal defects, densification and mechanical properties obtained in the materials manufactured with both virgin and reused powders, up to three times/cycles for the super duplex and for the cobalt-based Stellite® 21 type alloy.

3.1. Reused Powders—Chemical Composition and Particle Morphology Characterization

The first study conducted was the compositional element analysis in virgin and reused powder particles, following the methodologies and using the techniques stated in Section 2.1. In Tables 2 and 3, the chemical composition measurements are compiled for Stellite® 21 and super duplex SS, respectively; it is shown that as the number of cycles increases, so does the oxygen in the powder samples collected. The rest of the chemical elements maintain stable values despite the reuse of the powder during laser processing. This effect has been observed in other studies related to the sustainability of the LMD process, for example using AISI 316L reused powder alloys [2,3].

Table 2. Chemical composition measured (wt.%) in Stellite® 21 powder in virgin state and after three reuses.

Powder Sample	C	Si	Mn	Cr	Ni	Mo	Al	Fe	O	Co
Virgin	0.23	0.74	0.69	27.20	3.13	5.65	0.11	<0.10	0.017	Balance
1 use	0.24	0.74	0.69	27.50	3.16	5.60	0.10	0.19	0.230	Balance
2 uses	0.24	0.74	0.69	27.50	3.15	5.60	0.10	<0.10	0.180	Balance
3 uses	0.24	0.82	0.68	27.70	3.11	5.64	0.12	0.15	0.320	Balance

Table 3. Chemical composition measured (wt.%) in Super Duplex powder in virgin state and after three reuses.

Powder Sample	C	Si	Mn	P	S	Cr	Ni	Mo	Cu	V	Co	O	Fe
Virgin	0.017 ± 0.004	0.37 ± 0.03	0.44 ± 0.02	0.023 ± 0.002	<0.005	25.9 ± 0.4	6.49 ± 0.11	3.77 ± 0.08	0.15 ± 0.01	0.053 ± 0.004	0.060 ± 0.005	0.025	Bal.
1 use	0.02 ± 0.005	0.41 ± 0.04	0.53 ± 0.02	0.021 ± 0.002	<0.005	25.7 ± 0.4	6.62 ± 0.12	3.82 ± 0.08	0.12 ± 0.01	0.044 ± 0.003	0.048 ± 0.004	0.120	Bal.
2 uses	0.022 ± 0.005	0.43 ± 0.04	0.55 ± 0.02	0.018 ± 0.002	<0.005	26.0 ± 0.4	6.62 ± 0.12	3.80 ± 0.08	0.12 ± 0.01	0.042 ± 0.003	0.052 ± 0.005	0.330	Bal.
3 uses	0.023 ± 0.005	0.42 ± 0.04	0.55 ± 0.02	0.021 ± 0.002	<0.005	26.0 ± 0.4	6.59 ± 0.12	3.80 ± 0.08	0.12 ± 0.01	0.043 ± 0.003	0.051 ± 0.004	0.660	Bal.

On the other hand, the super duplex stainless steel powder particles are more reactive than the cobalt-based alloy studied, according to the oxygen content measured in the reused powder samples.

As a feedstock material quality check procedure, the powder particle virgin batch and reused samples were characterized and analysed using LOM and FESEM images. Figure 5 shows the FESEM images of super duplex SS virgin powder particles, with one use cycle, with two use cycles and with three use cycles.

In the case of the Stellite® 21 powder, particles were observed and analysed through images and semi-quantitative chemical compositions. Figure 6 shows the FESEM images of the powder particles, with 0 (virgin), 1, 2 and 3 use cycles respectively.

As for the external appearance of the powder particles, when observed in the FESEM, no significant differences were observed; however, differences in the coloration of these particles were observed after the visual inspection, becoming darker in the samples with more than one use, which suggests a burning effect that could lead to surface oxidation of the same.

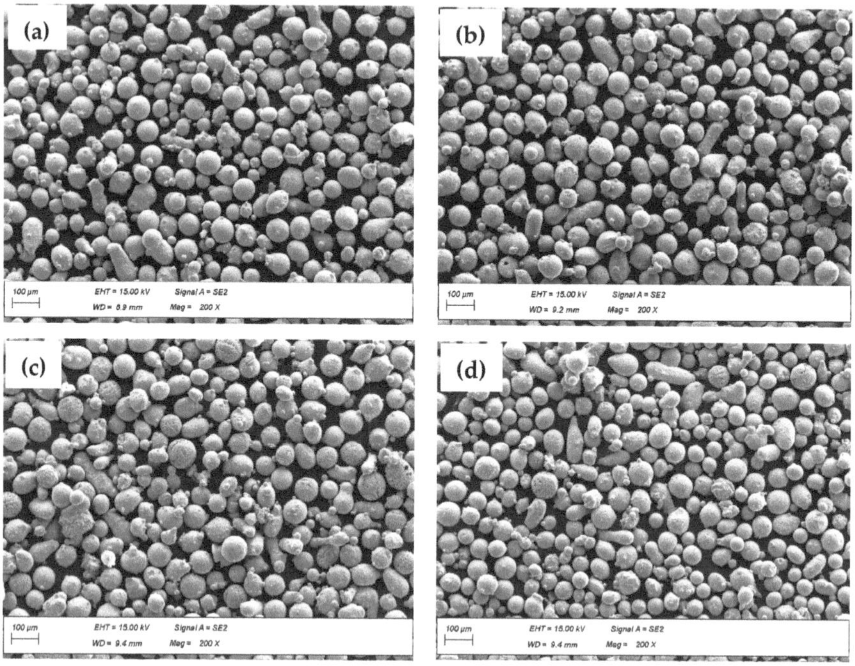

Figure 5. Micrographs of powder particles by FESEM (200× and SE mode) of super duplex stainless steel 2507: (**a**) virgin powder, (**b**) once-used powder, (**c**) twice-used powder and (**d**) thrice-used powder.

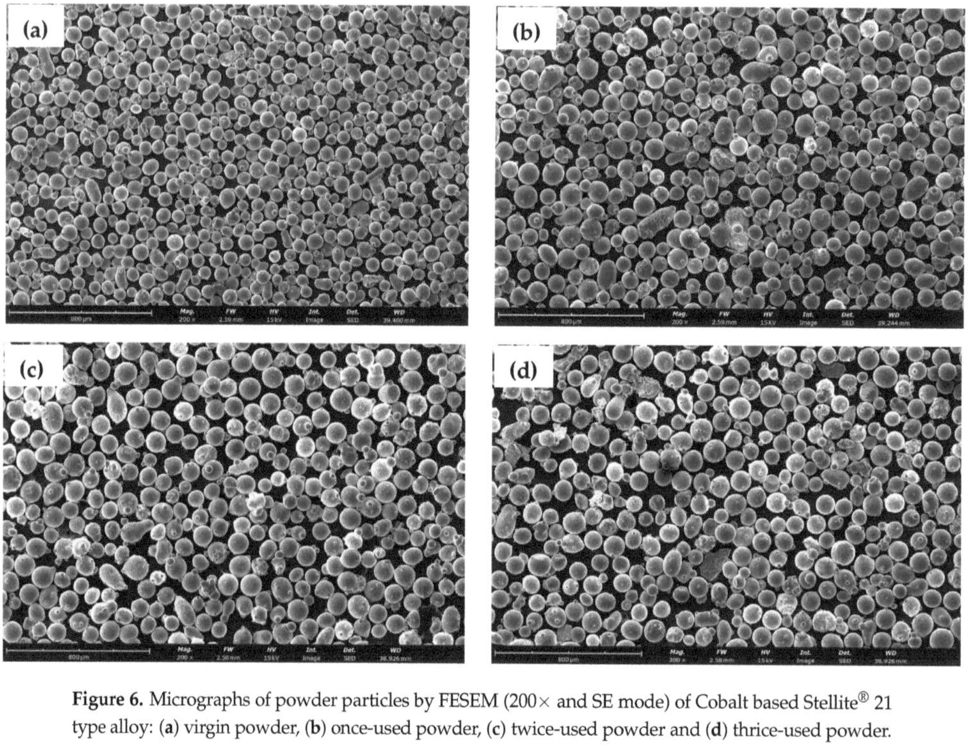

Figure 6. Micrographs of powder particles by FESEM (200× and SE mode) of Cobalt based Stellite® 21 type alloy: (**a**) virgin powder, (**b**) once-used powder, (**c**) twice-used powder and (**d**) thrice-used powder.

In terms of particle size distribution and morphology, there is a variation in the particle size distribution, with the virgin powder having a greater amount of large particles, measured as equivalent diameter; its particle size frequency distribution is not completely normal or Gaussian, with a tendency to skew to the right. This result may be due to the fact that in the direct laser deposition process, the smaller particles tend not to be part of the molten bath and, after successive uses of the powder, the proportion of small particles in the powder collected after the process increases compared to the larger ones, shifting the bias to the left. This can be seen for super duplex stainless steel particles in Figure 7 and for Stellite® 21 type alloy in Figure 8. The characteristic particle size parameters such as D10, D50 and D90 have also been calculated for both powders and the metrics derived from the particle's morphology study are summarized in Tables 4 and 5, respectively.

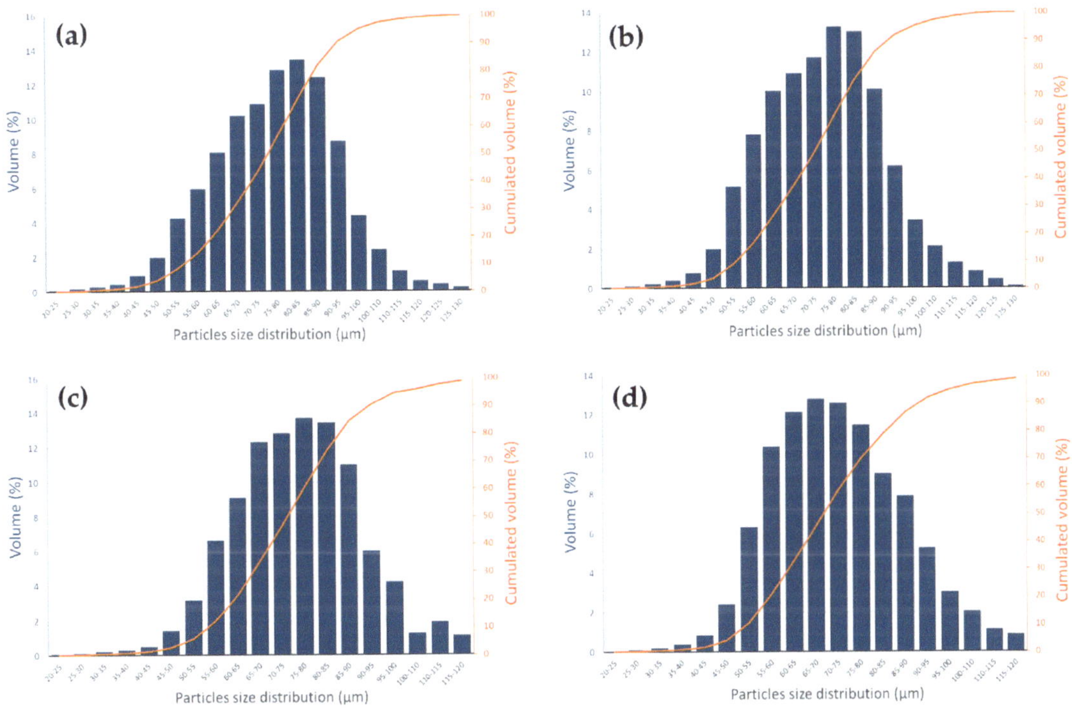

Figure 7. Particle size distribution of SS super duplex powder sample analysed with SEM images. (**a**) Virgin powder, (**b**) once-used powder, (**c**) twice-used powder and (**d**) thrice-used powder.

Table 4. Particle size distribution (PSD) in μm for materials studied.

Powder Sample	PSD (μm) for Stellite® 21 Powder			PSD (μm) for Super Duplex Powder		
	D10	D50	D90	D10	D50	D90
Virgin	77.65	103.43	122.44	56.51	77.60	94.57
1 use	72.38	98.39	121.17	55.78	75.29	93.61
2 uses	86.70	109.61	124.36	58.24	76.29	94.54
3 uses	81.67	105.36	122.37	54.76	71.76	93.36

Concerning the morphological analysis of the powder particles, it was observed that after the collection and sieving process of the particles, the circularity improves (one reuse powder), but as they are reused again (2nd and 3rd cycles), their circularity decreases. It has also been observed that the SDSS powder particles are smaller in diameter and more circular than the Stellite® 21 particles, which leads to the inference that they will have better flowability than the latter.

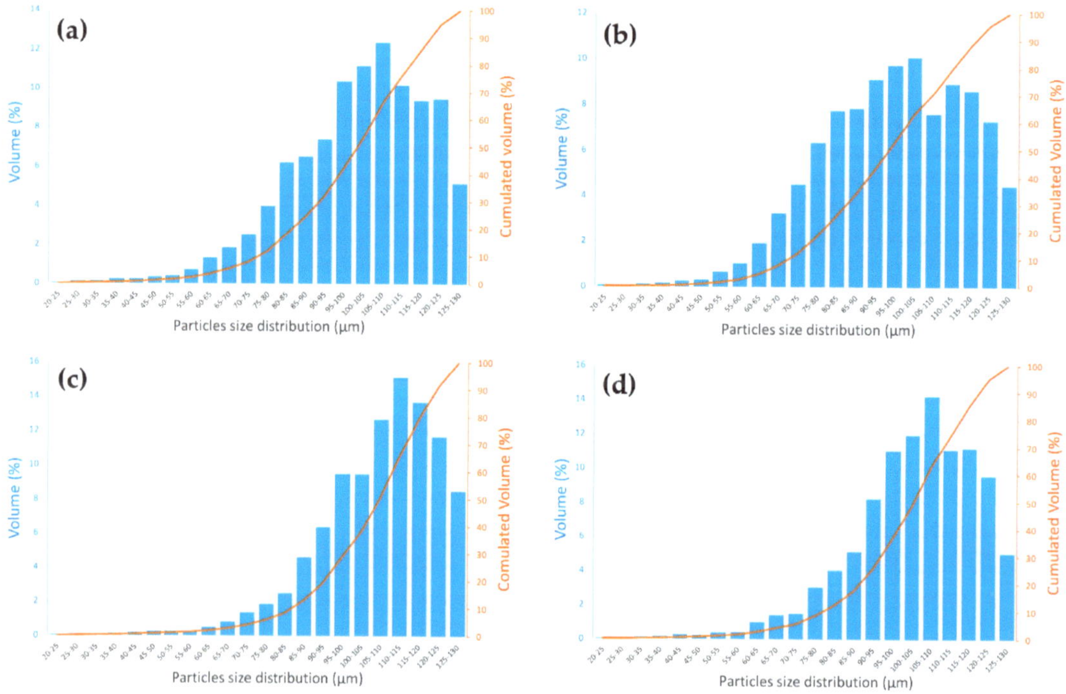

Figure 8. Particle size distribution of Stellite® 21 type alloy powder sample analysed with SEM images. (**a**) Virgin powder, (**b**) once-used powder, (**c**) twice-used powder and (**d**) thrice-used powder.

Table 5. Particle morphology study in powder samples studied.

Powder Sample	Stellite® 21 Type Particles				Super duplex Particles			
	Total	Circular	Circular with Ar < 0.4	% Circularity	Total	Circular	Circular with Ar < 0.4	% Circularity
Virgin	2757	2166	1761	63.9	4530	4138	3288	72.6
1 use	2909	2339	1929	66.3	4739	4318	3584	75.6
2 uses	2693	2067	1577	58.6	4452	3961	3266	73.4
3 uses	3291	2539	1854	56.3	5192	4587	3826	73.7

3.2. Manufacturing Trials—Geometry Selected and Process Efficiency

The depositions were applied on a rectangular substrate of grade C45 with dimensions of $100 \times 200 \times 30$ mm^3. Square multi-layer depositions of 15 mm width and 20 mm for Stellite® 21 and SDSS, respectively, were performed on the substrate, aiming to obtain 10 mm height cubes (see Figure 9), keeping the same deposition process parameters for each reuse cycle of the powder (nothing was changed in the successive rounds of cube manufacturing). Bidirectional deposition strategy was used at every layer in alternate perpendicular directions. The most relevant LMD process parameters employed for cuboids manufacturing, such as the laser power (P), linear deposition speed (S), powder flow rate (F), stepover distance (d) and layer thickness (h) are displayed in Table 6. These process parameters for cube geometries were developed from previously published research work by the authors [24] for super duplex stainless steel, and from the experience of DED-LB process technicians and the literature [11,12] for Stellite 21® type alloy. For the studies of chemical composition and microstructure analysis in the bulk material (in as-built condition), the cubes were cut from the substrate, removing the first deposited layers, to minimize the influence of the substrate material dilution with the additive material. The dilution effect has not been studied in this work.

(a) (b)

Figure 9. Manufactured LMD Cubes: (**a**) Example of 15 × 15 × 10 mm^3 Stellite® 21 cube manufactured with virgin powder, and (**b**) 20 × 20 × 10 mm^3 SAF 2507 cubes manufactured with one use (one cycle) powder.

Table 6. DED-LB process parameters for cubes manufacturing trials.

Process Parameter	Value for Each Material	
	Stellite® 21	**SAF 2507**
Laser power (W)	1800	1000
Laser spot diameter (mm)	2.7	2.1
Speed (mm/s)	10	15
Powder feed (g/min)	10	7.4
Stepover distance	1.0	1.3
Layer thickness (mm)	0.71	0.65

Prior to the laser metal deposition process, the build plate substrates were ground and cleaned with acetone. In addition, the substrates were weighed before and after production of each manufacturing batch to calculate the amount of material deposited in each case. The powder catchment efficiency was calculated for each sample to evaluate process effectiveness. It was obtained by dividing the amount of deposited powder by the amount of supplied powder during laser-on in the deposition trajectory path. The above steps were repeated four times to obtain the deposited samples of Stellite® 21 type alloy and SDSS obtained by LMD after reusing zero to three times. The efficiency of the catchment process did not exhibit a clear correlation with the number of powder reuse cycles (Table 7), but seems to be linked to the circularity of the particles and maybe the flowability index of them. Despite this, the process has not in fact been optimized for powder catchment efficiency. This was primarily due to the process parameters not being fine-tuned for maximum efficiency. Additionally, the cube construction process lacked a control mechanism for a constant layer height, further impacting the stand-off distance and overall efficiency of the deposition process.

Table 7. Powder catchment efficiency of powders.

Cube Manufactured with Powder	Process Efficiency (%)	
	Stellite® 21	**SAF 2507**
Virgin	40	73.2
1 use	36	58.6
2 uses	45	63.4
3 uses	41	61.1

3.3. Manufacturing Trials—Bulk Material Densification, Defects and Chemical Composition

After powder reuses, the Stellite® 21 samples did not show significant internal defects. Actually, no major defects were seen throughout the analysed cross-sections of samples in the early recycling cycles (see Figure 10). However, in the third reuse, porosity appeared in the inter-diffusion zone (see Figure 11). Besides the change in geometrical shape, this could happen due to the increase of the temperature in the part during the process, but more statistics are needed to test this hypothesis. It is worth noting that the superficial colour of the deposition darkens when recycled powder is used, as reported by Terrassa et al. with reused austenitic stainless-steel powders [2].

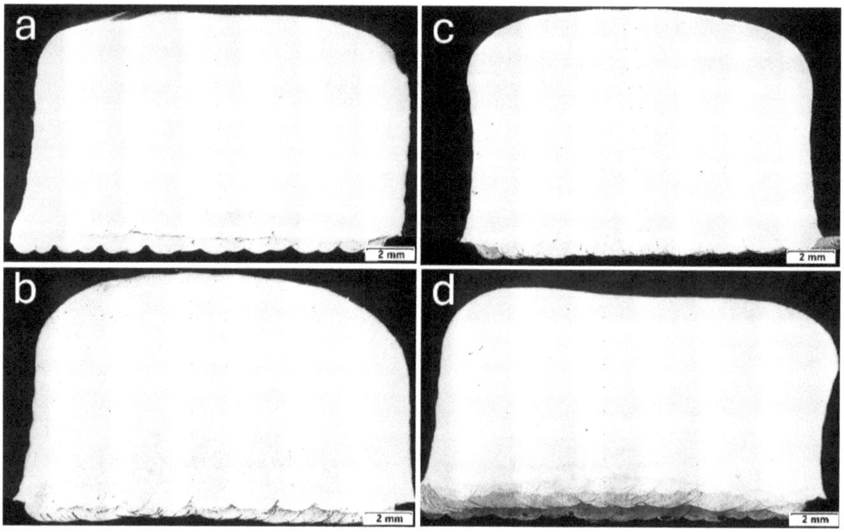

Figure 10. Macrographs of cube cross-sections, manufactured with Stellite® 21 (**a**) virgin powder, (**b**) once-used powder, (**c**) twice-used powder and (**d**) thrice-used powder.

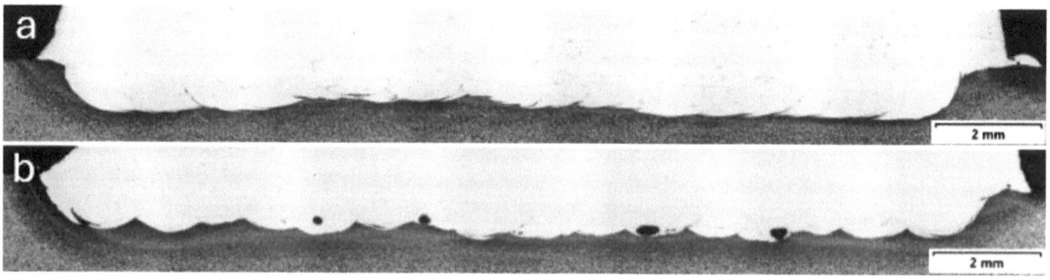

Figure 11. Dilution area in manufactured cubes with Stellite® 21 (**a**) using once-used powder, and (**b**) with powder reused three times.

After manufacturing and cutting the bulk cubes obtained by the DED-LB (LMD) process with the super duplex stainless steel composition grade, and to evaluate their cross-section, a metallographic preparation of the samples was carried out. It is observed that the surface appearance of the cubes follows the same trend, with the cubes made from virgin powder having the best appearance. Figure 12 shows the cross section of the fabricated cubes, where the level of defects is clearly observed and more evident in the samples with two and three cycles of powder reuse.

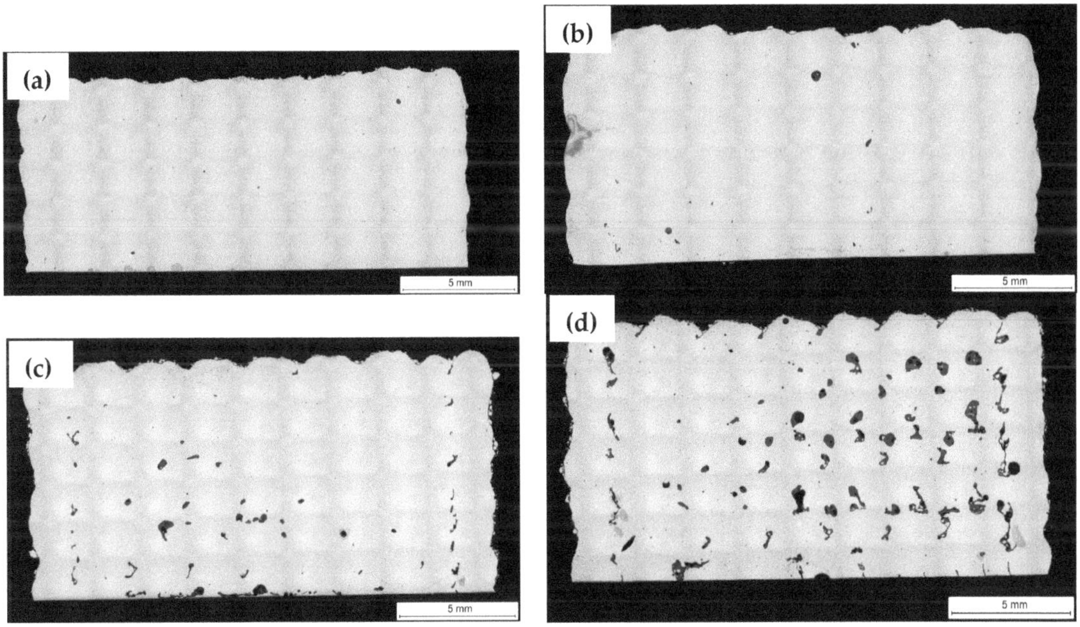

Figure 12. Macrographs of cubes cross-sections, manufactured with SAF 2507 super duplex steel with (**a**) virgin powder, (**b**) once-used powder, (**c**) twice-used powder and (**d**) thrice-used powder.

The internal porosity of the DED samples was measured using the LOM image analysis. Table 8 shows the measured values. In general, higher densification was obtained with virgin powders, but in the case of SDSS powder this decreased after two cycles of reuse.

Table 8. Densification values in bulk cubes manufactured with SAF 2507 powder.

Cube Manufactured with Powder	Material Densification (%)	
	Stellite® 21	SAF 2507 Super Duplex
Virgin	99.97 ± 0.03	99.95 ± 0.02
1 use	99.92 ± 0.05	99.80 ± 0.36
2 uses	99.93 ± 0.05	99.42 ± 0.47
3 uses	99.77 ± 0.14	92.99 ± 3.89

The chemical composition of the bulk cubes obtained was analysed in the same way as the powder samples collected after each manufacturing cycle. The aim was to analyse how the main elements present vary, but particularly oxygen content. Tables 9 and 10 compile the results of the measurements obtained.

Table 9. Chemical composition measured (wt.%) in manufactured Stellite®21 LMD cubes.

Powder Type	C	Si	Mn	Cr	Ni	Mo	Al	Fe	O	Co
Virgin	0.20	0.70	0.65	27.9	3.13	5.61	0.11	0.14	0.019	Balance
1 use	0.20	0.54	0.52	27.7	3.12	5.63	0.11	0.36	0.042	Balance
2 uses	0.21	0.48	0.49	27.7	3.17	5.56	0.11	0.24	0.052	Balance
3 uses	0.16	0.51	0.52	27.8	3.09	5.59	0.11	0.74	0.051	Balance

Table 10. Chemical composition measured (wt.%) in manufactured Super Duplex LMD cubes.

Powder Type	C	Si	Mn	P	S	Cr	Ni	Mo	Cu	V	Co	O	Fe
Virgin	0.018 ± 0.005	0.42 ± 0.04	0.42 ± 0.02	0.022 ± 0.002	<0.005	25.5 ± 0.4	6.53 ± 0.11	3.76 ± 0.08	0.15 ± 0.01	0.051 ± 0.004	0.062 ± 0.005	0.130	Bal.
1 use	0.02 ± 0.005	0.40 ± 0.04	0.55 ± 0.02	0.015 ± 0.002	<0.005	25.1 ± 0.4	6.74 ± 0.12	3.78 ± 0.08	0.10 ± 0.01	0.039 ± 0.003	0.048 ± 0.004	0.150	Bal.
2 uses	0.018 ± 0.005	0.36 ± 0.03	0.45 ± 0.02	0.017 ± 0.002	<0.005	25.2 ± 0.4	6.73 ± 0.12	3.79 ± 0.08	0.11 ± 0.01	0.038 ± 0.003	0.054 ± 0.005	0.320	Bal.
3 uses	0.018 ± 0.005	0.32 ± 0.03	0.42 ± 0.02	0.018 ± 0.002	<0.005	25.1 ± 0.4	6.73 ± 0.12	3.80 ± 0.08	0.12 ± 0.01	0.036 ± 0.003	0.054 ± 0.004	0.490	Bal.

As observed, there is no appreciable volatilization of chemical elements after laser metal deposition of the materials, perhaps because none of them are light elements. What is observed is the influence of the use of powder with high surface oxidation in the particles, i.e., in the SDSS samples there is a higher oxygen content in the bulk material (as-built condition), which in the case of this super duplex grade composition can have a significant impact on the formation of oxides and on the balance of austenite/ferrite phases that form the dual-phase microstructure.

3.4. As-Built Material—Microstructure Analysis

The microstructure of deposited samples is composed of dendrites (see Figures 13 and 14). As the number of cycles of powder reuse increases, the deposited layers still showed good metallurgical bonding and adequate adhesion between layers. The dendrite growth direction is not continuous because of the building strategy, which alternates perpendicularly between odd and even layers. This strategy alters the thermal dissipation of each layer that is deposited, thereby inhibiting the continuous growth of columnar crystals. This behaviour was observed in both materials.

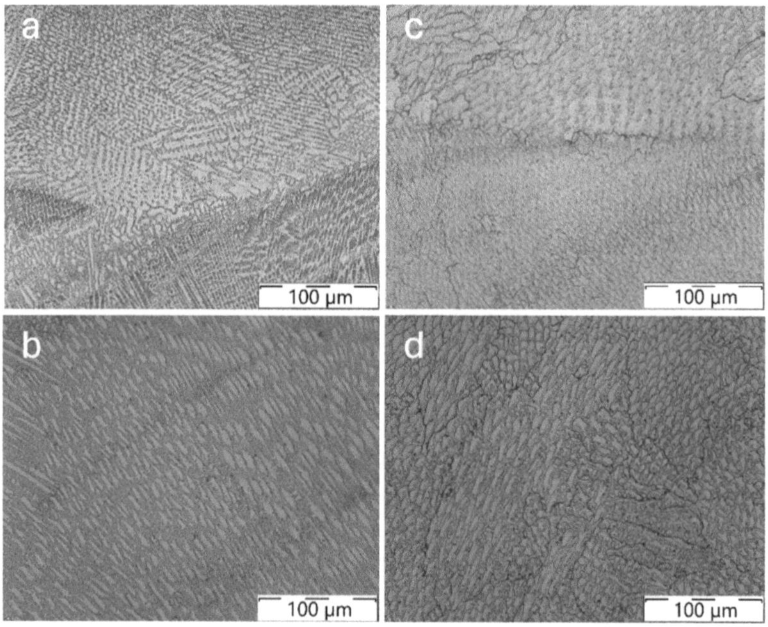

Figure 13. Detail of dendritic structure in Stellite® 21 samples manufactured with: (**a**) Virgin powder (zero cycles), (**b**) one-reuse powder (one cycle), (**c**) two-reuses powder (two cycles), and (**d**) three-reuses powder (three cycles).

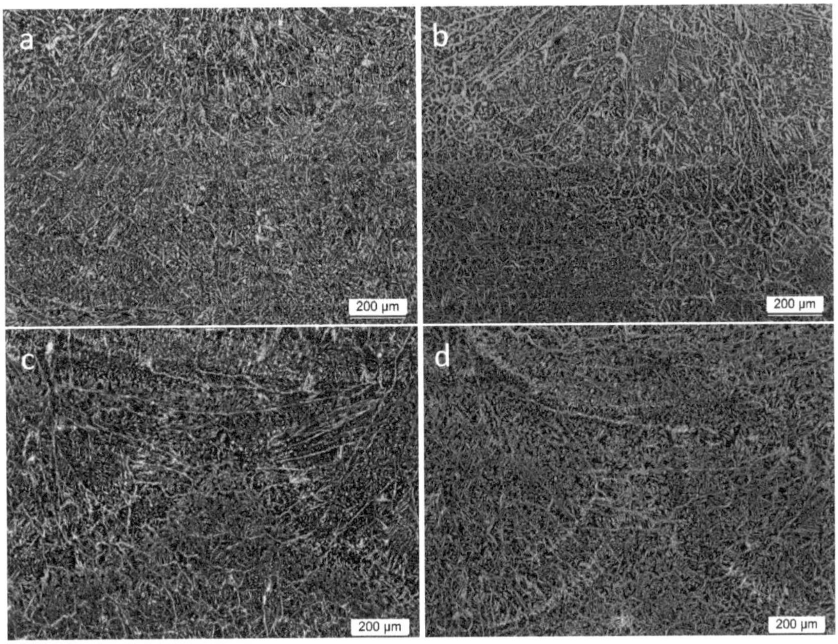

Figure 14. Detail of austenite-ferrite dendritic structure (LOM micrograph, 100×, etched) in SAF 2507 samples manufactured with: (**a**) Virgin powder (zero cycles), (**b**) one-reuse powder (one cycle), (**c**) two-reuses powder (two cycles), and (**d**) three-reuses powder (three cycles).

In the etched condition for the SDSS sample (chemical etching by immersion in Beraha's reagent) the dendritic microstructure is a bit coarser than the dendritic microstructure observed in the cobalt-based material. In Figure 14, the optical micrographs are compiled for SAF 2504 cubes. The microstructure of as-built material is composed of austenite-ferrite in this case.

The SDSS alloy seems to be more sensitive to the reuse of the powder than the cobalt-based alloy. At higher magnification, the unbalance of the austenite and ferrite phases is visible (see quantification values compiled in Table 11) and internal defects appear after solidification, such as pores and lack of fusion; also, oxides inclusions that usually surround these defects are more evident in the SAF 2507 samples manufactured with two- and three-reuses powder (cycles two and three), see the oxides inclusion identification and EDS maps in Figures 15 and 16, respectively.

Table 11. Phase quantification in the microstructure of bulk cubes manufactured with SAF 2507 powder.

Cube Manufactured with Powder	Phase Quantification (%)		Ferrite Ratio
	Austenite	Ferrite	
Virgin	49.9 ± 2.8	50.1 ± 2.8	0.50
1 use	53.6 ± 4.3	46.4 ± 4.3	0.46
2 uses	57.4 ± 4.7	42.6 ± 4.7	0.43
3 uses	63.0 ± 6.0	37.0 ± 6.0	0.37

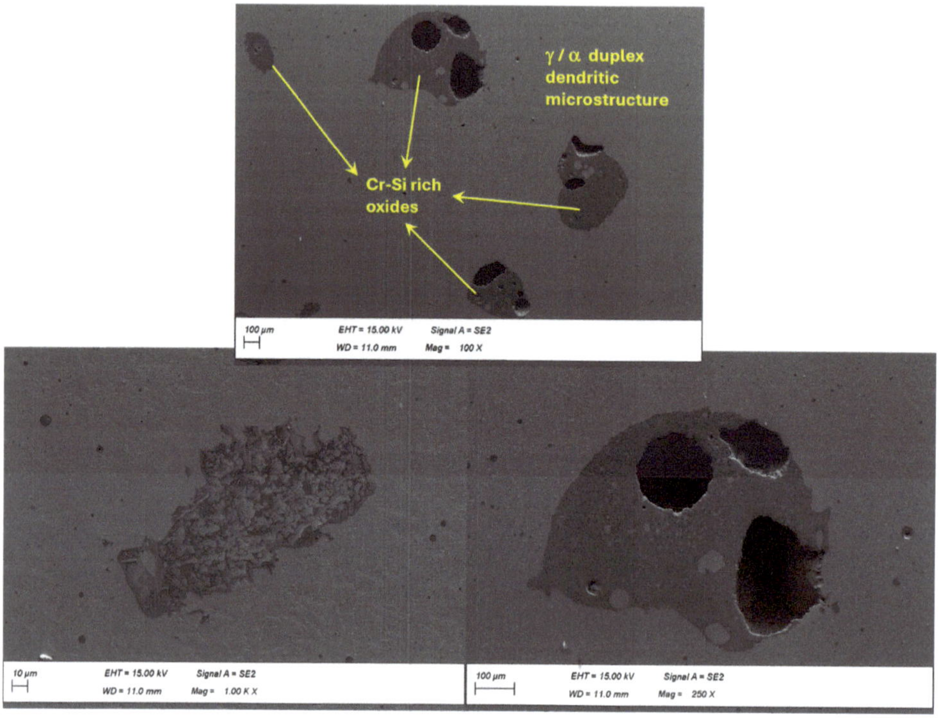

Figure 15. Detail of Cr-Si rich oxides present in a SAF 2507 cube manufactured with thrice-used powder (FESEM micrographs at 250× and 1000×, BSE mode, sample polished).

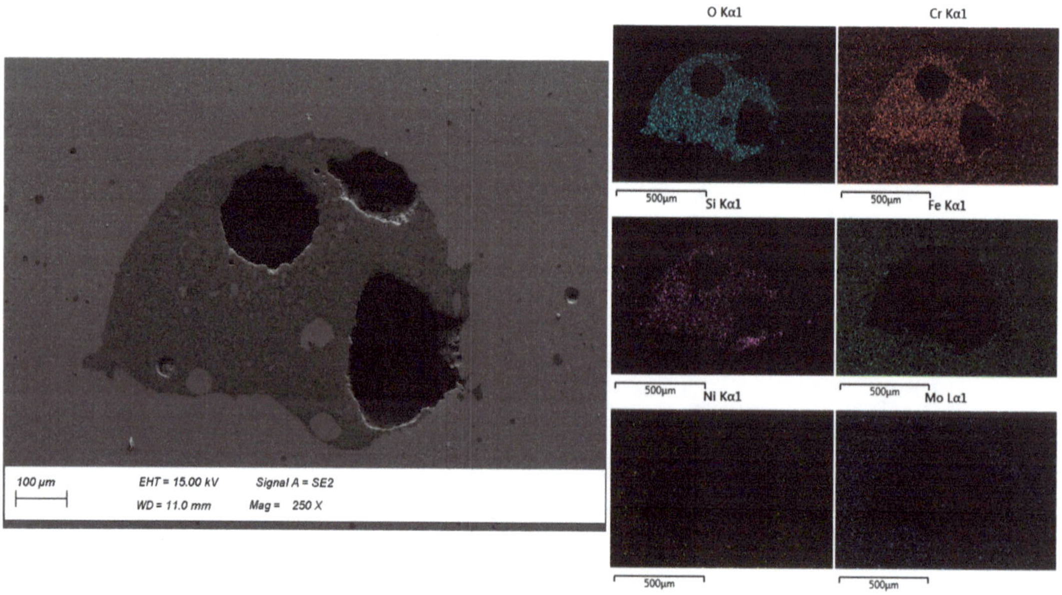

Figure 16. FESEM micrograph (BSE mode, 250×) and EDS maps of Cr-Si rich oxide present in a SAF 2507 cube manufactured with thrice-used powder (three cycles).

The relation between oxygen content and the stabilization of austenite and Cr-Si rich oxides formation in the super duplex stainless steel is evident. As can be observed in Figure 17, the higher the oxygen content in the bulk material (present mainly in the form of oxides around the pores and fusion faults), the greater the stabilization of the austenite phase, so that after solidification of the material a lower proportion of the ferrite phase is obtained. This is because oxygen is itself a gamma-magnetic element, and this promotes the stabilization of the austenite phase in SS. The ferrite ratio drops from 0.50 using virgin powder to 0.37 using reused powder in three processing cycles by LMD.

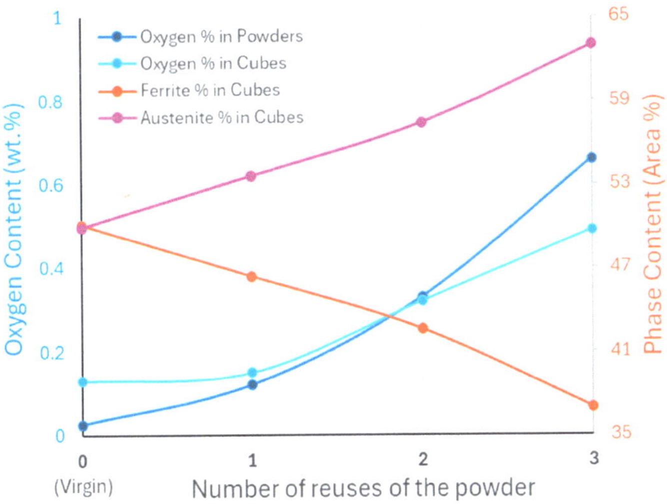

Figure 17. Evolution of oxygen content in powders and cubes and main phases present in the microstructure of SAF 2507 super duplex stainless steel obtained by LMD (DED-LB) process.

3.5. As-Built Material—Mechanical Properties/Hardness

Vickers micro-hardness measurements of every Stellite® 21 sample was obtained from the surface of the cube to the interface between the deposited and substrate materials every 1 mm with a load of 0.5 kg. The resulting average micro-hardness value for each sample is represented in Figure 18. The hardness of the samples decreases as the powder is reused. This can be caused by the reduction of carbon and manganese and increment of oxygen content in the cubes manufactured with reused powder, considering that the C content of a Stellite® 21 alloy determines the volume fraction of carbides precipitated in the alloy and hence its hardness, as reported previously by Liu et al. [25].

Hardness measurements were also performed in super duplex cubes cross-sections, making a sweep with indentations from the surface of the cube; the first indentation was made at 250 μm from the surface and then the following ones separated 500 μm between them, reaching the first layer. In a similar way, the hardness of SAF 2507 LMD samples decreases as the powder is reused (see the graph of Figure 18), because the amount of oxygen increases and more oxides and austenite are formed, lowering the strength due to reduced ferrite content in comparison with samples produced with virgin powder.

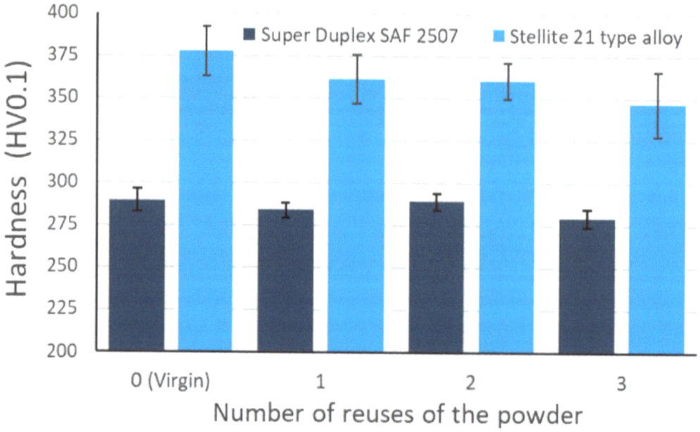

Figure 18. Average microhardness measured in cross-section of cubes manufactured by LMD with virgin and reused powder.

4. Conclusions

The main conclusions based on the results obtained and their analysis in this research work are as follows:

- A methodology has been proposed for the reuse of unmelted metallic powder after the DED-LB (LMD) manufacturing process. Two different alloys, one cobalt-based and one iron-based, have been evaluated and processed at two LMD robotic stations in different locations for the evaluation of the reuse of powders.
- In the present study, cobalt-based alloys such as the Stellite® 21 type are less reactive and less sensitive to powder reuse in the DED-LB (LMD) process than super duplex stainless steel type SAF 2507 (similar to UNS S32750/DIN 1.4410). This is because cobalt-based alloys are a less reactive material and not prone to oxides formation and austenitic phase stabilisation in their microstructure.
- In general, higher densification was obtained in as-built samples manufactured with virgin powders, but in the case of SDSS powder the densification decreases after two cycles of reuse. No loss of densification was observed in the case of Stellite® 21 type alloy, in fact the densification after three reuses in this powder did not drop below 99.74%.
- The ferrite ratio (proportion of ferrite in the duplex microstructure) drops from 0.50 using virgin powder to 0.37 after three reuse cycles in the case of super duplex stainless steel. This can have a strong influence on the performance of components manufactured by DED-LB and subjected to corrosion and loading.
- The average micro-hardness of the deposited samples of Stellite® 21 and SAF 2507 super duplex stainless steel decreases as the powder reuse increases due to the fact that the oxygen content increases, and the reduction of carbon and manganese in the cobalt-based alloy and the formation of oxides and austenite for the SDSS alloy.
- Powder particles of super duplex stainless steel alloy can be reused only one time without an increase of internal porosity and oxides formation within deposited layers in the additive process, while powder particles of the Stellite® 21 type alloy can be reused up to three times without an increase in internal porosity or lack of fusion in the deposited material.

Author Contributions: Conceptualization, J.C.P.; Data curation, U.I. and D.G.; Formal analysis, J.C.P., A.S. and D.G.; Funding acquisition, J.C.P., C.S. and A.L.; Investigation, U.I., A.S. and J.E.R.; Methodology, A.S.; Resources, J.E.R.; Supervision, A.L.; Writing—original draft, U.I., A.S. and D.G.; Writing—review & editing, J.C.P., C.S. and A.L. All authors have read and agreed to the published version of the manuscript.

Funding: This research work was supported by the Basque Government (Eusko Jaurlaritza, Department of Economic Development and Infrastructure, Programme ELKARTEK) through EDISON project (grant KK-2022/00070) and through ECOFAST project (grant KK-2024/00018).

Data Availability Statement: The data presented in this study are available on request from the corresponding author due to technical or time limitations.

Conflicts of Interest: The authors declare no conflicts of interest.

References

1. Gutjahr, J.; Pereira, M.; de Sousa, J.M.S.; Ferreira, H.S.; Júnior, A.T. Powder degradation as a consequence of laser interaction: A study of SS 316L powder reuse on the laser directed energy deposition process. *J. Laser Appl.* **2024**, *36*, 12022. [CrossRef]
2. Terrassa, K.L.; Haley, J.C.; MacDonald, B.E.; Schoenung, J.M. Reuse of powder feedstock for directed energy deposition. *Powder Technol.* **2018**, *338*, 819–829. [CrossRef]
3. Li, S.; Chen, B.; Tan, C.; Song, X. Study on Recyclability of 316L Stainless Steel Powder by Using Laser Directed Energy Deposition. *J. Mater. Eng. Perform.* **2022**, *31*, 400–409. [CrossRef]
4. Verdi, D.; Yang, S.; Soh, N.; Tay, G.; Patran, A. *The Role of Powder Feedstock in Directed Energy Deposition Sustainability*; ASTM International: West Conshohocken, PA, USA, 2022. [CrossRef]
5. Joju, J.; Verdi, D.; Han, W.S.; Hang, L.Y.; Soh, N.; Hampo, C.C.; Liu, N.; Yang, S.S. Sustainability assessment of feedstock powder reuse for Directed Laser Deposition. *J. Clean. Prod.* **2023**, *388*, 136005. [CrossRef]
6. Warner, J.H.; Ringer, S.P.; Proust, G. Strategies for metallic powder reuse in powder bed fusion: A review. *J. Manuf. Process.* **2024**, *110*, 263–290. [CrossRef]
7. Li, J.; Liu, W.; Shen, J.; Zhang, X.; Li, S.; Wang, Z. Research progress of the metal powder reuse for powder bed fusion additive manufacturing technology. *Powder Technol.* **2024**, *441*, 119815. [CrossRef]
8. Santecchia, E.; Spigarelli, S.; Cabibbo, M. Material Reuse in Laser Powder Bed Fusion: Side Effects of the Laser—Metal Powder Interaction. *Metals* **2020**, *10*, 341. [CrossRef]
9. Lanzutti, A.; Marin, E. The Challenges and Advances in Recycling/Re-Using Powder for Metal 3D Printing: A Comprehensive Review. *Metals* **2024**, *14*, 886. [CrossRef]
10. Yao, J.; Zhang, Q.; Liu, R.; Wu, G. (Eds.) *Laser Cladding of Stellite Alloys BT—Laser Applications in Surface Modification*; Springer: Singapore, 2022; pp. 11–57. [CrossRef]
11. Díaz, E.; Tobar, M.J.; Yáñez, A.; García, J.; Taibo, J. Laser Powder Welding with a Co-based alloy for repairing steam circuit components in thermal power stations. *Phys. Procedia* **2010**, *5*, 349–358. [CrossRef]
12. Díaz, E.; Amado, J.M.; Montero, J.; Tobar, M.J.; Yáñez, A. Comparative Study of Co-based Alloys in Repairing Low Cr-Mo steel Components by Laser Cladding. *Phys. Procedia.* **2012**, *39*, 368–375. [CrossRef]
13. Kendall, O.; Abrahams, R.; Paradowska, A.; Reid, M.; Qiu, C.; Mutton, P.; Schläfer, T.; Yan, W. Influence of multi-layer laser cladding depositions and rail curvature on residual stress in light rail components. *Eng. Fail. Anal.* **2023**, *150*, 107330. [CrossRef]
14. Foster, J.; Cullen, C.; Fitzpatrick, S.; Payne, G.; Hall, L.; Marashi, J. Remanufacture of hot forging tools and dies using laser metal deposition with powder and a hard-facing alloy Stellite 21®. *J. Remanuf.* **2019**, *9*, 189–203. [CrossRef]
15. Tavares, S.S.M.; Pardal, J.M.; Lima, L.D.; Bastos, I.N.; Nascimento, A.M.; de Souza, J.A. Characterization of microstructure, chemical composition, corrosion resistance and toughness of a multipass weld joint of superduplex stainless steel UNS S32750. *Mater. Charact.* **2007**, *58*, 610–616. [CrossRef]
16. Tan, H.; Jiang, Y.; Deng, B.; Sun, T.; Xu, J.; Li, J. Effect of annealing temperature on the pitting corrosion resistance of super duplex stainless steel UNS S32750. *Mater. Charact.* **2009**, *60*, 1049–1054. [CrossRef]
17. Jiang, D.; Gao, X.; Zhu, Y.; Hutchinson, C.; Huang, A. In-situ duplex structure formation and high tensile strength of super duplex stainless steel produced by directed laser deposition. *Mater. Sci. Eng. A* **2022**, *833*, 142557. [CrossRef]
18. Iams, A.D.; Keist, J.S.; Palmer, T.A. Formation of Austenite in Additively Manufactured and Post-Processed Duplex Stainless Steel Alloys. *Metall. Mater. Trans. A* **2020**, *51*, 982–999. [CrossRef]
19. Iams, A.D.; Keist, J.S.; Giannuzzi, L.A.; Palmer, T.A. The Evolution of Oxygen-Based Inclusions in an Additively Manufactured Super-Duplex Stainless Steel. *Metall. Mater. Trans. A* **2021**, *52*, 3401–3412. [CrossRef]
20. Brázda, M.; Salvetr, P.; Rzepa, S.; Melzer, D.; Vavrik, J. Effect of heat treatment on mechanical properties of duplex steel SAF 2507 manufactured by DED. *IOP Conf. Ser. Mater. Sci. Eng.* **2021**, *1178*, 12008. [CrossRef]
21. *ASTM E562-19e1*; Standard Test Method for Determining Volume Fraction by Systematic Manual Point Count. ASTM International: West Conshohocken, PA, USA, 2020; pp. 1–7. [CrossRef]
22. *ASTM E1019-18*; Standard Test Methods for Determination of Carbon, Sulfur, Nitrogen, and Oxygen in Steel, Iron, Nickel, and Cobalt Alloys by Various Combustion and Inert Gas Fusion Techniques. ASTM International: West Conshohocken, PA, USA, 2018. [CrossRef]
23. *UNE-EN 10361:2016*; Aceros Aleados. Determinación del Contenido de Níquel. Método por Espectrometría de Emisión Óptica con Fuente de Plasma Inducido. ASTM International: West Conshohocken, PA, USA, 2019.

24. Pereira, J.C.; Aguilar, D.; Tellería, I.; Gómez, R.; Sebastian, M.S. Semi-Continuous Functionally Graded Material Austenitic to Super Duplex Stainless Steel Obtained by Laser-Based Directed Energy Deposition. *J. Manuf. Mater. Process.* **2023**, *7*, 150. [CrossRef]
25. Liu, R.; Ding, Y.; Kamal, K. Advanced Super Alloys with High Performance for Severe Operation Environment Applications. Published in International Journal of Trend in Research and Development (IJTRD), ISSN: 2394-9333, Conference Proceeding | IPMESS-19, June 2019. Available online: http://www.ijtrd.com/papers/IJTRD20589.pdf (accessed on 23 July 2024).

metals

Article

Material Model Fidelity Comparison for the Efficacy of Predicting Residual Stresses in L-PBF Additively Manufactured IN718 Components

David P. Failla, Jr. [1,2,3], Matthew J. Dantin [4], Chuyen J. Nguyen [3,5] and Matthew W. Priddy [1,3,*]

1 Michael W. Hall School of Mechanical Engineering, Mississippi State University, Mississippi State, MS 39762, USA; dpf39@msstate.edu
2 U.S. Army Engineer Research and Development Center, Vicksburg, MS 39180, USA
3 Center for Advanced Vehicular Systems, Mississippi State University, Mississippi State, MS 39759, USA
4 Naval Surface Warfare Center Carderock Division, West Bethesda, MD 20817, USA
5 Department of Computer Science & Engineering, Mississippi State University, Mississippi State, MS 39762, USA
* Correspondence: mwpriddy@me.msstate.edu

Abstract: Internal state variable models are well suited to predict the effects of an evolving microstructure as a result of metal-based additive manufacturing (MBAM) processes in components with complex features. As advanced manufacturing techniques such as MBAM become increasingly employed, accurate methods for predicting residual stresses are critical for insight into component performance. To this end, the evolving microstructural model of inelasticity (EMMI) is suited to modeling these residual stresses due to its ability to capture the evolution of rate- and temperature-dependent material hardening as a result of the rapid thermal cycling present in MBAM processes. The current effort contrasts the efficacy of using EMMI with an elastic–perfectly plastic (EPP) material model to predict the residual stresses for an Inconel 718 component produced via laser powder bed fusion (L-PBF). Both constitutive models are used within a thermo-mechanical finite element framework and are validated by published neutron diffraction measurements to demonstrate the need for higher-fidelity models to predict residual stresses in complex components. Both EPP and EMMI can qualitatively predict the residual stresses trends induced by the L-PBF local raster scanning effects on the component, but the influence of the temperature-dependent yield and lack of plastic strain hardening allowed EPP to perform similar to EMMI away from free surfaces. EMMI offered the most insight at the free surfaces and around critical component features, but this work also highlights EMMI as a process–property-dependent model that needs be calibrated to specimens produced with a similar reference structure for microstructure evolution effects to be accurately predicted.

Keywords: additive manufacturing; thermomechanical modeling; internal state variable models; residual stress; finite element method; inconel alloys

1. Introduction

Metal-based additive manufacturing (MBAM) is an advanced manufacturing process by which a component is fabricated through adding metal material to build a component, rather than by removing or casting material [1,2]. MBAM commonly consists of depositing material in a layer-by-layer fashion that is then fused together through melting the deposited material with a moving heat source. This methodology is executed with a variety of different mechanisms, including laser powder bed fusion (L-PBF) [3], directed energy deposition (DED) [1], and wire-arc additive manufacturing (WAAM) [4]. The process of melting and solidifying between each layer occurs rapidly, which induces residual stresses that can impact component performance in terms of fatigue life [5] and can cause part cracking [6] and part deformation [4,7,8]. For example, as the initial layer is deposited, it solidifies and achieves mechanical equilibrium via expansion. After the next layer is deposited and subsequently melted, the surface between the two layers begins to interact. As the current layer material is melted, heat is conducted through it to the previous layers, causing them to expand. However, the previously deposited layer and current layer heat to different temperatures, and therefore have different cooling rates. This mismatch in cooling rates causes the parts to shrink non-uniformly, which induces a combined tension and compression state in the component. These effects are further exacerbated by the addition of more layers and subsequent thermal cycling. These residual stresses are strongly influenced by the machine process parameters used [9–12]. The machine process parameters include, but are not limited to, laser power, laser scan speed, layer thickness, and scan strategy [13]. Therefore, designers must perform numerous experiments to determine optimal machine process parameter sets for a given material for each MBAM technique. To aid in this endeavor, numerical methods have been developed to model MBAM processing effects on fabricated components. Amongst the most common methods employed for part-scale models are finite element (FE) sequentially coupled thermomechanical simulations that model the response of the part during fabrication [14–17]. For L-PBF, the FE method can be employed by using a moving heat source to represent the laser combined with an element birth and death technique [18,19] through a progressive element activation scheme to mimic the powder deposition for each layer [7]. This allows for the nodal temperature histories to be used as initial conditions to solve for the thermally induced stresses in the mechanical model. For material response, a number of modeling efforts employ an elastic-perfectly plastic (EPP) material model [14] with temperature-dependent properties [20], while others attempt to account for the inherent anisotropy induced as an artifact of the MBAM process or stress relaxation in the form of a relaxation temperature [21]. A higher fidelity material modeling approach can be accomplished through leveraging internal state variable (ISV) models, which attempts to predict the effects of the lower-length scales on the continuum scale. However, little research has been done to predict the thermomechanical effects of the MBAM process with ISV models outside of an implementation of a modified Bammann–Chiesa–Johnson (BCJ) material model [22] and the Evolving Microstructural Model of Inelasticity (EMMI) for DED modeling [23].

The cyclic temperature history observed during the L-PBF process results in a continually changing microstructure. The resultant microstructure and its associated features such as dislocation movement, phase transformation, and hardening, directly contribute to the as-built component's residual stresses, distortion, and mechanical properties. Thus, accounting for the evolutionary history of a component's features with a physically based ISV modeling methodology will better connect the machine process parameters with the component's mechanical response compared to modeling approaches that do not account for the microstructural effects of processing history. The proposed ISV model of interest in this work is EMMI. Although EMMI was primarily developed to capture the effects of high strain rate finite deformation of metals [24], the large fluctuations in temperature and thermal cycling along with resultant distortion and residual stresses make EMMI an ideal candidate to model the process due to its temperature dependence and ability to capture the evolution of hardening. While this work focuses on L-PBF, EMMI has been used to model DED in previous work with success [23]. The primary features of EMMI used in this work are as follows: (i) the elastic modulus is temperature- and damage-dependent; (ii) plasticity is described by isotropic and kinematic hardening variables and models both the hardening and recovery mechanisms that characterize the dislocations and cell structures formed during deformation.

The constitutive models employed in this work are temperature-dependent EPP and EMMI to represent low- and high-fidelity modeling approaches, respectively. The implementation of EMMI in this work evaluates whether more physically informed material models are needed for more accurate residual stress predictions. The aim for this effort is to contrast predictions for the residual stresses experienced in an L-PBF produced part of Inconel 718 (IN718) with complex geometry features such as fillets and holes by contrasting with neutron diffraction results [25] for validation to further develop an understanding of material modeling fidelity limitations and needs for MBAM.

2. Materials and Methods

All simulations leveraged Abaqus/Standard 2019 [26]. The thermal simulation was conducted on a local machine using a 32-core AMD Threadripper. The thermal results were then transferred to a Cray CS300-LC cluster, where 3 nodes with 20 threads per node were used to solve for the mechanical response. No graphical processing unit (GPU) acceleration was used for these simulations [27]. Both the EPP and EMMI implementations used the same thermal history input for the FE sequentially coupled thermomechanical framework to predict the mechanical response for an accurate comparison [28]. The computer aided designs (CAD) were developed using Solid Edge 2023 [29], and meshing was accomplished using Coreform Cubit 2022.6 [30].

2.1. L-Shape Geometry

To contrast the efficacy of EMMI with an EPP model, a component with complex features such as curved edges and holes was needed. The adopted geometry as shown in Figure 1 is replicated from [25], where neutron diffraction was performed on the beamline KOWARI with the Australian Nuclear Science and Technology Organization (ANSTO) to determine internal residual stresses on the L-PBF produced component. To mimic the L-PBF post-processing, the part was extruded an additional 0.4 mm down to capture the stress relief and material loss in the part from wire electrical discharge machine (EDM) removal. The part was modeled on square substrate with a depth of 12.7 mm. The entire substrate was not modeled to reduce computational costs, and the size was reduced to a square cross-sectional area of 65 mm × 65 mm to allow for there to be at least 10 mm all around the component. Both the substrate and the L-shape part were assumed to be fully dense, and any porosity was assumed to be negligible. The substrate is not shown in Figure 1 for clarity. The modeled materials for the component and substrate were IN718 and 316 L, respectively. The cross-sections labeled C2 and C3 intersecting the hole shown in Figure 1 will be of focus for characterizing residual stresses around complex geometries.

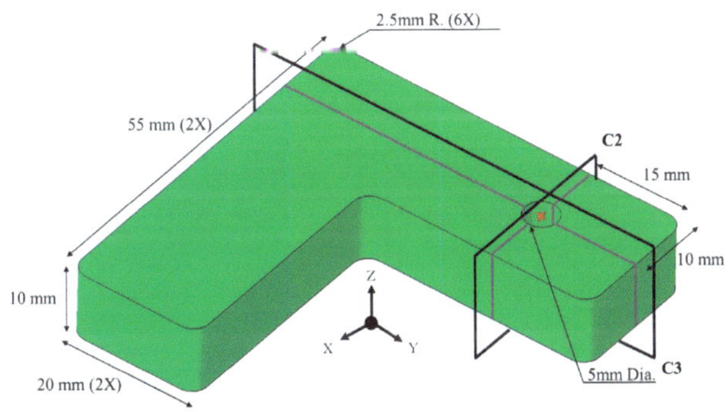

Figure 1. Image of L-shape part showing the C2 and C3 cross–sections intersecting the hole of interest to compare with experimental measurements.

2.2. Mesh

Partitions were made to the geometry along the cross-sections C2 and C3, as shown in Figure 1, to ensure nodes and element faces were on the cross-sections to match experimental measurement locations without interpolating. An additional partition was added between the part and the substrate for sectional property assignments. A non-uniform mesh with a static adaptive mesh refinement was leveraged for element mesh generation [31], where the element size increased through the powder and substrate to reduce computational cost but maintain accuracy for the component. The thermal mesh formulated consisted of the part on the substrate surrounded by powder, as illustrated in Figure 2a, with a total of 688,636 DC3D8 elements. For the mechanical analysis, the element type was changed to C3D8R, and the elements associated with powder were removed since the powder will have a negligible contribution to the mechanical response of the part [32]. The edges of the part were seeded at 0.4 mm, whereas the rest of the part had a global seed size of 2.5 mm using a sweep meshing scheme. The seed size of 0.4 mm was used to allow for 25 element layers to be progressively activated to represent the L-PBF process. The mesh consisted of a single part, and therefore, tie constraints between the substrate, part, and powder were not needed. Furthermore, this approach uses a lumped-mass assumption, as each element layer represents 10 layers of powder to reduce computational costs [15]. Mesh quality was determined through the aspect ratio as illustrated in Figure 2b.

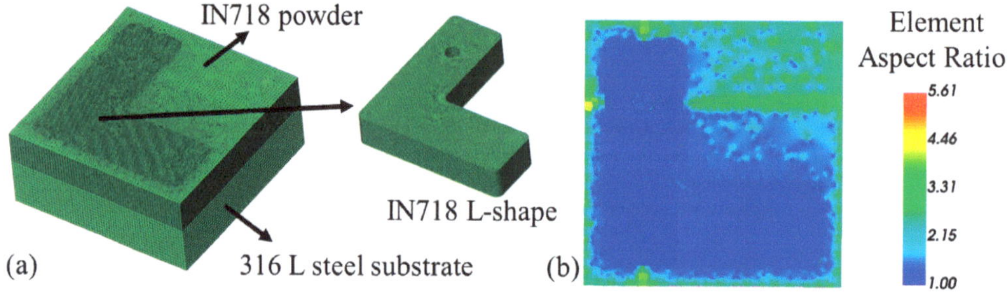

Figure 2. (**a**) The mesh used for the thermal model was partitioned into three sections for applying different materials models to each: the part, the powder, and the substrate. The elements associated with the powder were removed for mechanical analysis. (**b**) Mesh element aspect ratio contour to illustrate quality of mesh from the top-down view. Element density was coarsened in the powder and substrate since they were not the area of interest, but were still needed to capture their effects on the part.

The average aspect ratio for the L-shape part was 1.04, while the average aspect ratio was allowed to change and deviate more for the powder and substrate, as shown in Table 1.

Table 1. Mesh quality metrics.

Aspect Ratio	L-Shape	Powder	Substrate
Average	1.04	1.64	2.29
Minimum	1.00	1.00	1.00
Maximum	1.66	4.52	5.61

2.3. Loads and Boundary Conditions

For the thermal model, an initial and two heat-transfer simulation steps were defined for the printing and cooling processes. As outlined in Table 2, the substrate was assigned an initial temperature to mimic the heated chamber, while the elements associated with the deposited material were initialized to room temperature upon activation. The deposited material represents both the L-shape part and the un-melted powder, as illustrated in Figure 2a.

For the mechanical model, the bottom surface nodes of the substrate were fixed and the elements associated with the powder were removed to reduce computational time. As each layer of elements was activated in the mechanical model, a relaxation temperature of 750 °C [33] was assigned to all elements, and then each element was prescribed a set of nodal temperatures as determined from the output of thermal analysis. Following the cooling step in the mechanical model, a cut removal step via *MODEL CHANGE in Abaqus was leveraged to remove the elements that made up the additional 0.4 mm of thickness added to separate the part from the substrate. Separate steps were used to mimic actual post-processing conditions. The mechanical output at the end of each step is shown in Figure 3.

Table 2. Thermal loading steps and boundary conditions.

Thermal Loading and Boundary Conditions			
Step	Description	316L steel substrate	IN718 L-Shape and Powder
1	Initialization	Substrate initialized to 200 °C	Deposited material progressively initialized to 23 °C
2	Material Deposition/Printing	Nodes on substrate bottom surface fixed to 200 °C	Constant convection and radiation applied to exterior surfaces with sink temperature of 200 °C Concentrated moving heat source dictated by event series applied to L-shape part
3	Cooling	Nodes on substrate bottom surface fixed to 23 °C	Constant convection and radiation applied to exterior surfaces with sink temperature of 23 °C

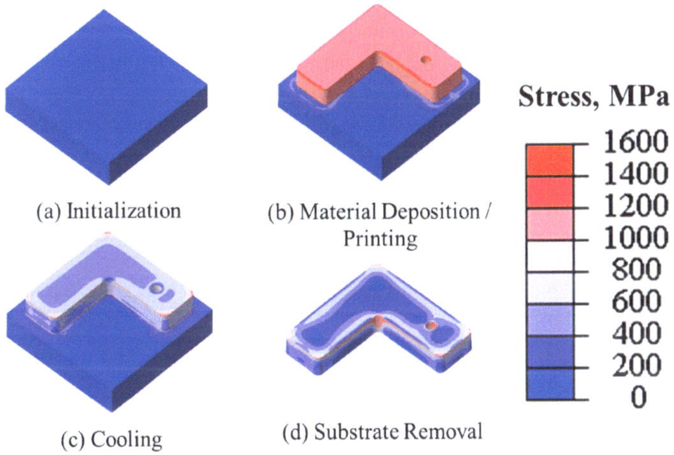

Figure 3. The mechanical simulation showing von Mises stresses consists of 4 steps: (**a**) initialization of substrate and initial boundary conditions, (**b**) predictions of stresses as a result of the layer-wise printing, (**c**) stress relaxation predictions from cooling, (**d**) predictions of stress as part achieves equilibrium after being removed from the substrate.

Automatic incrementation was used to determine the stable time increment [26] with the maximum for the thermal and mechanical models set to 20 s and 50 s, respectively, during printing. The minimum stable time increment was set to 1×10^{-5} s for both the thermal and mechanical modes. In addition, the thermal simulation used an output time series based on the generated event series to force Abaqus to adjust the increment size to solve for the temperatures at key event times during the printing process [34]. These key event times were defined as before element layer activation, immediately before raster scanning, at 5 interval time points during the raster scanning, and at the end of raster scanning. The output cycle was repeated for all layers during the thermal model and was used to provide the key points of the thermal history to the mechanical model. This significantly reduced the computational time of using a specified increment size and allowed for a larger time increment to be used.

2.4. Thermal Modeling

The governing equation for the transient heat conduction that drives and dominates the thermal model is portrayed in Equation (1):

$$\rho C_p \frac{\partial T}{\partial t} = \nabla \cdot (-\kappa \nabla T) + Q \quad (1)$$

where ρ is the material density; C_p is the specific heat capacity; T is the temperature; t is the time; κ is the thermal conductivity; and Q represents the thermal load from the heat source. The thermal load consists of the heat contributions from the conduction via the concentrated moving heat source, convection, and radiation. The convection heat transfer is determined using Newton's law of cooling as illustrated in Equation (2):

$$q_{conv} = h(T_S - T_\infty) \quad (2)$$

where h is the convection heat transfer coefficient; T_S is the surface temperature of the part; and T_∞ is the sink temperature. The convection heat transfer coefficient for the whole model was assumed constant at 15 W/m^2K from [5]. The radiation effects are accounted for with the Stefan–Boltzmann law as shown in Equation (3):

$$q_{rad} = \sigma\varepsilon\left(T_S^4 - T_\infty^4\right) \quad (3)$$

where ε is the emissivity and σ is the Stefan–Boltzmann constant. The emissivity for the whole model was assumed constant at 0.3 from [5]. A concentrated moving heat source is used to represent the conduction from the laser. This strategy was implemented over other models such as a Goldak Ellipsoidal model [35] due to the elements being much larger than the laser spot size and for computational efficiency [36]. This assumption is reasonable due to the high speed of the laser with respect to the size of the elements and the overall size of the part, and the cross-section in the build-direction has minimal variance throughout the build [10]. The laser process parameters for laser power and speed were 200 W and 900 mm/s for the infill and a 100 W and 450 mm/s for the contour, respectively, to reflect the build from [25]. The scan strategy, or heat source path, via the event series selected was meander. The original component was printed using a stripe scan strategy, but due to event series generation limits, meander was used. The difference in the effects of these scan strategies for this component are then assumed to be negligible. The CAD file used for mesh generation was also used to develop a g-code file via Slic3r [37] as shown in Figure 4.

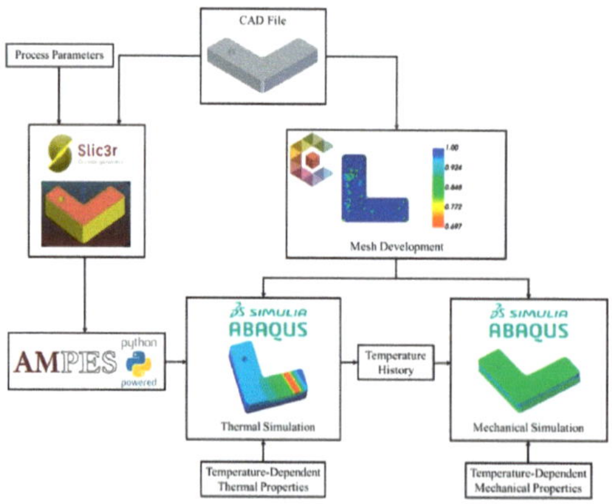

Figure 4. Thermomechanical modeling flow showing transition from CAD to mesh and process parameters and from CAD to g-code to the AMPES-generated event series to create inputs for the thermal and mechanical models.

The machine process parameters derived from [25] alongside the g-code files generated by Slic3r were provided as inputs to AMPES [34], a Python event series generation pre-processing tool that uses a RepRap flavored g-code file to create print-path event series for use with numerical solvers. The resulting input power and theoretical volumetric energy density, as estimated from the generated event series are shown in Figure 5a and Figure 5b, respectively. The input volumetric energy density was approximated using Equation (4) from [13].

$$ED_{est} = \frac{P\left(\frac{1}{hatch \cdot v}\right)}{t} \tag{4}$$

where ED_{est} is the estimated input volumetric energy density; P is the input power; *hatch* is the hatch spacing; v is the laser speed; and t is the layer thickness. The meshed files designed in Cubit paired with the generated event series were used as the input to Abaqus to implicitly simulate the thermal and mechanical response of the component. See Appendices A and C for thermal material constants for IN718 and 316L, respectively. Approximating an interlayer dwell time [38] of 180 s with 347 layers, the total step time for printing was computed by AMPES to be 73,249 s, or 20.3 h, which is assumed to be close to how long the L-shape parts from [25] took to fabricate with the selected L-PBF machine process parameters.

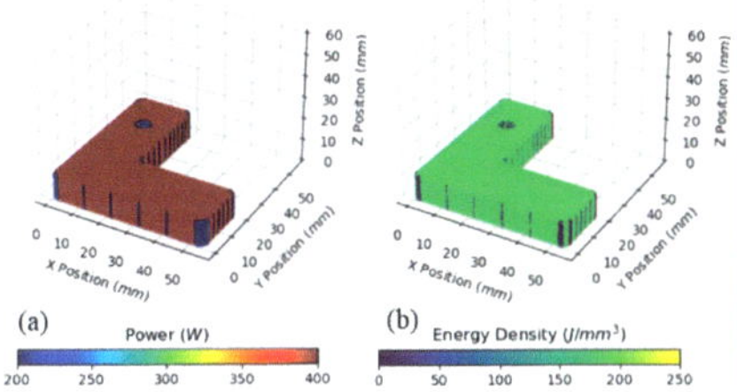

Process Parameters	Value	Units
Layer Height	30	μm
Laser Power (Perimeter)	100	W
Laser Power (Infill)	200	W
Laser Speed (Perimeter)	450	mm/s
Laser Speed (Infill)	900	mm/s

Figure 5. (**a**) Power and (**b**) theoretical input volumetric energy density plotted in 3D space as estimated by the generated event series with associated machine process parameters from [25].

2.5. Mechanical Modeling—Elastic–Perfectly Plastic

The EPP model assumes that no hardening occurs in the material system shown in Equation (5):

$$\varepsilon = \begin{cases} \frac{\sigma}{E} & \sigma < \sigma_y \\ \varepsilon_{el} + \varepsilon_{pl} + \varepsilon_{th} & \sigma \geq \sigma_y \end{cases} \tag{5}$$

where ε is the total strain; ε_{el} is the elastic strain contribution; ε_{pl} is the plastic strain contribution; ε_{th} is the thermal strain contribution; σ is the nominal stress; σ_y is the yield stress of the material; and E is the Young's Modulus. Though the actuality of a material foregoing any form of strain hardening or thermal softening is unrealistic, the amount of hardening can be small enough in ductile metals for the use of the EPP model as an approximation. For the purposes of demonstrating the robustness of EMMI to capture flow stress, the limitations of EPP serve to stress the need for a physically based ISV. The mechanical strains are interpreted from the prescribed nodal temperatures via thermal expansion given as:

$$\varepsilon^{th} = \alpha(T, f_\beta)\left(T - T^0\right) - \alpha\left(T^I, f_\beta^I\right)\left(T^I - T^0\right) \tag{6}$$

where α thermal expansion coefficient; f_β are the current values for the predefined field variables; T^0 is the reference temperature; T^I is the initial temperature; f_β^I are the initial value of the of the field variables [26]. These strains are related to stress via Hooke's law:

$$\sigma = C\varepsilon \tag{7}$$

where C is a rank four material stiffness tensor. Mechanical properties for IN718 and 316L are listed in Appendices B and D, respectively.

2.6. Mechanical Modeling—Evolving Microstructural Model of Inelasticity

EMMI was developed to predict the inelastic response of metals in environments containing high strain rate and large fluctuations in temperature. Thus, EMMI is an ideal candidate to model the L-PBF process due to the aforementioned rapid thermal cycling observed. EMMI is a dislocation mechanics-based ISV model and is the successor to the widely used BCJ plasticity model [39]. Relevant updates to EMMI from BCJ include updated equations representing changes to strain rate effects and recovery mechanisms. These updates position EMMI as more physically based than BCJ, particularly in the formulation of plasticity. Furthermore, as an ISV model, EMMI captures state variables to track the cyclic temperature history, hardening, and recovery effects observed in the L-PBF process. The version of EMMI used in this work was modified to include the relaxation of deviatoric stresses, hardening variables, and plastic strains above 80% of the melting temperature of the metal [23]. The assumption that justifies this modification lies in the physical basis of increased dislocation motion that occurs above 80% of the melting temperature resulting in annihilation of work hardening which is caused by the metal behaving as a linear viscous material [40]. Furthermore, the implemented version of EMMI was based upon three features: (i) the temperature dependence of yield strength and Young's Modulus; (ii) isotropic and kinematic hardening variables to track the evolution of plasticity; and (iii) hardening and recovery mechanisms that represent the cell structures and dislocations created under deformation.

The finite strain deformation gradient, F, is multiplicatively decomposed into the following: (i) the thermal deformation gradient, F_θ, (ii) the deviatoric plastic deformation gradient, F_p, (iii) the volumetric deformation gradient, F_d, and (iv) the elastic recoverable deformation gradient, F_e [24]. Thus, the total deformation gradient is shown in Equation (8). Damage is neglected in this implementation due to the relatively small strain nature of L-PBF.

$$F = F_e F_d F_p F_\theta \tag{8}$$

The plastic strain is represented by three equations: (i) the inelastic flow rule, (ii) isotropic hardening, and (iii) kinematic hardening. The inelastic flow rule is shown in Equation (9), where $\overline{\sigma}_{eq}$ is the equivalent stress, $\overline{\kappa_s}$ is internal stress due to isotropic hardening, $Y(\theta)$ is the temperature-dependent yield function, and $f(\theta)$ is a temperature-dependent material parameter. The strain rate due to isotropic hardening, $\dot{\overline{\varepsilon}}_s$, is shown in Equation (10), where H is a hardening material constant, $R_D(\theta)$ is a dynamic recovery constant, and $R_s(\theta)$ and $Q_s(\theta)$ are temperature-dependent static recovery constants. The strain rate due to kinematic hardening, $\dot{\overline{\beta}}$, is shown in Equation (11), where h is a hardening material constant, $r_d(\theta)$ is a temperature-dependent dynamic recovery constant, and $\overline{\beta}$ is isotropic hardening. The calibration constants used in this study can be found at reference [41]. Further documentation for EMMI can be found in Marin et al. [24]. See Appendix E for EMMI material parameters used.

$$\dot{\overline{\varepsilon}}^p = f(\theta)\left[sinh\left(\langle\frac{\overline{\sigma}_{eq}}{\overline{\kappa_s} + Y(\theta)} - 1\rangle\right)\right]^n \tag{9}$$

$$\dot{\overline{\varepsilon}}_s = [H - R_D(\theta)\overline{\varepsilon}_s]\dot{\overline{\varepsilon}}^p - R_s(\theta)\overline{\varepsilon}_s sinh[Q_s(\theta)\overline{\varepsilon}_s] \tag{10}$$

$$\dot{\overline{\beta}} = hF^{e^T}d^pF^e - r_d(\theta)\dot{\overline{\varepsilon}}\sqrt{\frac{2}{3}}||\overline{\beta}||\overline{\beta} \tag{11}$$

3. Results

The computational resources used for the thermal model and mechanical models with each material model are listed in Table 3. The total runtime for all jobs was about

100 h, or just over 4 days. This framework then offers significant computational time savings over using a smaller increment size and no initial temperature in the mechanical model. Through preliminary runs using a fixed time increment size of 0.4 s with similar mass lumping assumptions, the thermal model was estimated to have taken 3 to 4 months to finish, which was not feasible. This also did not account for the mechanical model predictions, which usually take much longer on account of the added degrees of freedom. It is also important to note that the thermal model took longer to solve due to its much smaller time increment scheme, as noted in Section 2.3. This resulted in the thermal model having 3191 increments, as opposed to the EPP and EMMI mechanical model predictions having 498 and 504 increments, respectively. Despite having fewer degrees of freedom, the smaller increment size combined with fewer computing threads resulted in a much higher runtime than the mechanical model predictions. Lastly, the thermal predictions were solved on a different thread count than the mechanical predictions due to resource availability.

Table 3. Computational resources used.

	Thermal	EPP	EMMI
Thread Count	32	60	60
Element Count	688,636	540,176	540,176
Runtime (h)	62.6	17.2	20

Figure 6 shows the moving heat source at the final increment of the final layer. The expected thermal gradation outlined in the literature is qualitatively present [42], as shown in Figure 6a, and the laser radiation penetration depth is shallow, as expected for IN718 being printed with L-PBF [43], as shown in Figure 6b. The temperatures experienced by the part due to the heat source were approximately 350 °C, and it provided those thermal histories around the hole. This is the rationale for using an initial temperature of 750 °C [33] in the mechanical models as a relaxation temperature [21]. This approach provides more information on the local raster scanning effects on residual stresses at complex features than just layer heating by combining the initialized temperature in the mechanical model with the lower temperatures from the local raster scanning at key time points.

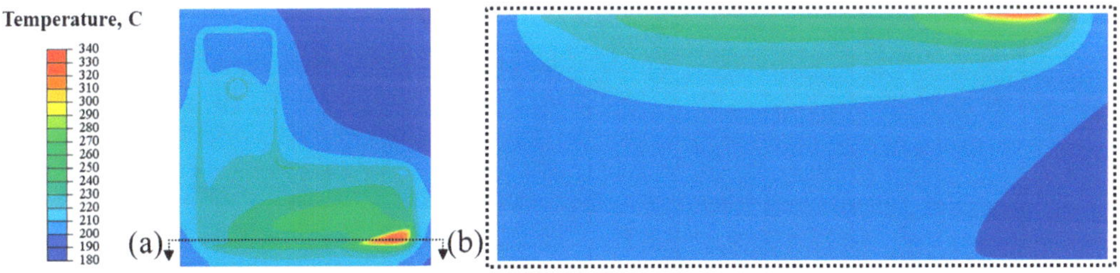

Figure 6. Predicted thermal history dictated by moving heat source at final increment of a layer during printing of (**a**) top view and (**b**) front view cross-section.

3.1. Mechanical Model Results—Surface C2

The mechanical models show the general qualitative trend of AM produced parts, with tensile stress states at the free surfaces and compression internally, as shown in the literature [25,44]. Contrasting EMMI with EPP, EMMI tends to yield much higher residual stresses as opposed to EPP in Figure 7, with the greatest deviation being along the sides of the hole in the build direction.

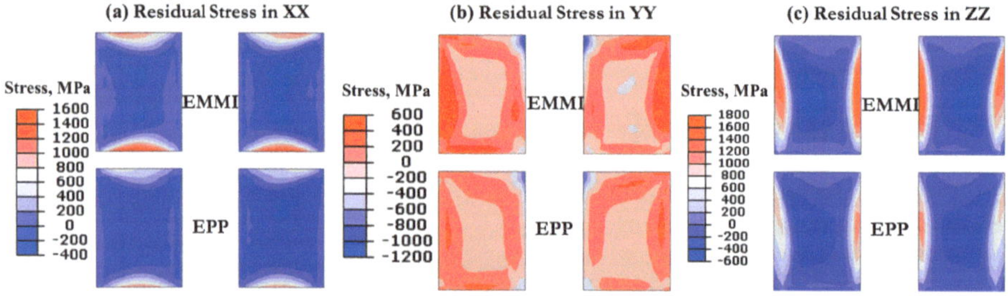

Figure 7. Predicted EPP and EMMI residual stress contours for stresses in (**a**) XX, (**b**) YY, and (**c**) ZZ directions for cross–section C2.

By observing the residual stresses along the line of measurement in Figure 8 on cross-section C2, it is determined that both EPP and EMMI capture the inflection of tensile to compressive to tensile again for all principal stress directions, showing that both models can qualitatively capture the stress response. However, the compressive stresses predicted by both EPP and EMMI are much stiffer than the experimental observation by as much as 200 MPa. Lastly, both models show higher extrema closer to free surfaces, but due to the implementation of the neutron diffraction measurements not approaching the free surface, these simulation points cannot be validated.

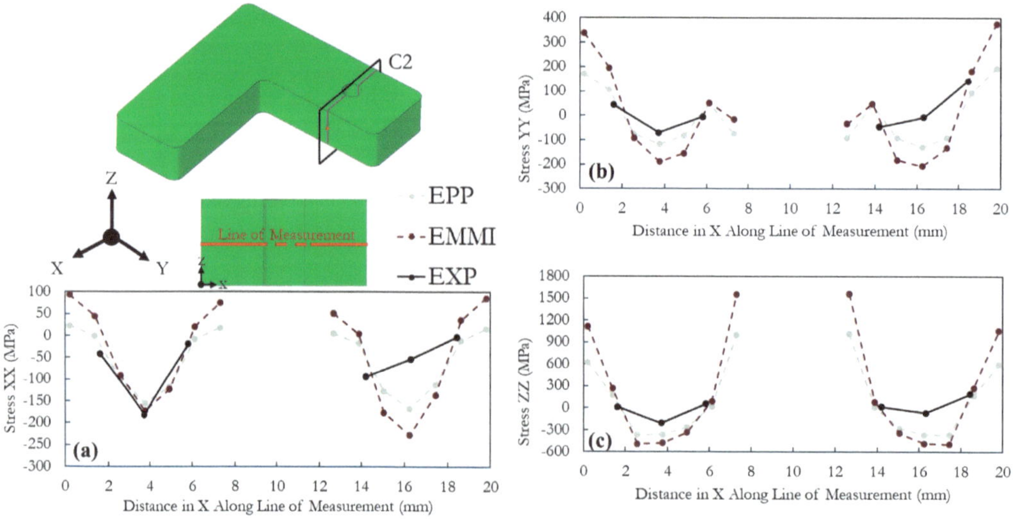

Figure 8. The predicted EPP and EMMI residual stresses (dashed lines) contrasted with the experimentally measured (EXP, solid lines) residual stresses [5] in the (**a**) XX, (**b**) YY, and (**c**) ZZ along the line of measurement on cross–section C2.

3.2. Mechanical Model Results—Surface C3

Similar findings can be found with cross-section C2 for cross-section C3 when contrasting EPP and EMMI in Figures 9–11. The stress extrema predicted by EMMI are higher for both tensile and compressive states, with the stresses in the YY and ZZ at the surface of the hole being the most notable.

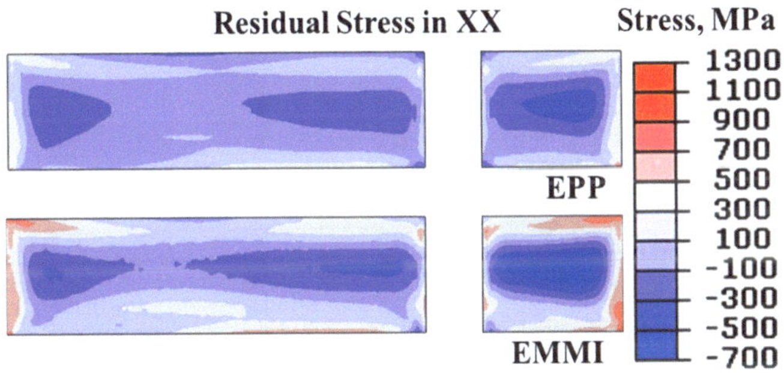

Figure 9. Predicted EPP and EMMI residual stress contours for stress in XX for cross–section C3.

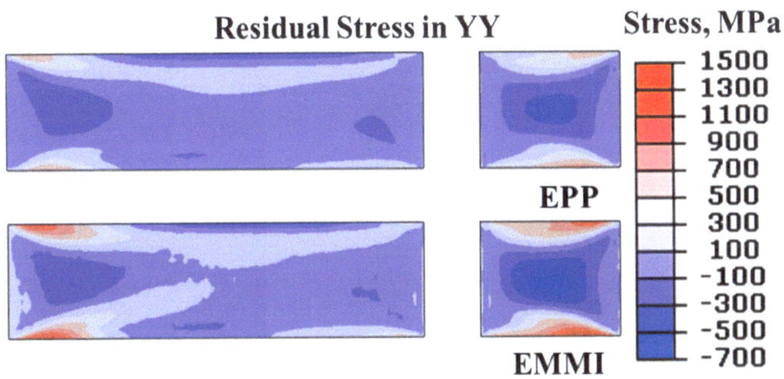

Figure 10. Predicted EPP and EMMI residual stress contours for stress in YY for cross–section C3.

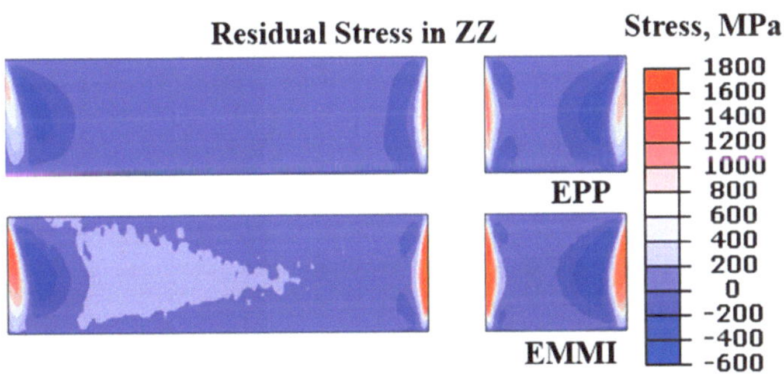

Figure 11. Predicted EPP and EMMI residual stress contours for stress in ZZ for cross–section C3.

Qualitatively, both models yield a similar stress contour to the neutron diffraction results. Interestingly, EPP outperforms EMMI for the stress measurements along the line of measurement from Figure 12 for both the YY and ZZ principal stress directions, highlighting that there is artificial hardening being induced by EMMI for the stress predictions. Furthermore, based on the performance of EPP and the minimal plastic stresses observed, it would show that temperature dependent yield is what drives the residual stresses in the selected L-PBF component produced with the given machine process parameters.

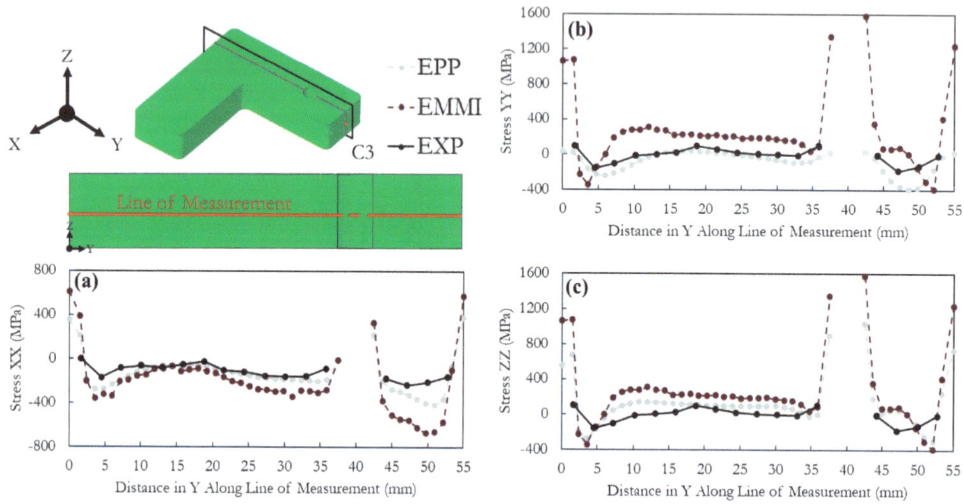

Figure 12. The predicted EPP and EMMI residual stresses (dashed lines) contrasted with the experimentally measured (EXP, solid lines) residual stresses [5] in the (**a**) XX, (**b**) YY, and (**c**) ZZ along the line of measurement on cross–section C3.

3.3. Modeling Limitations

Due to the size of the mesh and nature of the L-PBF simulation process, an h-refinement was not feasible for this effort, and the mesh became dependent on aspect ratio for quality assurance. Furthermore, due to the lack of experimental in-situ thermal history data, quantitative thermal model validation via heat source spot size or thermal gradation through the build height was not possible. Also, due to larger time increments used, the peak temperatures above melt expected from L-PBF are not captured in the thermal predictions and far-field temperatures are used. Lastly, both the L-shape part and substrate were assumed to be fully dense since information on part density was not provided with the neutron diffraction data from [5]. However, from the literature characterizing similar IN718 L-shape parts [45] and determining that porosity was present, it is likely that the referenced L-shape part modeled in this effort also had porosity throughout. By not accounting for this porosity in the current work, stress predictions are likely to over-predict where porosity is dense in the L-shape part due to the inability of the L-shape part to carry a high load in these regions.

4. Conclusions

A comparison between neutron diffraction experimental results and numerical predictions produced through FE methods leveraging EMMI and EPP material models of the induced residual stresses from L-PBF was made. Key takeaways from the comparison are as follows:

- Leveraging lower raster scanning temperatures at key event points with an initialized temperature equivalent to the relaxation of the given material in the mechanical model provides a reasonable amount of far-field thermal history information to predict residual stresses accurately.
- Stresses were the highest at the free surfaces, shown both experimentally and numerically.
- The localized stresses at regions of complex features such as holes demonstrate the need to account for local raster scanning effects in numerical models.
- Plastic hardening appears to have little effect on the L-PBF response away from the free surfaces of the L-shape part, as determined experimentally.

- Both EPP and EMMI qualitatively agree with the neutron diffraction measurements for stress on the C2 and C3 surfaces.
- The residual stresses in the L-shape part are strongly influenced by the temperature-dependent properties of the IN718 material.
- The influence of the temperature-dependent yield on the solution and lack of plastic strain induced hardening in areas away from free surfaces allowed for EPP to more closely match the neutron diffraction measurements than EMMI in some cases, specifically on the cross-section C3 for XX and YY.
- Though not validated by experimental data due to neutron diffraction depth penetration limitations, the stresses at the hole surfaces predicted by EMMI are as much as 67% higher than EPP, which could be attributed to plastic hardening at the free surfaces.
- Both EMMI and EPP struggled to predict maximum compressive stresses in the XX, YY, and ZZ directions on the C3 surface from 45 mm to 52 mm. One explanation for the deviation could be the presence of porosity between the hole and the free surface.

It can be concluded that both EMMI and temperature-dependent EPP material models can provide accurate predictions of residual stresses for complex geometries on the part-scale. However, for both this geometry and material, EPP outperformed EMMI in most cases. One theory for the difference in EPP and EMMI results is attributed to EMMI being calibrated to wrought IN718 [41] as opposed to MBAM-produced IN718 specimens, which have different microstructures. This work therefore highlights the need for future-work where microstructurally sensitive ISV models are calibrated to a reference condition of the selected material system to determine if this could enable EMMI to better predict the effects of the L-PBF process–structure relationship on the desired component. Lastly, the presence of such low experimentally measured maximum stresses from [25] that this study is based on could indicate high amounts of porosity [46,47]. Novel scan strategy work completed for this same L-shape part printed out of IN718 in a different orientation and with different scan strategies [45] further supports this theory. This would also explain why both the EPP and EMMI predictions trended to higher maxima at some locations than the neutron diffraction measurements in Figures 8 and 12, since porosity was assumed to be negligible.

Author Contributions: D.P.F.J.: Writing—review and editing, Writing—original draft, Software, Project administration, Methodology, Investigation, Conceptualization. M.J.D.: Writing—original draft, Writing—review and editing, Investigation, Conceptualization. C.J.N.: Writing—review and editing, Software, Methodology. M.W.P.: Writing—review and editing, Supervision, Project administration, Conceptualization. All authors have read and agreed to the published version of the manuscript.

Funding: This research received no external funding.

Data Availability Statement: The original contributions presented in the study are included in the article, further inquiries can be directed to the corresponding author.

Acknowledgments: The authors would like to thank the Center for Advanced Vehicular Systems for assistance with computational resources used for this effort.

Conflicts of Interest: The authors declare that they have no known competing financial interests or personal relationships that could have appeared to influence the work reported in this paper.

Appendix A. IN718 Thermal Material Properties

Temperature-dependent conductivity and specific heat in Table A1 were derived from [5], which were adapted from [48]. Density was taken as 8193 kg/m^3 assumed to be constant and was derived from [49]. The latent heat of fusion of 210 J/g, solidus temperature of 1260 °C, and liquidus temperature of 1336 °C were derived from [50].

Table A1. Temperature-dependent thermal properties of IN718.

Temperature (°C)	Conductivity (W/mK)	Specific Heat (J/kgK)
20	11.4	427
100	12.5	442
300	14	482
500	15.5	522
700	21.5	562
727	21	-
900	-	602
927	25	-
1227	30	-
1350	-	692

Appendix B. IN718 Mechanical Material Properties

Temperature-dependent Young's modulus, yield strength, and coefficient of thermal expansion in Table A2 were derived from [5], which were adapted from [48]. Poisson ratio was taken as 0.3 from [48] and assumed constant.

Table A2. Temperature-dependent mechanical properties of IN718.

Temperature (°C)	Elastic Modulus (GPa)	Yield Strength (MPa)	Coefficient of Thermal Expansion
21	208	1172	-
93	205	1172	1.28×10^{-5}
204	202	-	1.35×10^{-5}
316	194	-	1.39×10^{-5}
427	186	1089	1.42×10^{-5}
538	179	1068	1.44×10^{-5}
649	172	1034	1.51×10^{-5}
760	162	827	1.60×10^{-5}
871	127	286	-
954	17.8	138	-

Appendix C. 316L Thermal Material Properties

Temperature-dependent conductivity and specific heat in Table A3 were derived from [14]. Density was assumed to be constant and taken as 8.0 kg/m^3 from [20]. The latent heat of fusion was taken as 207 J/g [51] and the solidus temperature of 1375 °C and liquidus temperature of 1400 °C were derived from [52].

Table A3. Temperature-dependent thermal properties of 316L.

Temperature (°C)	Conductivity (W/mK)	Specific Heat (J/kgK)
0	12.76	4.40×10^8
159	14.94	5.10×10^8
317	17.18	5.45×10^8
476	19.3	5.60×10^8
634	21.48	5.85×10^8
793	23.66	6.20×10^8
951	25.84	6.50×10^8
1110	28.02	6.80×10^8
1268	30.2	7.13×10^8
1377	-	7.34×10^8
1387	-	6.19×10^9
1417	-	6.19×10^9
1427	32.38	7.44×10^8

Appendix D. 316L Mechanical Material Properties

Temperature-dependent Young's modulus, yield strength, and coefficient of thermal expansion in Table A4 were derived from [14]. Poisson ratio was taken as 0.294 from [20] and assumed constant.

Table A4. Temperature-dependent mechanical properties of 316L.

Temperature (°C)	Elastic Modulus (GPa)	Yield Strength (MPa)	Coefficient of Thermal Expansion
0	200.8	529	1.51×10^{-5}
159	188.9	402.04	1.61×10^{-5}
317	176.3	322.69	1.70×10^{-5}
476	163.1	280.37	1.77×10^{-5}
634	149.1	232.76	1.83×10^{-5}
793	134.6	179.86	1.87×10^{-5}
951	119.3	137.54	1.91×10^{-5}
1110	103.4	89.93	1.92×10^{-5}
1268	86.8	47.61	1.92×10^{-5}
1427	69.5	0.001	1.92×10^{-5}

Appendix E. IN718 EMMI Parameters

The EMMI parameters used were derived from [41] and were used as calibrated. As mentioned before, damage was neglected in this work, and all associated parameters were set to 0. The constants driving the EMMI response in MBAM IN718 are the m_1, m_2, m_3, m_4, and m_5 constants used for the temperature dependent yield function for IN718 from [41] and listed in Table A5.

Table A5. EMMI parameters for temperature-dependent yield function.

Material	m_1	m_2	m_3	m_4	m_5
IN718	1.2321	0.4508	0.14395	11.49	0.67071

References

1. Frazier, W.E. Metal Additive Manufacturing: A Review. *J. Mater. Eng. Perform.* **2014**, *23*, 1917–1928. [CrossRef]
2. Baumers, M.; Dickens, P.; Tuck, C.; Hague, R. The cost of additive manufacturing: Machine productivity, economies of scale and technology-push. *Technol. Forecast. Soc. Chang.* **2016**, *102*, 193–201. [CrossRef]
3. Dunbar, A.J.; Denlinger, E.R.; Gouge, M.F.; Simpson, T.W.; Michaleris, P. Comparisons of laser powder bed fusion additive manufacturing builds through experimental in situ distortion and temperature measurements. *Addit. Manuf.* **2017**, *15*, 57–65. [CrossRef]
4. Wu, B.; Pan, Z.; Ding, D.; Cuiuri, D.; Li, H.; Xu, J.; Norrish, J. A review of the wire arc additive manufacturing of metals: Properties, defects and quality improvement. *J. Manuf. Process.* **2018**, *35*, 127–139. [CrossRef]
5. Pant, P.; Sjöström, S.; Simonsson, K.; Moverare, J.; Proper, S.; Hosseini, S.; Luzin, V.; Peng, R. A Simplified Layer-by-Layer Model for Prediction of Residual Stress Distribution in Additively Manufactured Parts. *Metals* **2021**, *11*, 861. [CrossRef]
6. Holfelder, P.; Brenner, F.; Rund, M.; Witte, A.; Junghans, S.; Seyfert, C.; Richter, M.; Dell, H.; Koukolikova, M.; Gese, H.; et al. Finite element simulation of plasticity and fracture for Inconel 718 deposited by laser powder bed fusion—Chances, use and challenges. *Addit. Manuf.* **2022**, *56*, 102888. [CrossRef]
7. Yang, Y.P.; Jamshidinia, M.; Boulware, P.; Kelly, S.M. Prediction of microstructure, residual stress, and deformation in laser powder bed fusion process. *Comput. Mech.* **2018**, *61*, 599–615. [CrossRef]
8. Wang, D.; Song, C.; Yang, Y.; Bai, Y. Investigation of crystal growth mechanism during selective laser melting and mechanical property characterization of 316L stainless steel parts. *Mater. Des.* **2016**, *100*, 291–299. [CrossRef]
9. Bugatti, M.; Semeraro, Q. Limitations of the inherent strain method in simulating powder bed fusion processes. *Addit. Manuf.* **2018**, *23*, 329–346. [CrossRef]
10. Gouge, M.; Denlinger, E.; Irwin, J.; Li, C.; Michaleris, P. Experimental validation of thermo-mechanical part-scale modeling for laser powder bed fusion processes. *Addit. Manuf.* **2019**, *29*, 100771. [CrossRef]
11. Gockel, J.; Sheridan, L.; Koerper, B.; Whip, B. The influence of additive manufacturing processing parameters on surface roughness and fatigue life. *Int. J. Fatigue* **2019**, *124*, 380–388. [CrossRef]

12. Stokes, R.M.; Yadollahi, A.; Priddy, M.W.; Bian, L.; Hammond, V.H.; Doude, H.R. Effects of Build Interruption and Restart Procedure on Microstructure and Mechanical Properties of Laser Powder Bed Fusion Al-Si-10Mg. *J. Mater. Eng. Perform.* **2022**, *32*, 1576–1588. [CrossRef]
13. Cherry, J.A.; Davies, H.M.; Mehmood, S.; Lavery, N.P.; Brown, S.G.R.; Sienz, J. Investigation into the effect of process parameters on microstructural and physical properties of 316L stainless steel parts by selective laser melting. *Int. J. Adv. Manuf. Technol.* **2015**, *76*, 869–879. [CrossRef]
14. Hodge, N.E.; Ferencz, R.M.; Solberg, J.M. Implementation of a thermomechanical model for the simulation of selective laser melting. *Comput. Mech.* **2014**, *54*, 33–51. [CrossRef]
15. Ganeriwala, R.K.; Strantza, M.; King, W.E.; Clausen, B.; Phan, T.Q.; Levine, L.E.; Brown, D.W.; Hodge, N.E. Evaluation of a thermomechanical model for prediction of residual stress during laser powder bed fusion of Ti-6Al-4V. *Addit. Manuf.* **2019**, *27*, 489–502. [CrossRef]
16. Lindgren, L.-E.; Lundbäck, A.; Fisk, M.; Pederson, R.; Andersson, J. Simulation of additive manufacturing using coupled constitutive and microstructure models. *Addit. Manuf.* **2016**, *12*, 144–158. [CrossRef]
17. Dantin, M.J.; Furr, W.M.; Priddy, M.W. Toward a Physical Basis for a Predictive Finite Element Thermal Model of the LENS^{TM} Process Leveraging Dual-Wavelength Pyrometer Datasets. *Integrating Mater. Manuf. Innov.* **2022**, *11*, 407–417. [CrossRef]
18. Soundararajan, B.; Sofia, D.; Barletta, D.; Poletto, M. Review on modeling techniques for powder bed fusion processes based on physical principles. *Addit. Manuf.* **2021**, *47*, 102336. [CrossRef]
19. Baiges, J.; Chiumenti, M.; Moreira, C.A.; Cervera, M.; Codina, R. An adaptive Finite Element strategy for the numerical simulation of additive manufacturing processes. *Addit. Manuf.* **2021**, *37*, 101650. [CrossRef]
20. Hodge, N.E.; Ferencz, R.M.; Vignes, R.M. Experimental comparison of residual stresses for a thermomechanical model for the simulation of selective laser melting. *Addit. Manuf.* **2016**, *12*, 159–168. [CrossRef]
21. Denlinger, E.R.; Michaleris, P. Effect of stress relaxation on distortion in additive manufacturing process modeling. *Addit. Manuf.* **2016**, *12*, 51–59. [CrossRef]
22. Johnson, K.L.; Emery, J.M.; Hammetter, C.I.; Brown, J.A.; Grange, S.J.; Ford, K.R.; Bishop, J.E. Predicting the reliability of an additively-manufactured metal part for the third Sandia fracture challenge by accounting for random material defects. *Int. J. Fract.* **2019**, *218*, 231–243. [CrossRef]
23. Dantin, M.J.; Priddy, M.W. A Rate- and Temperature-Dependent Thermomechanical Internal State Variable Model of the Directed Energy Deposition Process. *J. Mater. Eng. Perform.* **2024**, *33*, 4051–4064. [CrossRef]
24. Marin, E.B.; Bammann, D.J.; Regueiro, R.A.; Johnson, G.C. *On the Formulation, Parameter Identification and Numerical Integration of the EMMI Model: Plasticity and sotropic Damage*; Sandia National Laboratories: Albuquerque, NM, USA, 2006.
25. Pant, P.; Proper, S.; Luzin, V.; Sjöström, S.; Simonsson, K.; Moverare, J.; Hosseini, S.; Pacheco, V.; Peng, R.L. Mapping of residual stresses in as-built Inconel 718 fabricated by laser powder bed fusion: A neutron diffraction study of build orientation influence on residual stresses. *Addit. Manuf.* **2020**, *36*, 101501. [CrossRef]
26. Hibbett, Karlsson and Sorensen. *ABAQUS/Standard: User's Manual*; Hibbitt, Karlsson and Sorenson Incorporated: Pawtucket, RI, USA, 1998; Volume 1.
27. Betts, J.L.; Downs, W.; Dantin, M.J.; Priddy, M.W. Examining the Gpu Acceleration Speed-Up for Finite Element Modeling of Additive Manufacturing. Available online: https://repositories.lib.utexas.edu/server/api/core/bitstreams/c004ad59-d8d5-4f4b-abf5-92eea1458740/content (accessed on 26 July 2024).
28. An, N.; Yang, G.; Yang, K.; Wang, J.; Li, M.; Zhou, J. Implementation of Abaqus user subroutines and plugin for thermal analysis of powder-bed electron-beam-melting additive manufacturing process. *Mater. Today Commun.* **2021**, *27*, 102307. [CrossRef]
29. Solid Edge Documentation. Siemens. Available online: https://docs.sw.siemens.com/en-US/documents/246738425/PL20190201104957790?start=0 (accessed on 26 July 2024).
30. Skroch, M. Coreform CubitTM 2024.7 User Documentation. Coreform. Available online: https://coreform.com/cubit_help/cubithelp.htm#t=cubit_users_manual.html (accessed on 26 July 2024).
31. Luo, Z.; Zhao, Y. A survey of finite element analysis of temperature and thermal stress fields in powder bed fusion Additive Manufacturing. *Addit. Manuf.* **2018**, *21*, 318–332. [CrossRef]
32. Song, X.; Feih, S.; Zhai, W.; Sun, C.-N.; Li, F.; Maiti, R.; Wei, J.; Yang, Y.; Oancea, V.; Romano Brandt, L.; et al. Advances in additive manufacturing process simulation: Residual stresses and distortion predictions in complex metallic components. *Mater. Des.* **2020**, *193*, 108779. [CrossRef]
33. Rahimi, S.; King, M.; Dumont, C. Stress relaxation behaviour in IN718 nickel based superalloy during ageing heat treatments. *Mater. Sci. Eng. A* **2017**, *708*, 563–573. [CrossRef]
34. Failla, D.P., Jr.; Nguyen, C.J. AMPES: Additive Manufacturing Process Event Series Generator. Available online: https://github.com/Computational-Mechanics-Materials-Lab/AMPES (accessed on 8 June 2024).
35. Goldak, J.; Chakravarti, A.; Bibby, M. A new finite element model for welding heat sources. *Metall. Trans. B* **1984**, *15*, 299–305. [CrossRef]
36. Kiran, A.; Li, Y.; Hodek, J.; Brázda, M.; Urbánek, M.; Džugan, J. Heat Source Modeling and Residual Stress Analysis for Metal Directed Energy Deposition Additive Manufacturing. *Materials* **2022**, *15*, 2545. [CrossRef]
37. Hodgson, G. Slic3r Manual. Available online: https://manual.slic3r.org/ (accessed on 10 May 2024).

38. Mohr, G.; Altenburg, S.J.; Hilgenberg, K. Effects of inter layer time and build height on resulting properties of 316L stainless steel processed by laser powder bed fusion. *Addit. Manuf.* **2020**, *32*, 101080. [CrossRef]
39. Bammann, D.J. An internal variable model of viscoplasticity. *Int. J. Eng. Sci.* **1984**, *22*, 1041–1053. [CrossRef]
40. Goldak, J.A.; Akhlagi, M. *Computational Welding Mechanics*; Springer: New York, NY, USA, 2005.
41. Marin, E.; Bammann, D.; Brown, A.; El Kadiri, H.; Costley, R.; Wang, P.; Horstemeyer, M. Modeling the Hot Forging of Nickel-Based Superalloys: IN718 and Alloy 718Plus®. In Proceedings of the 7th International Symposium on Superally 718 and Derivatives, Pittsburgh, PA, USA, 10–13 October 2010; pp. 331–342.
42. Lane, B.; Jacquemetton, L.; Piltch, M.; Beckett, D. *Thermal Calibration of Commercial Melt Pool Monitoring Sensors on a Laser Powder Bed Fusion System*; US Department of Commerce, National Institute of Standards and Technology: Gaithersburg, MD, USA, 2020; p. 20.
43. Ma, L.; Bin, H. Temperature and stress analysis and simulation in fractal scanning-based laser sintering. *Int. J. Adv. Manuf. Technol.* **2007**, *34*, 898–903. [CrossRef]
44. Phan, T.Q.; Strantza, M.; Hill, M.R.; Gnaupel-Herold, T.H.; Heigel, J.; D'Elia, C.R.; DeWald, A.T.; Clausen, B.; Pagan, D.C.; Peter Ko, J.Y.; et al. Elastic Residual Strain and Stress Measurements and Corresponding Part Deflections of 3D Additive Manufacturing Builds of IN625 AM-Bench Artifacts Using Neutron Diffraction, Synchrotron X-Ray Diffraction, and Contour Method. *Integrating Mater. Manuf. Innov.* **2019**, *8*, 318–334. [CrossRef]
45. Pant, P.; Salvemini, F.; Proper, S.; Luzin, V.; Simonsson, K.; Sjöström, S.; Hosseini, S.; Peng, R.L.; Moverare, J. A study of the influence of novel scan strategies on residual stress and microstructure of L-shaped LPBF IN718 samples. *Mater. Des.* **2022**, *214*, 110386. [CrossRef]
46. Dejene, N.D.; Lemu, H.G. Current Status and Challenges of Powder Bed Fusion-Based Metal Additive Manufacturing: Literature Review. *Metals* **2023**, *13*, 424. [CrossRef]
47. Guillen, D.; Wahlquist, S.; Ali, A. Critical Review of LPBF Metal Print Defects Detection: Roles of Selective Sensing Technology. *Appl. Sci.* **2024**, *14*, 6718. [CrossRef]
48. Denlinger, E.R.; Gouge, M.; Irwin, J.; Michaleris, P. Thermomechanical model development and in situ experimental validation of the Laser Powder-Bed Fusion process. *Addit. Manuf.* **2017**, *16*, 73–80. [CrossRef]
49. ASM Aerospace Specification Metals Inc. Special Metals INCONEL® Alloy 718. Available online: https://asm.matweb.com/search/SpecificMaterial.asp?bassnum=NINC34 (accessed on 22 July 2024).
50. Knapp, G.L.; Raghavan, N.; Plotkowski, A.; DebRoy, T. Experiments and simulations on solidification microstructure for Inconel 718 in powder bed fusion electron beam additive manufacturing. *Addit. Manuf.* **2019**, *25*, 511–521. [CrossRef]
51. Stainless Steel-Grade 316 (UNS S31600). AZO Materials. Available online: https://www.azom.com/properties.aspx?ArticleID=863 (accessed on 22 July 2024).
52. ASM Aerospace Specification Metals Inc. AISI Type 316L Stainless Steel, Annealed Bar. Available online: https://asm.matweb.com/search/SpecificMaterial.asp?bassnum=MQ316Q (accessed on 22 July 2024).

metals

Article

Real-Time Modeling for Design and Control of Material Additive Manufacturing Processes

Amir M. Horr

LKR Light Metals Technologies, Austrian Institute of Technology, 1210 Vienna, Austria; amir.horr@ait.ac.at

Abstract: The use of digital twin and shadow concepts for industrial material processes has introduced new approaches to bridge the gap between physical and cyber manufacturing processes. Consequently, many multidisciplinary areas, such as advanced sensor technologies, material science, data analytics, and machine learning algorithms, are employed to create these hybrid systems. Meanwhile, new additive manufacturing (AM) processes for metals and polymers, based on emerging technologies, have shown promise for the manufacturing of sophisticated parts with complex geometries. These processes are undergoing a major transformation with the advent of digital technology, hybrid physical-data-driven modeling, and fast-reduced models. This study presents a fresh perspective on hybrid physical-data-driven and reduced order modeling (ROM) techniques for the digitalization of AM processes within a digital twin concept. The main contribution of this study is to demonstrate the benefits of ROM and machine learning (ML) technologies for process data handling, optimization/control, and their integration into the real-time assessment of AM processes. Therefore, a novel combination of efficient data-solver technology and an architecturally designed neural network (NN) module is developed for transient manufacturing processes with high heating/cooling rates. Furthermore, a real-world case study is presented, showcasing the use of hybrid modeling with ROM and ML schemes for an industrial wire arc AM (WAAM) process.

Keywords: additive manufacturing; real time modeling; machine learning; hybrid physical-data-driven modeling

1. Introduction

The role of real-time ROM technology, along with hybrid and smart ML schemes, has already revolutionized many industrial processes. The development of digital twin and digital shadow concepts has optimized process control and design. The introduction of the process-data paradigm and the use of ML technologies have significantly modernized process control concepts, enabling real-time prediction and correction for better control of industrial processes. Live sensor and remote sensing data, offline experimental and literature data, and numerical simulation predictions are all part of this processing paradigm, where data acquisition, handling, and training can greatly influence the manufacturing process. Furthermore, hybrid and smart data-handling technologies can help assess the performance of final parts regarding their service life. However, the introduction of these new technologies has not rendered conventional analytical, experimental, and numerical modeling redundant; rather, it has made their use smarter by integrating their results into evolving databases. Thus, dynamic and evolving databases can be accumulated for model training using data generated from analytical, experimental, and further offline detailed numerical simulation studies [1–3].

The systematic integration of these new technologies into AM processes can help establish faster and more accurate predictive models, avoiding long and expensive additional experimental and numerical efforts. Therefore, appropriate ROM techniques have been developed to enhance the digital twin and digital shadow concepts [4,5]. In this study, a combination of hybrid physical-data-driven and ROM techniques, along with ML modules, is employed for the WAAM deposition process. Additionally, a numerical simulation framework is set up to provide sufficient data for the ROM and ML schemes. Furthermore, a snapshot matrix is generated for the deposition process using various input variables (e.g., initial temperature, deposition speed, and torch power) to examine their effects on the final component. To ensure the ROM predictive model works effectively, some of the data are used to derive the reduced model (including its training), while the rest of the data are used to validate the performance of the ROM.

2. Simulation of AM Processes—Methodology

Different modeling techniques have been employed for the simulation of AM processes, considering many inherent thermal and mechanical characteristics. Although various AM processes have been developed for metallic and non-metallic components, many of the process functionalities are similar in terms of numerical modeling approaches [6–8]. The finite element (FE) method has been extensively used for simulating processes like AM, where different approaches have been developed to address the dynamic nature of these processes [9–12]. As AM parts are deposited and built up layer by layer, conventional FE methods, with their fixed domain size, can only simulate the final geometry without considering the progressive development of the layered structure. Although many alternative FE techniques have been developed to alleviate these shortcomings through dynamic mesh generation, deactivation/activation, and hybrid techniques, accurately modeling deposition processes remains a challenging task [13–16].

2.1. Dynamic FE Techniques

The application of alternative FE methods for dynamic manufacturing processes like casting and AM has recently been promoted to simulate progressively generated parts using both fixed-size and variable-size numerical domains. Since these processes inherently involve multi-physical thermal, fluid, and solid features and may also include multi-scale microstructural evolution phenomena, the proposed dynamic FE methods need to handle the different physical and dimensional aspects of these processes. This poses challenges for conventional FE methods, where geometries of domains are prepared and discretized at the start of the simulation, and all system matrices are initially assembled to calculate the full system responses. To alleviate these numerical complications, different simulation strategies, including deactivation and activation methods and novel dynamic mesh-generating frameworks, have been proposed [9,17].

For the so-called Deactivation and Activation Technique (DAT) for AM processes, the common practice is to deactivate almost all of the mesh related to the finished part at the start of the simulation (except initial base elements) and re-activate it (layer by layer or block-wise) as the simulation progresses [16,18,19]. Although this helps simulate the dynamic nature of the process using conventional FE methods, the deactivation process does not remove elements from assembled matrices; rather, it reduces their numerical impact to very small values (e.g., using very small multipliers). Additionally, any changes in material properties and system boundaries during the process cannot be simulated, as the system matrices are assembled at the start of the simulation. This means the solver needs to solve full system matrices from the start, which is computationally inefficient. Hence, dynamic mesh techniques (DMTs) have recently been proposed to overcome obstacles related to DAT by generating or splitting mesh during the simulation, where new mesh blocks or layers can be adapted into mainstream FE techniques for AM applications [4,12]. These techniques treat the progressive generation of AM parts during the simulation by generating and appending mesh blocks in a predefined or calculated manner, which are then attached to the main numerical domain.

2.2. Evolving Domain Technique

One of the recent DMT frameworks proposed for AM processes is the evolving domain technique. This technique considers the transient nature of the numerical domain and its continuous expansion throughout the processing time. The new technique, along with its accompanying block-mesh scheme, has been developed to address the numerical issues associated with domain expansion and the dynamic discretization of evolving domains. The expansion of the numerical domain is managed using a dynamic zone, where new layers of mesh (or mesh blocks) can be appended to the existing domain in a tempo-spatial manner. The spatial directions for the expansion of the numerical domain can be defined using a predefined path (e.g., material deposition direction) or by calculating the required growth during the transient simulation. This technique eliminates the need for generating a steady-state mesh with birth and death features and also removes the necessity for splitting element layers at the deposition front.

For AM process simulations, continuous deposition processes in various spatial directions can be simulated using a mesh evolution technique at the deposition front. In this research, an in-house code was developed to control and implement dynamic mesh generation and insertion (in Python) as follows:

- First, the initial geometry of necessary components, including the baseplate, is generated and meshed.
- In the second step, the initial thermal and mechanical boundaries are considered, and the initial system matrices are assembled.
- The first time-steps/iterations are subsequently solved using a thermal–mechanical solver, and the deposition front coordinates are updated.
- The domain geometry and mesh are later adapted by inserting a mesh block based on the deposition direction and speed.
- In the next step, the domain matrices are updated with new mesh entries, and the extra input energy is disseminated amongst the domain using the mapping technique.
- In the final step of the loop, after achieving thermal energy balance, the previous converged solution is used as a first step for a newly updated domain, and the simulation scheme continues with the new geometry/mesh till the next evolution step is triggered.

The computational performance of the technique for industrial casting applications has already been examined, and extensive numerical investigations have revealed that higher speed and accuracy can be achieved using the evolving domain technique [16]. Figure 1a,b show the flowchart for the evolving algorithm and a comparison study of the numerical performance of this technique versus conventional methods (for horizontal casting cases). Figure 1c illustrates the mesh-insertion technique for simulating the WAAM deposition process, where new meshes are appended to the existing domain to match the material deposition process [12]. More comprehensive discussions about the evolving domain technique can be found in [20].

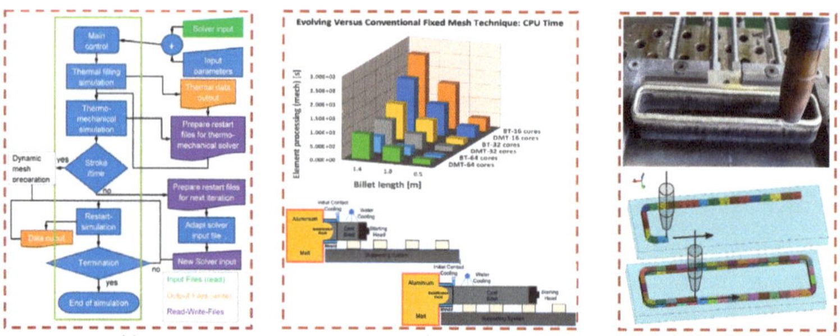

(a) (b) (c)

Figure 1. (**a**) Flowchart for evolving algorithm, (**b**) CPU time comparison for DAT and DMT for horizontal casting case using different numbers of computational cores. Reprinted from Ref. [12], and (**c**) dynamic WAAM deposition and its numerical mesh-insertion technique for DMT. Reprinted with permission from refs. [16]. 2020, ELSEVIER.

3. Methods for AM Reduced Models

Cyber–physical modeling, along with the new concepts of digital twins and shadows, has already found its way into industrial processes, revolutionizing optimized design and control through digitalization schemes. Cyber manufacturing processes, often described as the combination of physical and cyber tools/digital replicas and twinning systems, are well known. However, in this research, only the aspects of digital replicas and twinning systems were considered, with physical testing and design of experiments (DOEs) conducted solely for model calibration. Sensing technologies, data gathering, data filtering, and data-processing schemes, along with reduced-order models (ROMs) for fast predictive–corrective analyses, form a crucial part of the digitalization framework for a process-data paradigm. ROM technologies have also been rapidly evolving to achieve fast and reliable real-time predictions using new analytical and mathematical concepts. Although conventional reduced modeling techniques have been employed over the last fifty years to limit computational time and resources for detailed process simulations, they are unable to provide real-time predictions during processes. Consequently, recent attention has focused on the new generation of ROMs, which enable fast predictions and corrections in real-time during AM processes.

3.1. ROM Techniques for AM Processes

The increasing demand for real-time simulations of complex manufacturing processes has spurred numerous research activities aimed at developing fast ROMs using novel mathematical and analytical concepts [21,22]. Despite the exploitation of various techniques to create suitable ROMs for both transient and steady-state processes, challenges related to accuracy, speed, and solution stability remain unresolved for many multi-physical aspects of industrial processes. Consequently, numerous mathematical decomposition, projection, and dimensional reduction techniques have been developed to mitigate the extensive computational efforts required for modeling the multi-physical nature of these processes. In AM processes, detailed thermal–mechanical and/or fluid–thermal–mechanical simulations are routinely performed to optimize power, material delivery, deposition speed, heating, and cooling. However, the reduced versions of these simulations are cumbersome due to the transient nature of AM processes and their continuous domain expansion, which are difficult to predict using any simplification scheme.

The concepts of characteristic equations, eigenvalue analyses, and matrix factorization have been employed in this research to create a fast ROM framework for transient AM processes. This framework enables real-time predictive and corrective measures using previously measured or calculated data. To begin, let us consider a conventional second-order differential equation as follows:

$$[K][u] + [M][\ddot{u}] = 0 \tag{1}$$

where $[K]$, $[M]$, and $[u]$ are the stiffness, mass, and displacement matrices of the system. The simple solution, along with the eigenvalue model representation for the system, can be written as

$$u(t) = Ae^{i\omega t} \rightarrow \{[K] - \omega^2[M]\}[u] = 0 \tag{2}$$

where ω and A are eigenvalues (natural frequencies) and a constant multiplier, respectively. If the tempo-spatial characteristic of the system is decomposed, the characteristic equation of the system along with its solution for a point in space can then be written as

$$\left(\omega^2\right)^N + c_1\left(\omega^2\right)^{N-1} + c_2\left(\omega^2\right)^{N-2} + \ldots\ldots c_N = 0 \quad u(x,t) = \sum_n \hat{u}_n(x,\omega_n)e^{i\omega_n t} \tag{3}$$

where ω_n and $\hat{u}_n$ are the system's natural frequencies and Fourier coefficients, respectively. The characteristic equation of the system is calculated using properties such as geometry, material properties, and boundaries. However, solving the system equations using conventional time and frequency domain solvers requires substantial computational time and effort. This makes the calculation of system characteristics for engineering problems and processes lengthy and costly, rendering it unsuitable for real-time digital twinning. Consequently, much research has been conducted to reduce computational time and effort by developing specialized types of reduced-order models (ROMs) for industrial processes [3,23,24]. One interesting alternative technique, also used in the current research, is to estimate the characteristic equation of systems and processes using smart data-driven techniques instead of calculating them directly.

This ROM technology is initially based on the same mathematical concepts of eigenvalues and orthogonal eigenvectors to find the characteristic equation of the process. However, instead of using system properties to formulate the system matrices, it relies on calculated or measured responses of the system across a range of variables. If the responses of the system under varying parameters are known, a matrix of system responses under changing variables can be constructed, which holds the basic characteristic information about the system. The estimation of the system's characteristic equation is then performed using mathematical decomposition techniques traditionally used for eigenvalue analyses. To understand the basis of the mode decomposition technique for estimating the characteristic equation, let us consider the governing equation for a vibrational problem as follows:

$$[K][u(x,t)] + [M]\left[u(\ddot{x},t)\right] = 0 \rightarrow \ ([K] - \omega^2[M]).[u(x,t)] = 0 \rightarrow \ [K_d]^{-1}.[u(x,t)] = 0 \quad (4)$$

where $[K_d]^{-1} = [K] - \omega^2[M]$ is called the dynamic stiffness matrix. Now, we can consider the alternative governing equation for the response of the system as

$$[Y(x,t)] = [C].\ [X(x,t)] \quad (5)$$

where $[Y(x,t)]$ can be interpreted as the desired response at specific values for variables, $[C]$ is the estimated system characteristics (using data techniques), and $[X(x,t)]$ is the input values. In this research work, Singular Value Decomposition (SVD) is used to decompose the system response matrix, which gives the optimal expansion of the matrix as

$$[Y(x,t)] = U\sum V^T \quad (6)$$

where U represents the spatial eigenvector decomposition, $\sum$ is the singular value diagonal matrix (e.g., eigenvalues), and V^T represent temporal decomposition for transient system responses. The SVD technique is a popular reduction method based on linear algebra that can reduce the size of predictive models in terms of dimensions and time. A more detailed mathematical description of the SVD method can be found in [25,26]. Using these reduced models, the transient responses of the system can be estimated in real-time using a database of previous responses.

3.2. Hybrid ROM-ML Techniques

The advancement of ROM techniques for industrial process simulations has encountered challenges related to the accuracy of data extrapolation and rate dependency (e.g., for high cooling/heating rates). In many of these processes, the limited amount of available data, combined with the large number of influential parameters, has promoted the use of integrated ROM-ML hybrid schemes, where further data training can be performed. To establish a foundation for such a hybrid framework, appropriate ML and ROM techniques must be employed to enhance the predictive power of these models. This combined ROM-ML technique can facilitate the development of agile and efficient models for transient thermal–mechanical and/or fluid–thermal–mechanical WAAM process modeling.

For data training in this research, a neural network (NN) ML scheme was implemented using in-house code. This NN scheme can employ either Multi-Layer Perceptron (MLP) or Back-Propagation Neural Networks (BPNN) for training system response data. For data interpolation and extrapolation, data extracted from experimental and detailed simulations were stored in a file. During each exemplar epoch, a certain number of data points were selected randomly without repeating any learning vector. Since batching concurrent inputs is computationally more efficient than sequential inputs, epochs and total computation of errors were used as follows:

$$E_{epoch} = \frac{1}{n_{epoch}} \sum\nolimits_{n=1}^{n_{epoch}} \sum\nolimits_{u=1}^{U} \left(t_u - y_u\right)^2 \tag{7}$$

where t_u and y_u are desired and model outputs, and the network error is computed based on the RMS standard formulation, where it is propagated into the network until convergence is reached. In the back-propagating networks, the modification of weights is proportional to error differential ∂E with respect to weight ∂w as

$$\Delta w_{ij} \propto -\frac{\partial E_{epoch}}{\partial w_{ij}} \tag{8}$$

After data training, the resulting ROM-ML models can further update themselves using the NN scheme, enhancing thermal and mechanical predictions for AM processes. Additionally, these hybrid models can be used for modeling sub-processes such as heating and cooling during AM processes [4].

3.3. Case Study: Reduced Models for WAAM Process

Digital twin and cyber–physical modeling, along with real-time prediction and correction concepts, have already been employed in some AM industrial processes. The optimization and active control of these processes are among the main goals of twinning technology, where sensing technology, data generation, and data-processing schemes can be combined for greener production. In this case study, attention is focused on the WAAM process, a type of directed energy deposition (DED) scheme. In this technique, single or multiple plasma or electric arc torches are used to weld layers of wire feedstock onto a predefined geometry. The heat generated by the moving torch melts the wire, allowing layers of material to be built up through an inter-layer fusion process.

Although various sophisticated transient and steady-state numerical simulations of WAAM processes have already been developed, their use in digital twinning is limited due to the substantial computational time and resources required. Therefore, for real-time predictions and corrections, the predictive power of ROM and ML can be employed. We can consider Equation (6) to predict temperature (and stress) responses during a WAAM process as follows:

$$[Y(x,t)] = U\sum V^T \rightarrow T_k(x,y,z,t) = U_k \sum\nolimits_k V_k^T \quad \sigma_k(x,y,z,t) = U_{k\sigma} \sum\nolimits_{k\sigma} V_{k\sigma}^T \tag{9}$$

where T_k σ_k are temperatures and stresses at process time t, and the snapshot results can be interpolated using the radial basis function (RBF) or adaptive redial basis function (ARBF) as follows [27,28]:

$$f(T) = \sum\nolimits_{i=1}^{k} a_i \varphi(||T - T_i||) \tag{10}$$

where a_i and φ are the weighting coefficient for the RBF. Here, the temperature at the any location can be predicted during the WAAM process using a sum of k radial basis functions, which are weighted by appropriate coefficients. For the RBF definition of WAAM processes with their high heating rate (passing of a torch over a measuring point), the artificial network architecture developed earlier can be employed, where an input, hidden, and output layer can be defined as

$$\varphi_i = e^{\left(-\frac{||\overline{T}-\overline{T}_i||^2}{2\mu_i^2}\right)} \tag{11}$$

where $\overline{T}$, $\overline{T}_i$, μ, and φ_i are the temperature input vector, ith neuron sample vector, ith neuron bandwidth, and ith neuron output, respectively. The innovative combination of SVD and the neural network RBF for implementing a reduced model for WAAM can produce fast and reliable predictions, even in zones with high-gradient data (e.g., high heating rates). To investigate the accuracy and reliability of this reducing technique, a WAAM case study was conducted to simulate the welding of a single layer on top of a thermally pre-conditioned wall. The wall, with dimensions of 100 mm $\times$ 50 mm $\times$ 6 mm, was welded with aluminum 6061 alloy using a plasma technique. For the initial verification study, experimental work was conducted using thermocouples to measure the temperature at selected locations. The results of these studies were then compared to thermal–mechanical FE simulations for verification and calibration. After completing the initial verification study, a snapshot matrix was built using varying process parameters, such as torch power, deposition speed, and initial temperatures [5]. A series of FE simulations were then performed using these scenarios to generate a database for building the reduced model. Nine scenarios were initially used to build the model, while three DOE cases were later carried out for the validation of the reduced models. Figure 2 shows the snapshot matrix along with the FE mesh and its typical temperature contour results during the torch passage.

Scenario No.	ROM ID	Initial Temp [°]	Power [W]	Deposition Vel. [m/s]
1	B1	25	2200	0.007
2	B2	75	2000	0.0075
3	B3	115	1800	0.006
4	B4	160	1600	0.008
5	B5	200	1500	0.007
6	B6	230	1400	0.0075
7	B7	250	1300	0.007
8	B8	300	1150	0.0065
9	B9	350	1100	0.008
10	V1	90	1950	0.006
11	V2	180	1530	0.007
12	V3	320	1120	0.0065

(a)

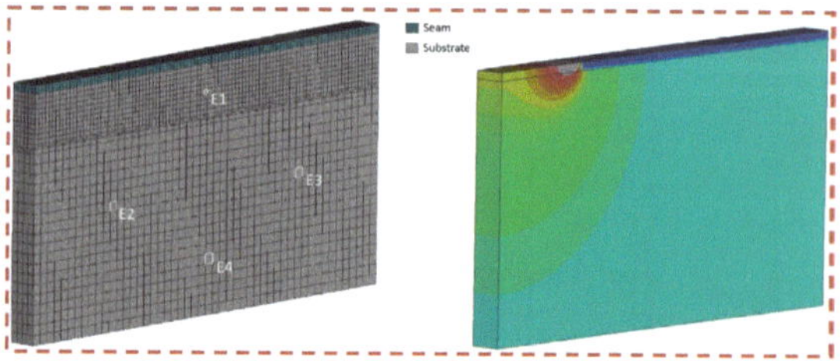

(b)

Figure 2. (**a**) Snapshot matrix for WAAM scenarios with varying process parameters; (**b**) FE mesh with selected measuring points and representative temperature contour during passage of torch. Reprinted with permission from ref. [5].

For data interpolation and training of the reduced models, two different techniques are used: genetic algorithm symbolic regression (GASR) and neural networks. The GASR technique is capable of data processing, handling, and fitting, which can be performed using available computer codes. Specifically, the academic code HeuristicLab, described in [29], was used in this research for data handling and training. GASR is a type of regression analysis that searches multi-dimensional space for mathematical expressions and operators using a genetic algorithm to find the most suitable model. For the RBF neural network, convergence is achieved using dynamic node creation [30] and an overlapping scheme for the test and training data. These techniques enhance the network's ability near the search space boundaries and even beyond (i.e., extrapolation). Since the RBF technique proved to be more practical and yielded more accurate results for this study, the main reduced model was built using a combination of SVD and RBF techniques. Figure 3 shows the neural network architecture and the GASR tree-like diagram, while Figure 4 illustrates their convergence and data training plots.

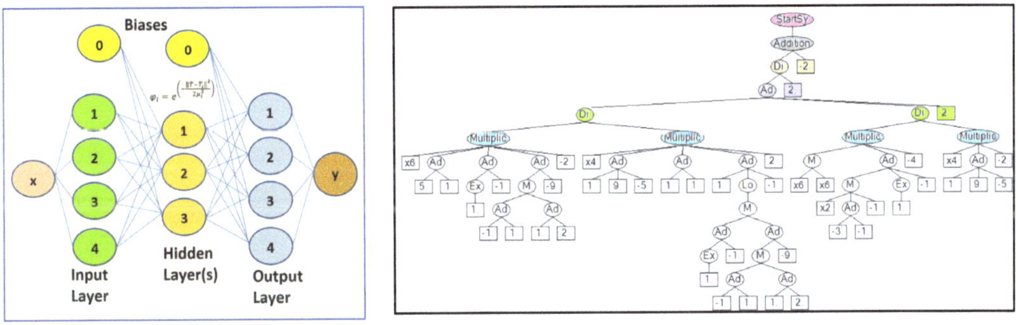

(**a**) (**b**)

Figure 3. (**a**) NN architecture for RBF interpolation; (**b**) tree-like GASR diagram for genetic algorithm fitting.

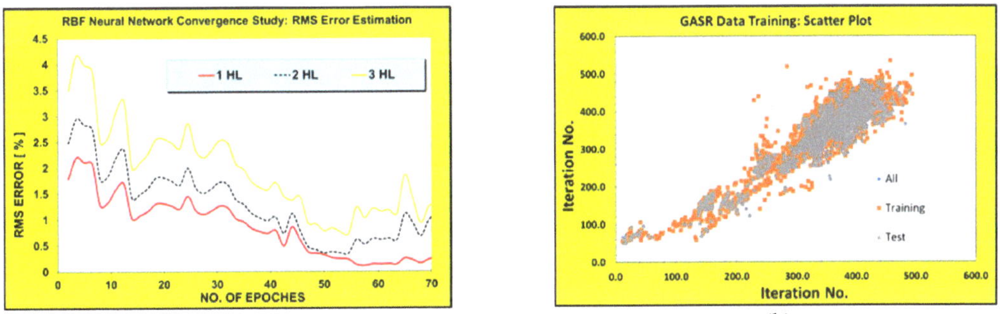

(**a**) (**b**)

Figure 4. (**a**) NN convergence of networks with different number of hidden layers, and; (**b**) scatter plot for training process of GASR.

In the final part of the study, the performance of the SVD-RBF reduced model was evaluated by comparing its results with other popular model-building techniques. Five different techniques—kriging, regression, support vector machine (SVM), clustering, and inverse distance (InvD)—were used to create alternative WAAM reduced models using the available data [31–33]. The performances of these reduced models were then compared to the trained SVD-RBF model using the results from three additional DOEs. Temperature time histories for the deposition process were collected for specific nodes along the wall, and the history results were compiled for the entire simulation time. The performances of these reduced models were then compared with the SVD-RBF model by calculating the normalized error histories.

4. Discussion

The development of accurate and real-time reduced models for metallic and polymeric AM processes, with their multi-physical aspects, is challenging and requires careful consideration of data training and testing. In this study, the eigen-based SVD technique was enhanced with the interpolation and training power of RBF and neural networks to establish an efficient and accurate reduced model. The verification of the proposed reduced model showed that it can reliably produce real-time predictions of temperature and stress/strain time histories. However, due to space constraints, only the temperature results are presented here, excluding mechanical deformations, stress, and strain results. At first glance, these models appear capable of predicting responses even with small and limited-size databases and a large range of process parameter variations. However, several challenges in generating and training these types of reduced models need to be addressed before they can be solidly employed within digital twin concepts, namely:

- ➢ For WAAM processes with thermal–mechanical and multi-physical aspects, reduced models need to cope with rapidly changing data, especially for processes with a high cooling and heating rate.
- ➢ The size and variation of data within the snapshot matrix can significantly affect the prediction power of these models. Different sampling techniques should be employed to cover the entire multi-dimensional search space (e.g., Sobol and Latin Hypercube).
- ➢ To carefully verify the performance of these models, a rigorous validation criterion is required, examining performance maps at internal, near-boundary, and extreme conditions (extrapolation) of the search space.
- ➢ Although the use of neural network and GASR techniques can greatly increase the predictive power of reduced models, customized training schemes are necessary for proper data interpolation and fitting.

Figure 5 shows the estimated time history temperature for a computational node along the deposited wall for the FE simulation, trained SVD-RBF, and SVD-Kriging reduced models.

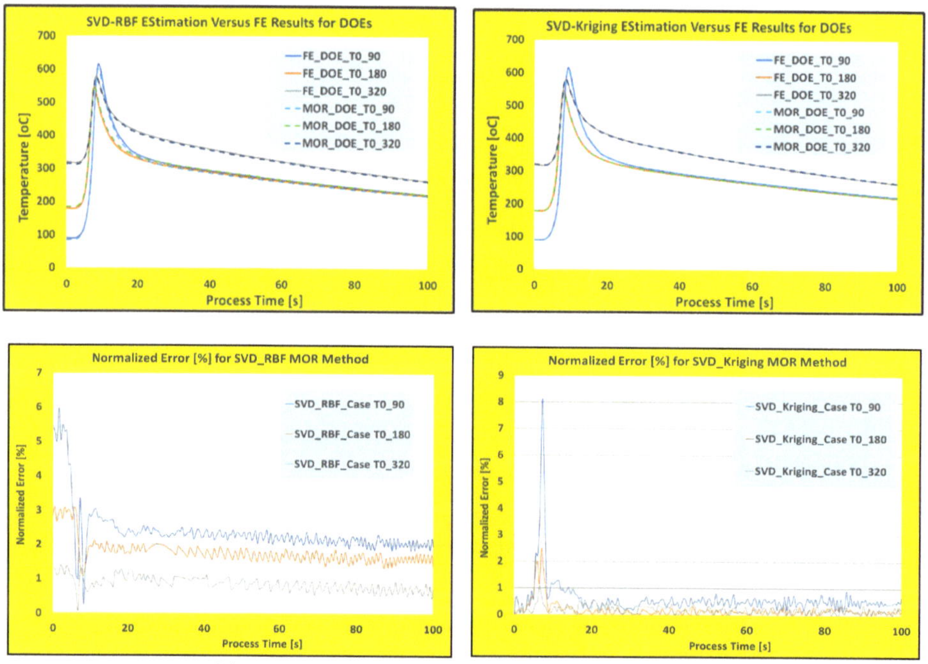

(a) (b)

Figure 5. Comparison of temperature time histories for FE versus ROMs along with their normalized errors for (**a**) SVB-RBF trained, and (**b**) SVD-Kriging ROM techniques.

The computational time for the FE simulation is approximately 2880 s per scenario (i.e., wall clock time), whereas the reduced models take only about 0.72 s to estimate the time history responses. Figure 5a compares the temperature predictions and normalized error graphs (relative to verified FE results) for the proposed SVD-RBF method across the three verification DOEs. Figure 5b presents these results for the popular SVD-Kriging reduced model. The Kriging method is one of the best interpolation techniques, using a limited set of sampled data points to evaluate spatio-temporal variables over a continuous search space.

At first glance, the calculated errors show that during periods of high heating rates (when the torch is passing over the measuring point), the proposed SVD-RBF-trained model exhibits lower error margins, indicating almost no data rate dependency. In contrast, the popular SVD-Kriging model shows higher error margins during these times but maintains a consistently lower error margin throughout the rest of the time history results. Further investigation of the data rate dependency for these two techniques is shown in Figure 6a,b, where the correlation of normalized error time histories with the rate of temperature changes (heating rate) is plotted for both techniques. Figure 6c,d illustrate the accumulated normalized error over the entire time history and the Pearson correlation index for both techniques.

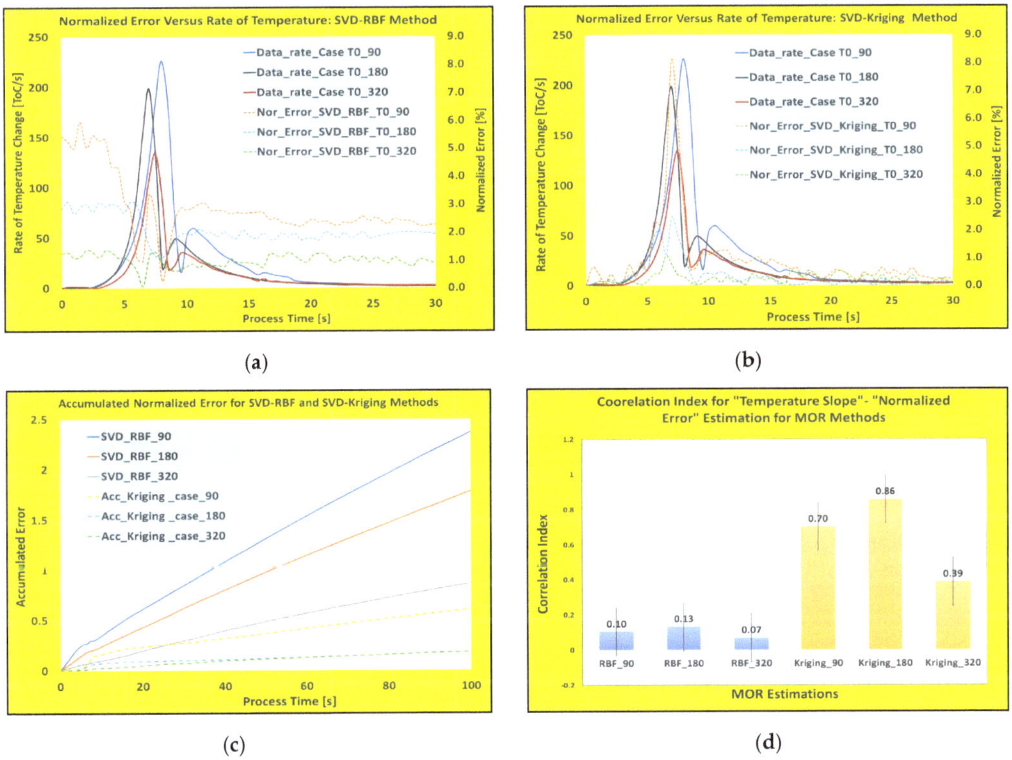

Figure 6. Rate of temperature changes (heating rate) versus normalized errors for (**a**) SVD-RBF trained, (**b**) SVD-Kriging, (**c**) accumulated normalized errors for SVD-RBF and SVD-Kriging, and (**d**) Pearson correlation index for heating rate with respect to normalized error for both techniques.

As these results indicate, while the SVD-RBF model predictions are more accurate in high-heating-rate zones (i.e., almost rate-independent), the SVD-Kriging normalized error results show strong heating rate dependency. However, the SVD-Kriging technique produces about 1% lower error margin after the torch passes over the measuring point, indicating better data fitting at low temperature changes. Therefore, due to their effective sampling and interpolation schemes, both SVD-RBF and SVD-Kriging techniques are suitable for WAAM processes with moderate heating/cooling rates. For high-heating/cooling-rate processes, it is recommended to use the trained SVD-RBF technique for model-building procedures. To demonstrate the superiority of these techniques for real-time modeling, the final part of the study is dedicated to showing the performance of other popular reduced model techniques for WAAM processes. Figure 7 presents the time history temperature predictions along with normalized error plots for clustering, regression, InvD, and SVM model-building techniques.

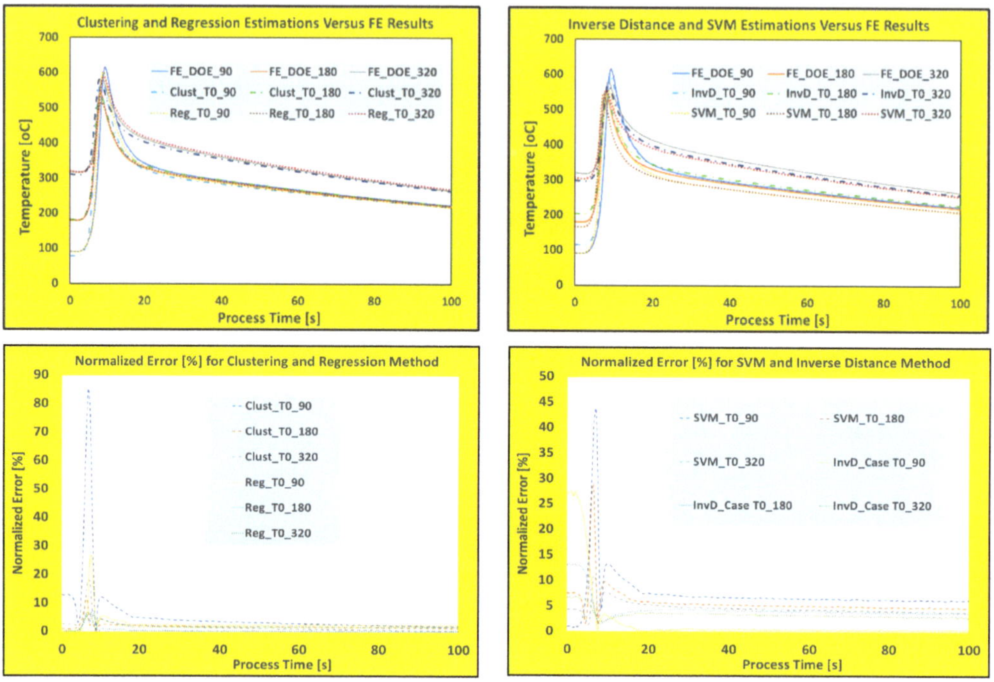

(a) (b)

Figure 7. Comparison of calculated temperature time histories for FE versus ROMs along with their normalized errors for (**a**) clustering and regression and (**b**) InvD and SVM techniques.

As clearly demonstrated in Figure 7, most of these popular reduced model techniques exhibit either heating rate dependency or an inability to fit the predefined initial conditions of the process (initial base temperature). The sampling rate for the thermal calculations of all these WAAM scenarios is half a second (2 Hz), with heating rates of up to 200 K/s, which may increase the error margin for real-time predictions. Furthermore, the heating rate dependency and deviation of initial temperatures are shown in Figure 8a, where the average maximum errors (over three DOEs) and the deviation of initial conditions (in percentages) are plotted for all six methods (including trained SVD-RBF and SVD-Kriging methods). Figure 8b shows the time history of the average normalized errors (over three DOEs) with the calculated heating rates for the four popular model-building techniques (for the first 25 s of the WAAM process).

As shown in these figures, the predictions based on these techniques exhibit either some degree of heating rate dependency or significant deviation in predicting the initial thermal conditions. Both trained SVD-RBF and SVD-Kriging models demonstrate superior overall performance, with low maximum errors and limited or zero deviation in the initial thermal conditions. Besides these two techniques, the best performance is achieved by the regression method, which shows relatively low rate dependency and no deviation in the initial conditions.

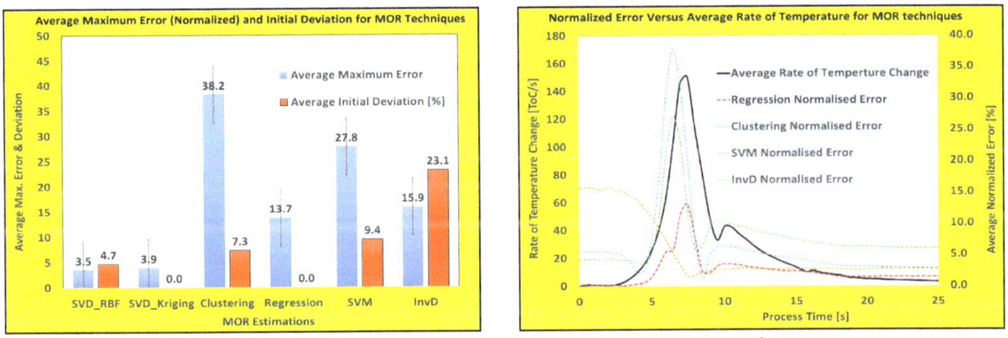

(**a**) (**b**)

Figure 8. (**a**) Average maximum errors (over three DOEs) and deviation of initial conditions (in percentages) for six ROM techniques; (**b**) correlation of heating rates with normalized errors for clustering, regression, InvD, and SVM techniques.

5. Conclusions

The reduced models and their fast and real-time prediction potentials can introduce opportunities for further developments toward the digitalization of additive manufacturing processes. In this study, different reduced model-building techniques were examined to evaluate their performance for optimizing and controlling WAAM processes. Initially, a brief description of the numerical simulation techniques for these dynamic processes was presented, along with some technical aspects of the new evolving-domain and dynamic-mesh technique. Furthermore, a brief mathematical description of popular reduced model-building techniques was provided, and the impacts of hybrid ML-ROM trends on the accuracy and reliability of the reduced models were discussed. In the following sections, the ROM description of a practical case study for the WAAM process was elaborated, examining its critical aspects of accuracy, heat rate dependency, and thermal initial boundary conditions.

There are challenges related to the accuracy of ROM techniques for industrial process modeling, including data extrapolation, initial boundary fitting, and rate dependency (e.g., for high cooling/heating rates). Additionally, for many of these processes with a large number of input parameters, the quality and size of the database can greatly influence the accuracy of predictions. The results presented here indicate that for the two developed SVD-RBF and SVD-Kriging methods, both the issues of heating/cooling rate dependency and modeling of initial boundary conditions can be partially resolved. However, with further ML training (e.g., NN data training), even better results can be achieved for high-gradient heating/cooling processes, where almost no rate dependency is observed in the predictions. Detailed investigations into the best performance "solver-interpolator" combinations have shown that some well-known solvers (e.g., SVD, POD) can produce more accurate results when used with the right interpolating schemes. Furthermore, there is potential to use ML for training and improving data representations within the hybrid ROM-ML framework to achieve superior accuracy and agility. This will promote the advancement of more adapted MOR-ML technologies for AM processes, where the features of both ML and MOR can be combined for more accurate reduced models.

As a final statement, this study aims to encourage the use of ROM-ML schemes for the real-time modeling of dynamic material processes like WAAM. Although these hybrid reduced models are not intended to replace other experimental or detailed numerical process simulations, they can be valuable assets within the digitalization framework for control and optimization. The opportunity to apply the hybrid modeling scheme to other multi-physical aspects of these processes, such as mechanical stress/strain and deformation/warping, will be explored in future publications.

Funding: This research was funded by the Austrian Federal Ministry for Climate Action, Environment, Energy, Mobility, Innovation and Technology; the Austrian Research Promotion Agency (FFG); and the Austrian Institute of Technology under the UF2023 scheme.

Data Availability Statement: The raw data supporting the conclusions of this article will partially be made available by the authors on request.

Acknowledgments: The author would like to thank the Austrian Federal Ministry for Climate Action, Environment, Energy, Mobility, Innovation and Technology; the Austrian Research Promotion Agency (FFG; and the Austrian Institute of Technology for the technical/financial support. The author would also like to thank Hugo Drexler and Johannes Kronsteiner, both from LKR, the Austrian Institute of Technology, for their contributions to the research work.

Conflicts of Interest: The author declares no conflicts of interest.

References

1. McBride, M.; Persson, N.; Reichmanis, E.; Grover, M.A. Solving Materials' Small Data Problem with Dynamic Experimental Databases. *J. Process.* **2018**, *6*, 79. [CrossRef]
2. Perno, M.; Hvam, L.; Haug, A. Implementation of digital twins in the process industry: A systematic literature review of enablers and barriers. *Comput. Ind.* **2022**, *34*, 103558. [CrossRef]
3. Favoretto, B.; de Hillerin, C.A.; Bettinotti, O.; Oancea, V.; Barbarulo, A. Reduced order modeling via PGD for highly transient thermal evolutions in additive manufacturing. *Comput. Methods Appl. Mech. Eng.* **2019**, *349*, 405–430. [CrossRef]
4. Horr, A.M. Optimization of Manufacturing Processes Using ML-Assisted Hybrid Technique. *J. Manuf. Lett.* **2021**, *31*, 24–27. [CrossRef]
5. Horr, A.M.; Drexler, H. On Reduced and Detailed Simulations of Additive Manufacturing Processes. In Proceedings of the MAMC2022, Graz, Austria, 26–28 September 2022. Available online: https://submit.asmet.org/event/63/book-of-abstracts.pdf (accessed on 1 November 2024).
6. Armstrong, M.; Mehrabi, H.; Naveed, N. An overview of modern metal additive manufacturing technology. *J. Manuf. Process.* **2022**, *84*, 1001–1029. [CrossRef]
7. Kanishka, K.; Acherjee, B. Revolutionizing manufacturing: A comprehensive overview of additive manufacturing processes, materials, developments, and challenges. *J. Manuf. Process.* **2023**, *107*, 574–619. [CrossRef]
8. Korkmaz, M.E.; Waqar, S.; Garcia-Collado, A.; Gupta, M.K.; Krolczyk, G.M. A technical overview of metallic parts in hybrid additive manufacturing industry. *J. Mater. Res. Technol.* **2022**, *18*, 384–395. [CrossRef]
9. Baiges, J.; Chiumenti, M.; Moreira, C.A.; Cervera, M.; Codina, R. An adaptive Finite Element strategy for the numerical simulation of additive manufacturing processes. *Addit. Manuf.* **2021**, *37*, 101650. [CrossRef]
10. Patil, N.; Pal, D.; Stucker, B. A New Finite Element Solver using Numerical Eigen Modes for Fast Simulation of Additive Manufacturing Processes. In Proceedings of the International Solid Freeform Fabrication Symposium, Austin, TX, USA, 12–14 August 2013. [CrossRef]
11. Lindgren, L.E.; Lundbäck, A. Approaches in computational welding mechanics applied to additive manufacturing: Review and outlook. *Comptes Rendus. Mécanique* **2018**, *346*, 1033–1042. [CrossRef]
12. Brötz, S.; Horr, A.M. Framework for progressive adaption of FE mesh to simulate generative manufacturing processes. *J. Manuf. Lett.* **2020**, *24*, 52–55. [CrossRef]
13. Setien, I.; Chiumenti, M.; Veen, S.; San Sebastian, M.; Garciandía, F.; Echeverría, A. Empirical methodology to determine inherent strains in additive manufacturing. *Comput. Math. Appl.* **2019**, *78*, 2282–2295. [CrossRef]
14. Huang, H.; Ma, N.; Chen, J.; Feng, Z.; Murakawa, H. Toward large-scale simulation of residual stress and distortion in wire and arc additive manufacturing. *Addit. Manuf.* **2020**, *34*, 101248. [CrossRef]
15. Michaleris, P. Modeling metal deposition in heat transfer analyses of additive manufacturing processes. *Finite Elem. Anal. Des.* **2014**, *86*, 51–60. [CrossRef]
16. Horr, A.M.; Kronsteiner, J. Dynamic material process simulations: Hybrid-evolving technique. *J. Met.* **2021**, *11*, 1884. [CrossRef]
17. Denlinger, E.R.; Irwin, J.; Michaleris, P. Thermomechanical Modeling of Additive manufacturing Large Parts. *ASME. J. Manuf. Sci. Eng.* **2014**, *136*, 061007. [CrossRef]
18. Gouge, M.; Michaleris, P.; Denlinger, E.; Irwin, J. The Finite Element Method for the Thermo-Mechanical Modeling of Additive Manufacturing Processes. In *Thermo-Mechanical Modeling of Additive Manufacturing*; Gouge, M., Michaleris, P., Eds.; Butterworth-Heinemann: Oxford, UK, 2018; pp. 19–38. [CrossRef]
19. Yang, Q.; Zhang, P.; Cheng, L.; Min, Z.; Chyu, M.; To, A.C. Finite element modeling and validation of thermomechanical behavior of Ti-6Al-4V in directed energy deposition additive manufacturing. *Addit. Manuf.* **2016**, *12 Pt B*, 169–177. [CrossRef]
20. Horr, A.M. Computational Evolving Technique for Casting Process of Alloys. *Math. Prob. Eng.* **2019**, *2019*, 6164092. [CrossRef]

21. Strobl, D.; Robens-Radermacher, A.; Ghnatios, C.; Pittner, A.; Rethmeier, M.; Unger, J.F. PGD model with domain mapping of Bead-on-Plate weld simulation for wire arc additive manufacturing. In Proceedings of 8th European Congress on Computational Methods in Applied Sciences and Engineering (ECCOMAS), Oslo, Norway, 5–9 June 2022; Available online: https://opus4.kobv.de/opus4-bam/frontdoor/index/index/docId/55111 (accessed on 1 November 2024).
22. Quaranta, G. Efficient Simulation Tools for Real-time Monitoring and Control Using Model Order Reduction and Data-Driven Techniques. Ph.D. Thesis, UPC, Departament d'Enginyeria Civil Ambiental, Barcelona, Spain, 2019. Available online: https://hdl.handle.net/2117/168567 (accessed on 1 November 2024).
23. Lu, Y.; Jones, K.K.; Gan, Z.; Liu, W.K. Adaptive hyper reduction for additive manufacturing thermal fluid analysis. *Comput. Methods Appl. Mech. Eng.* **2020**, *372*, 113312. [CrossRef]
24. Korneev, S.; Wang, Z.; Thiagarajan, V.; Nelaturi, S. Fabricated shape estimation for additive manufacturing processes with uncertainty. *Comput.-Aided Des.* **2020**, *127*, 102852. [CrossRef]
25. Brunton, S.L.; Kutz, J.N. *Data Driven Science & Engineering-Machine Learning, Dynamical Systems, and Control*; Cambridge University Press: Cambridge, UK, 2019. [CrossRef]
26. Qin, J.; Hu, F.; Liu, Y.; Witherell, P.; Wang, C.L.; Rosen, D.W.; Simpson, T.W.; Lu, Y.; Tang, Q. Research and application of machine learning for additive manufacturing. *Addit. Manuf.* **2022**, *52*, 102691. [CrossRef]
27. Nguyen, M.N.; Kim, H.G. An efficient PODI method for real-time simulation of indenter contact problems using RBF interpolation and contact domain decomposition. *Comput. Methods Appl. Mech. Eng.* **2022**, *388*, 114215. [CrossRef]
28. Farazandeh, E.; Mirzaei, D. A rational RBF interpolation with conditionally positive definite kernels. *Adv. Comput. Math.* **2021**, *47*, 74. [CrossRef]
29. Wagner, S.; Affenzeller, M. HeuristicLab: A generic and extensible optimization environment. In *Adaptive and Natural Computing Algorithms*; Springer: Berlin/Heidelberg, Germany, 2005; p. 538. [CrossRef]
30. Ash, T. Dynamic Node Creation in Backpropagation Networks. *Connect. Sci.* **1989**, *1*, 365–375. [CrossRef]
31. Kaushik, C.; McRae, A.D.; Davenport, M.; Muthukumar, V. New Equivalences between Interpolation and SVMs: Kernels and Structured Features. *SIAM J. Math. Data Sci.* **2024**, *6*, 761–787. [CrossRef]
32. Choi, K.; Chong, K. Modified Inverse Distance Weighting Interpolation for Particulate Matter Estimation and Mapping. *Atmosphere* **2022**, *13*, 846. [CrossRef]
33. Phan, D.T.; Nguyen, L.M.; Murali, P.; Pham, N.H.; Liu, H.; Kalagnanam, J.R. Regression Optimization for System-level Production Control. In Proceedings of the American Control Conference (ACC), New Orleans, LO, USA, 25–28 May 2021; pp. 5023–5028. [CrossRef]

metals

Article

Optimized Build Orientation and Laser Scanning Strategies for Reducing Thermal Residual Stress in Topology-Optimized Automotive Components

Jeongho Yang [1,2,†], Youngsuk Jung [3,†], Jaewoong Jung [4], Jae Dong Ock [5], Shinhu Cho [3], Sang Hu Park [2], Tae Hee Lee [4,*] and Jiyong Park [1,6,*]

1 Advanced Joining & Additive Manufacturing R&D Group, Korea Institute of Industrial Technology, Yeonsu-gu, Incheon 15014, Republic of Korea; junghotoo@kitech.re.kr
2 School of Mechanical Engineering, Pusan National University, Busan 46241, Republic of Korea; sanghu@pusan.ac.kr
3 Lightweight Materials Research Team, Hyundai Motor Group, Uiwang-si 16082, Republic of Korea; ys.jung@hyundai.com (Y.J.); s.cho@hyundai.com (S.C.)
4 Premium Vehicle R&D Center, Korea Automotive Technology Institute, Yeongam-gun 58463, Republic of Korea; jwjung@katech.re.kr
5 Woosin Industries Corporation, Gimhae-si 50877, Republic of Korea; dhrwoehd@nate.com
6 Department of Convergence Manufacturing System Engineering, University of Science and Technology (UST), 217 Gajeong-ro, Yuseong-gu, Daejeon 34113, Republic of Korea
* Correspondence: thlee@katech.re.kr (T.H.L.); j.park@kitech.re.kr (J.P.); Tel.: +82-61-813-3021 (T.H.L.); +82-32-850-0288 (J.P.); Fax: +82-61-813-3067 (T.H.L.); +82-32-850-0210 (J.P.)
† These authors contributed equally to this work.

Abstract: This study investigates the reduction in thermal residual stress during the powder bed fusion (PBF) process in a non-standardized shape generated by topology optimization method in lightweight automotive part of brake caliper. While the caliper of the braking system for reducing the CO_2 consumption in vehicle systems undergoes a redesign to increased strength and reduced weight, challenges arise due to biased melting area ratios in the topologically optimized design, causing the thermal deformation. To address this, our research proposes an efficient PBF scan strategy aimed at minimizing anisotropy and residual stress—a critical consideration for successful manufacturing. The effectiveness of the laser scan strategy is validated through testing on a brake dynamometer, following the JASO C406 test procedure, an authorized standard for commercial brake calipers. Furthermore, a comparative analysis between the conventional product and the proposed brake caliper highlights superior performance, particularly under lightweight conditions. This comprehensive approach contributes valuable insights to the field, offering a potential solution for overcoming manufacturing challenges associated with topologically optimized designs in automotive components.

Keywords: additive manufacturing (AM); powder bed fusion (PBF); topology optimization; brake caliper

1. Introduction

Additive manufacturing (AM) process technology developed rapidly in recent years, and the powder bed fusion (PBF) process is utilized in industrial applications for metal parts among AM processes [1–3]. The PBF process, which is primarily used in industrial applications, manufactures more sophisticated and complex geometries than conventional machining processes such as casting, forging, and rolling [4–6]. In addition to the advantages and significant potential of PBF process, including design freedom, it enables the manufacture of complex shapes and safety-critical products. In particular, structural lightweight design through topology optimization is applied to defense, aerospace, and automotive products [7–10]. The high degree of design freedom enables topology optimization to achieve optimized lightweight structures that not only fulfill specified requirements, but can also be optimally manufactured, making them attractive for automotive applications where lightweight components for CO_2 reduction are required [11–14]. The braking system included in the unsprung mass serves as a safety component, and all vehicles are equipped with their own safety devices for stopping them, and it also affects the drive comfort, road holding, and acceleration [15–17]. Efficient topology optimization via PBF process is vital to meet heightened demands for lighter materials, enhanced brake performance, and lightweight calipers, especially in complex structures and tight spatial constraints [18–20].

Certainly, the PBF process is not devoid of potential drawbacks to apply the automotive field; it possesses the capability to produce an array of product defects, encompassing surface inconsistencies, porosity, residual stress, cracks, and thermal deformations [21–24]. Especially, the thermal deformation defects that occur during the PBF process are caused by residual stress resulting from thermal accumulation [25]. Energy density and scan pattern can be applied to control the magnitude of residual stress and thermal accumulation [26]. Since the directionality of residual stress is generated according to the direction of the laser scan, when the laser scan is used for a long time in one direction, an anisotropic characteristic is generated, and a melted area with a long aspect ratio also causes anisotropic residual stress [27]. The chessboard scan strategy demonstrates a reduction in anisotropic residual stress compared to the strip pattern [28]. More research of the PBF scan pattern is needed for effective reduction in residual stress and for achieving isotropic characteristics.

In this study, an effective PBF process was analyzed to reduce anisotropic residual stress caused by irregular aspect ratio melting areas generated from a topology optimization brake caliper of an automotive part. The PBF scan strategy that reduces residual stress and ensures isotropic properties was analyzed and applied and this PBF scan strategy was used to successfully fabricate complex shapes with varying melting area ratios. To manufacture a high-strength, lightweight brake caliper, a high-strength lightweight model was designed using Ti-6Al-4V material and topology optimization. The stiffness of the proposed model was analyzed using the FEM simulation. The thermal deformation simulation was conducted to evaluate and improve manufacturability by analyzing the build direction that minimizes thermal deformation during the additive manufacturing process. The new brake caliper model, which has been precisely additively manufactured using the effective PBF process, can reduce weight while improving maximum stiffness and brake performance compared to the commercial product.

2. Laser Scan Strategy for Reducing Thermal Stress and Residual Stress

2.1. Material and Equipment

The conventional brake caliper is made from aluminum alloy. While performing topology optimization with aluminum alloy can result in a lighter design, it also leads to a reduction in stiffness, necessitating the use of high-strength materials. To produce a high-strength, a lightweight brake caliper, Ti-6Al-4V, known for its high stiffness-to-weight ratio, was used and manufactured through PBF process. Typically, the PBF process involves layer-by-layer manufacturing and can easily manufacture complex shapes with internal structures [29]. The PBF process repeats the transfer of powder from the powder platform to the plate via a recoater and melting the powder with a laser. In addition, since laser melting is applied in the PBF process, repeated overheating and rapid solidification result in high thermal stress. Process analysis is required to reduce excessive thermal residual stress. In this study, DMP Flex 350 (3D SYSTEMS, Rock Hill, SC, USA) was used with a printing volume of 275 mm × 275 mm × 420 mm. We used a laser power of 110 W, a scan speed of 1000 mm/s, a hatch spacing of 80 μm, and a layer thickness of 30 μm as the PBF process conditions. Argon gas was used as the shielding gas during the PBF process. The spherical powders with an average diameter of 32 μm were randomly distributed in the PBF process.

2.2. PBF Scan Strategy Method to Reduce Residual Stress

The topology optimization model has an irregularly shaped region, and different melting regions occur in each layer, which may lead to excessive thermal deformation depending on the residual stress characteristics. The effective laser scan strategy is required in the PBF process to reduce thermal residual stress. In this chapter, the residual stress of parts manufactured using the laser scanning process was analyzed using X-ray measurement (μ-X360 portable, Pulstec Industrial Co., Shizuoka, Japan). As shown in Figure 1, 20 × 20 × 20 mm^3 specimens were fabricated with strip, hexagonal and island pattern (Figure 1a). The residual stress in a specimen deposited on the plate tends to increase with height, with the highest residual stress typically measured at the top surface if part cutting was not performed [30]. The hatching length of the scan strategy was compared at 5 and 10 mm and the residual stress was averaged by measuring 16 points on the top surface.

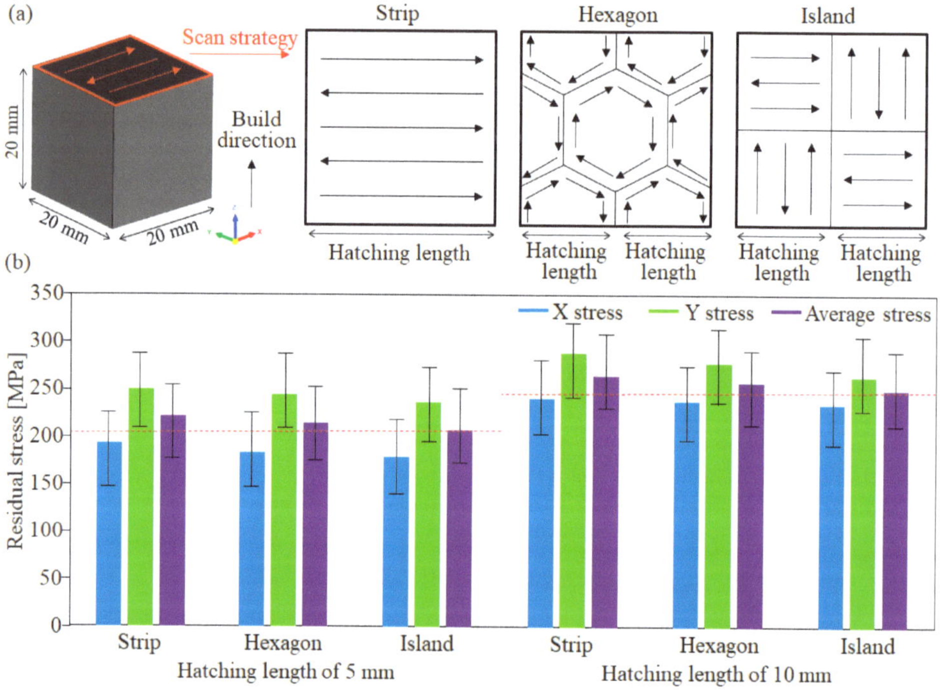

Figure 1. Measurement of residual stress, (**a**) laser scan strategies; and (**b**) measurement of residual stress according to scan pattern and hatching length.

Given the residual stress along scan directions, measurements were taken in both the X and Y directions of the specimen and then the residual stresses in the two directions were averaged. As shown in Figure 1b, the X direction of residual stress of 192.8, 183.3, and 178.8 MPa, the Y direction of residual stress of 249.4, 244.2, and 236.1 MPa, and average residuals stress of 221.1, 213.8, and 207.4 MPa were measured for the strip, hexagon, and island patterns in the hatching length of 5 mm cubic specimens. In addition, the X direction of the residual stress of 241.1, 238.2, and 234.5 MPa and the Y direction of residual stress of 289.2, 277.9, and 264.1 MPa and average residuals stress of 265.1, 257.8, and 249.3 MPa were measured for the strip, hexagon, and island pattern in the hatching length of 10 mm cubic specimens. The error bars of residuals stress are caused by the difference in thermal stress values according to position. This is due to the localized heating that induces a rapidly thermal gradient along the laser track, generating thermal stress in the subsequent track and layer. In the case of the laser scan pattern, the boundaries of overlapping hatching lengths lead to local heat accumulation, which affects the size of the error bars. Consequently, residual stresses with deviations, as indicated by the error bars, arise at certain points. The residual stress values were averaged to effectively compare the differences in error values that occurred for each scan pattern. By averaging the measurements, the average residual stress of the island pattern was found to be lower than other scan strategies such as strip or hexagon patterns. The residual stress was lower at a 5 mm hatching length than at a 10 mm hatching length. In the case of strip and hexagon, the anisotropic residual stress was high because they had a scan pattern in only one direction pattern. Furthermore, as the hatching length increases, the resulting residual stress tends to be higher due to the increasing anisotropic property of the part.

Figure 2 shows the measurement of the residual stress and cantilever deformation to analyze the isotropic properties according to scan strategy. As shown in Figure 2, the residual stress and thermal deformation characteristics were analyzed using specimens with sizes of 30 × 30 × 20 and 40 × 40 × 20 mm^3 and a cantilever with a size of 12 × 72 × 9 mm^3.

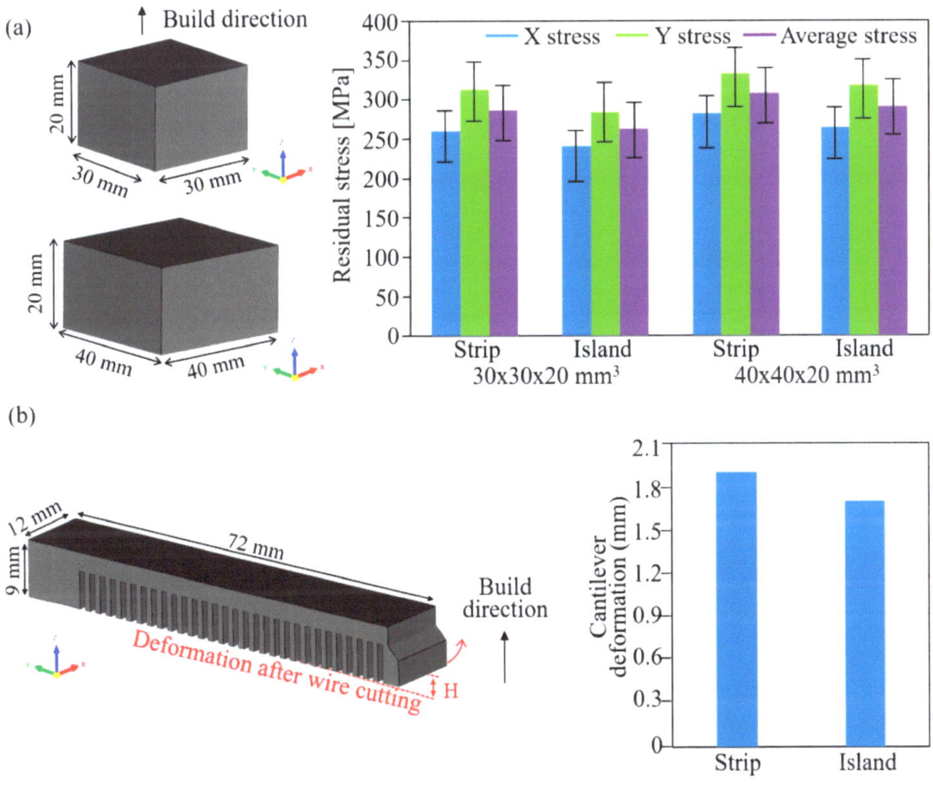

Figure 2. Measurement of residual stress and thermal deformation, (**a**) image of specimen model and measurement of residual stress; and (**b**) image of specimen model and thermal deformation of cantilever.

The X direction of the residual stress of 260.1 and 241.2 MPa, and the Y direction of residual stress of 312.7 and 283.5 MPa, and average residuals stress of 286.4 and 262.3 MPa were measured for the strip and island pattern in 30 × 30 × 20 mm^3 specimens. The X direction of residual stress of 282.2 and 263.8 MPa, and the Y direction of residual stress of 332.7 and 317.3 MPa, and average residuals stress of 307.4 and 290.6 MPa were measured for the strip and island pattern in 40 × 40 × 20 mm^3 specimen (Figure 2a). After additive manufacturing the cantilever, the bottom surface was cut using wire cutting, and the height from the bottom surface was measured to compare the thermal deformation. The deformation of the cantilever when using the strip and island patterns was 1.92 and 1.76 mm, respectively (Figure 2b). As the area of the specimen increases, the cumulative heat capacity of the part increases and the residual stress increases. By applying the island pattern, the residual stress was reduced by 5.47–8.41% in 30 × 30 × 20 mm^3 and 40 × 40 × 20 mm^3 specimens. In addition, the island pattern can reduce the cantilever deformation by 8.33%, reducing the anisotropic property.

2.3. Mechanical Property According to Build Orientation

The mechanical strength produced varies based on the build direction, owing to the layer-by-layer additive manufacturing process. When additively manufacturing a product, it is necessary to select the build direction considering the mechanical properties. The effect of melting area and cooling rate, thus the effect of the microstructure of additively manufactured specimens, affect the strength and strain [31,32]. Usually, the fabricated Ti–6Al–4V specimens using the PBF process have high ultimate tensile strength and lower elongation due to rapid cooling rate compared to conventionally produced titanium alloy [32,33]. As a low elongation rate can lead to the formation of cracks or defects within components during the PBF process, it is imperative to enhance the elongation rate corresponding to the build orientation. As shown in Figure 3a, in order to determine the build direction of the model (horizontal (0°), diagonal (45°), and vertical (90°) orientation.), improving the machinability and structure property, as-built specimens were manufactured according to the build direction and evaluated using tensile tests. The manufactured specimen was machined based on the ASTM E8 standard (Standard Test Methods for Tension Testing of Metallic Materials [34]), and a tensile test was performed. As shown in Figure 3b and Table 1, the tensile strength and elongation measured from 1333 MPa and 5.8% in the horizontal direction structure, 1289 MPa and 5.4% in the diagonal direction structure, and 1307 MPa and 6.7% in the vertical direction specimens.

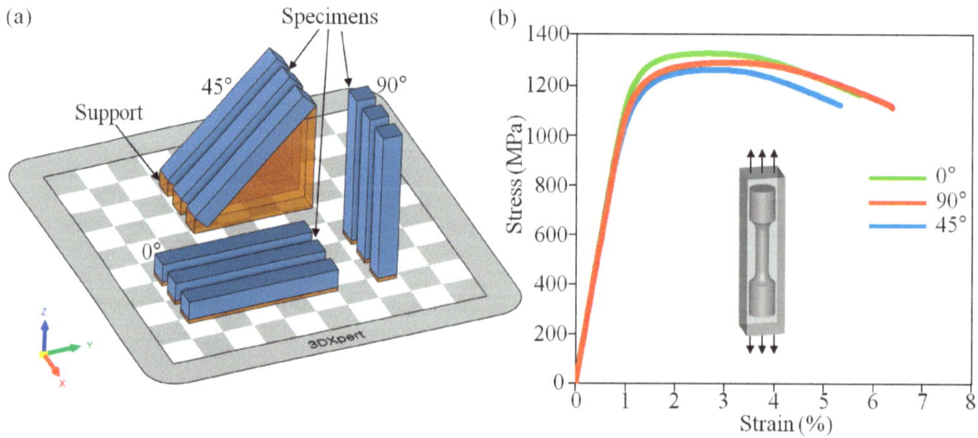

Figure 3. Stress–strain graph and tensile specimens along different directions. (**a**) direction of manufactured specimens; and (**b**) stress–strain curve.

Based on these experiments, low-ductility tensile strength was highest in horizontal orientation and elongation was highest in vertical orientation. The PBF process in the vertical direction with high elongation can effectively reduce defects attributed to low heat accumulation of a small melting area and thermal deformation.

Table 1. Result of tensile experiment according to building direction.

Parameters	Tensile Stress (MPa)	Elongation (%)
Horizontal orientation	1333 ± 23	5.8 ± 0.3
Diagonal orientation	1289 ± 27	5.4 ± 0.4
Vertical orientation	1307 ± 30	6.7 ± 0.4

3. Additive Manufacturing Strategy of Brake Caliper Model

3.1. Topology Optimization of Brake Caliper Model

The weight of the brake caliper model was reduced via topology optimization to determine the full potential of AM. The topology optimization was performed according to the following processes. First, the design domain was constructed based on the dimensions from the reference design (size of 300 × 174 × 110 mm^3, Woosin Industries Corporation, Gimhae-si, Republic of Korea) shown in Figure 4a. Oil channels and assembling holes were set as non-design areas, and unnecessary areas for machining and bolting were removed in advance. The main bridge of the reference design was replaced by bolting assemblies, and two side bridge areas were added to determine a new load path. Second, loading and boundary conditions were set for performance evaluation during the topology optimization. A hydraulic pressure of 13.7 MPa was applied to the inner wall of the oil channel of the non-design domain, and the jig-fixing regions of the two cylinder shapes were fixed, excluding the cylindrical rotation degree of freedom. Additionally, a force in the braking direction was applied to each center of the oil cylinder to consider the braking condition. Third, an optimization problem was formulated to minimize the weight while satisfying the deformation constraints of the reference design. The average deformation values of both inner and outer surfaces extending the oil cylinder centerlines were constrained for hydraulic pressure, and the average deformation values of both oil cylinder centers were constrained for the braking force. Finally, topology optimization design was performed with the solid isotropic material with penalization (SIMP) model using Altair OptiStruct [35] based on the material properties of Ti–6Al–4V. The strength was not directly constrained during topology optimization; however, the maximum stress was maintained within the allowable range because the deformation conditions were dominant.

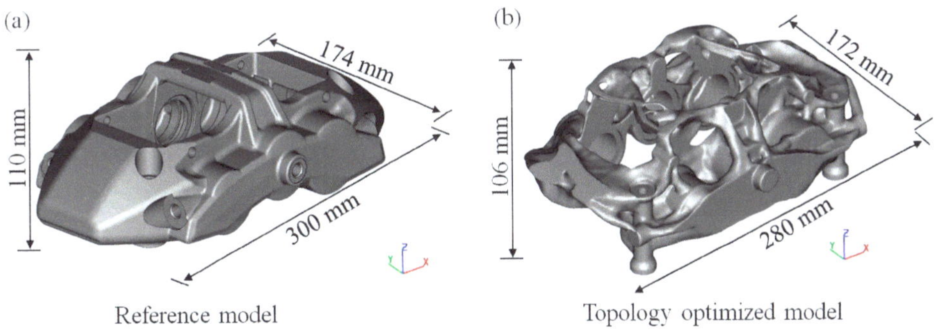

Figure 4. Brake caliper model; (**a**) reference model; and (**b**) topology-optimized model.

The result obtained from topology optimization shown in Figure 4b and the model has a structural shape that cannot be made by conventional fabrication methods such as casting or milling. Despite using titanium material, which has a 38% higher density than aluminum, a weight reduction of 20% was obtained while satisfying both deformation and stress conditions.

The topology optimized design shown in Figure 4b was smoothed using the OSSmooth function and exported to Altair Inspire for further modification. The PolyNURBS function in Inspire was used to modify the shape comprehensively. Extremely thin branch structures were removed or thickened, and the jig-fixing area was reinforced. Small holes were filled in to improve the AM quality but were eventually removed via the machining process.

The topology optimization method significantly impacts weight reduction in parts, but it can create a weak area due to thinning in unexpected places. To prevent this weak point, verification through FEM analysis is required. FEM was performed on the brake caliper components to evaluate the mechanical properties of the topology-optimized model using ABAQUS software (V6.5.1, Dassault Systemes, Vélizy-Villacoublay, France) [36]. The 3D CAD data, mesh type, complexity, mesh optimization, analysis reliability, and contact between caliper components were considered when constructing the FEM models. The FEM analysis of the topology-optimized model is shown in Figure 5. The boundary conditions were defined as shown in Figure 5a. The bolting position was fixed, and pistons in the caliper were applied at 13.7 MPa, which is the pressure level for the leaking test of the caliper. The contact between the pad faces and disc, as well as the contact between the plate of the outboard pad and caliper, were defined as surface-to-surface contact with friction-tangential behavior. The properties of Ti–6Al–4V material were applied to the topology optimization of the brake caliper model.

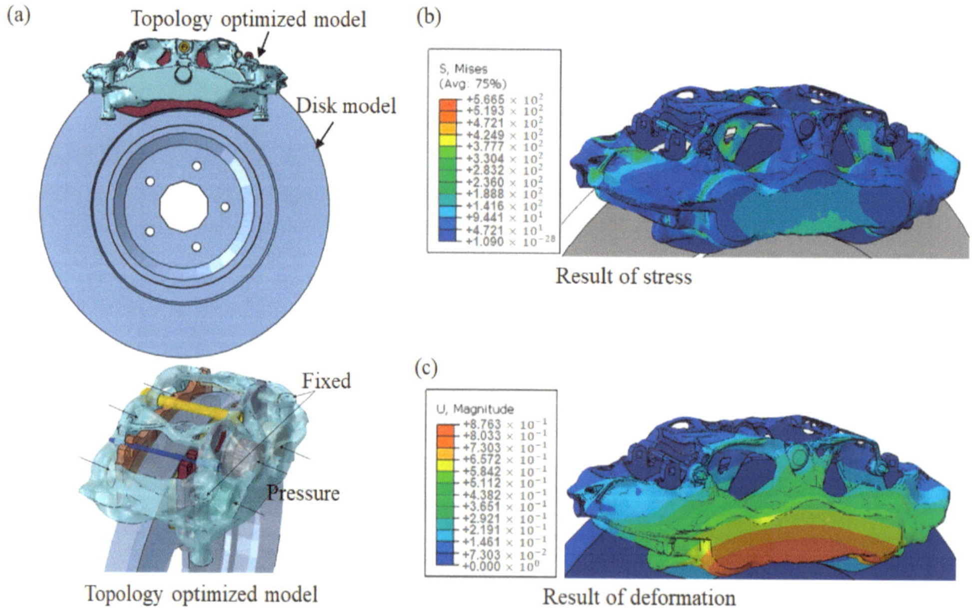

Figure 5. Boundary conditions and results of FEM analysis. (**a**) Topology-optimized model and boundary conditions; (**b**) result of stress; and (**c**) result of deformation.

The stress and deformation results of the brake calipers generated in the brake system are shown in Figure 5b,c, respectively. Regarding the topology optimization of the brake caliper, when a hydraulic pressure of 13.7 MPa was applied, the maximum stress was 566.5 MPa, which was 54.9% of the yield strength, and the maximum deformation was 0.88 mm in the end of caliper. As a result of FEM analysis, the yield strength of the brake system was lower than the yield strength of Ti–6Al–4V material, and the maximum stress generated was half of the yield strength. In addition, the topology model fabricated using titanium alloy has a higher mechanical strength than the existing caliper fabricated using aluminum alloy owing to its material characteristics; additionally, it is advantageous in terms of safety [37].

3.2. Design and Additive Manufacturing of Brake Caliper

The thermal conduction Ti–6Al–4V is lower than that of conventional materials. Therefore, Ti-6Al-4V material exhibits low heat dissipation, leading to heat accumulation, thermal deformation, and residual stress in the PBF process. To prevent excessive thermal deformation, the melt area must be divided and patterned through the PBF process strategy to reduce residual stress [38]. As shown in Figure 6, residual stress and melting area were analyzed according to the vertical and horizontal brake caliper model to reduce the thermal deformation of the PBF part. To analyze deformation caused by thermal stress during PBF, the thermal stress and deformation based on the building direction was compared via numerical analysis of the Amphyon in 3Dxpert software (V24.1.1, 3D Systems, Rock Hill, SC, USA) [39]. The material's mechanical properties and material characteristics suited to laser scanning were considered when constructing the thermal deformation simulation. As shown in Figure 6a, in order to analyze the thermal deformation of only additively manufactured part, the no-support design was generated, and the numerical analysis was performed. As a result of numerical analysis, the maximum thermal stress and deformation were 1124, 1034 MPa, and 13.7, 7.8 mm in the horizontal and vertical directions, respectively. As shown in Figure 6b, the vertical and horizontal melting areas are uneven, and in the case of the horizontal direction, the melting area per layer is higher. Despite the additional 33.5 h required for manufacture, the thermal stress and deformation in the vertical direction were 5.9 mm, 43.1% less than in the horizontal direction, due to the difference in the melting area of orientation.

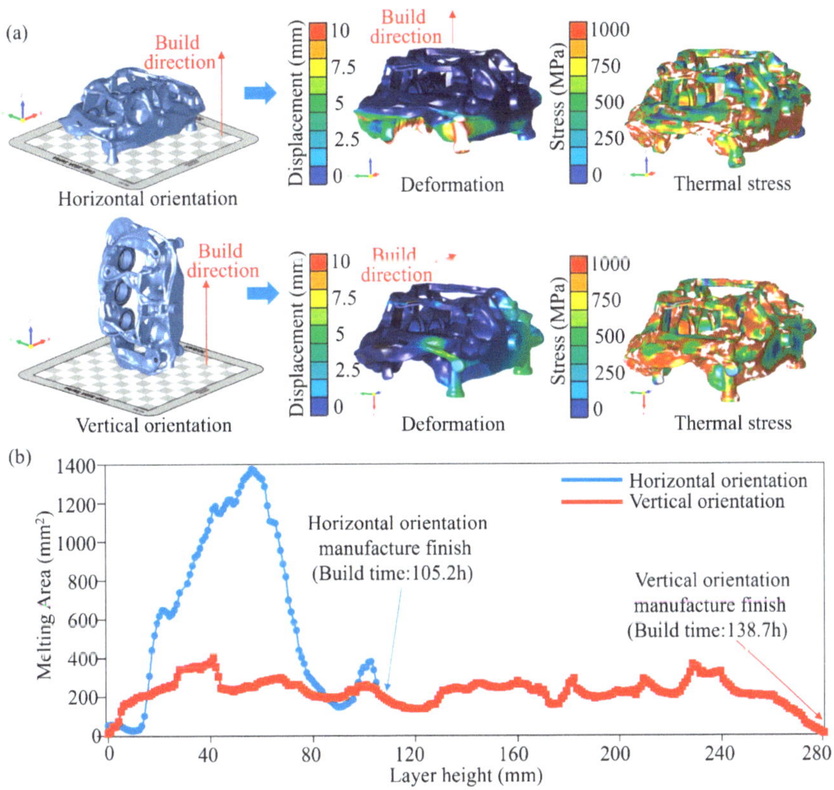

Figure 6. Considerations for AM orientation (**a**) thermal deformation of horizontal and vertical orientation (**b**) melting area on build direction according to orientation.

In the previous contents, the PBF process of reducing anisotropic residual stress and thermal deformation was analyzed. For high-quality manufacturing of additive manufacturing parts, a topology optimization brake caliper was manufactured by applying an island pattern with a hatching length of 5 mm and vertical orientation of the model. As shown in Figure 7a, the support design was carried out. To prevent excessive thermal deformation, the parts where thermal deformation is high in the analysis were designed with cone supports to prevent thermal deformation. As shown in Figure 7b,c, the topology optimization of the brake caliper was manufactured and the supports, which were manufactured alongside the brake caliper, have been processed, and the additively manufactured part has been finalized (part mass of 2.59 kg).

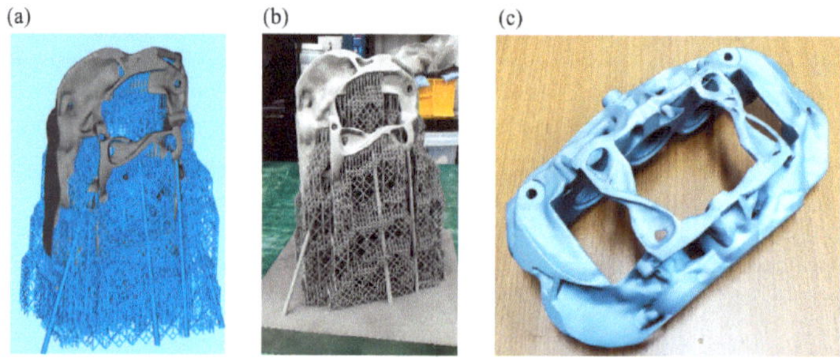

Figure 7. Additive manufacturing process of topology-optimized brake caliper. (**a**) Support design; (**b**) additively manufactured brake caliper; and (**c**) removing the support.

During the AM of large components, such as brake calipers, heat accumulation occurs owing to the large area involved, which causes thermal deformation [38,40]. Therefore, the manufactured caliper must be scanned and then compared with a CAD file to verify the dimensional error of the product. As shown in Figure 8, the error rate was analyzed by overlapping the two models based on the origin coordinates of the model scanned with the brake caliper using t Surveyor ZS-3040 (Laser Design Inc., Golden Valley, MN, USA) scanning equipment and CAD data for topology optimization. As a result of comparing the topology optimization model and scan data (additive manufacturing part), the maximum error rate was 97 μm and the topological brake caliper of the precise shape was manufactured using the effective PBF process.

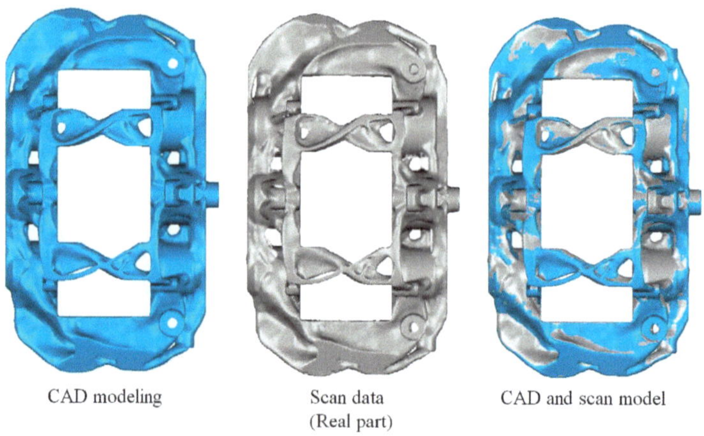

Figure 8. Evaluation of dimensional error via CAD modeling and scan data comparison.

3.3. Evaluation of Topology Optimized Brake Caliper

The performance of the topology-optimized brake caliper was evaluated based on comparison with the commercial product. As shown in Figure 9a, the topology-optimized brake caliper and the reference brake caliper were mounted on the disc. The test procedure of the brake caliper was evaluated based on the JASO C406 (Japanese Automobile Standards Organization C406) test procedure using a brake dynamometer [41]. The sequence of the JASO C406 test procedure is as follows: burnish, first effectiveness, first reburnish, first fade, recovery, second reburnish, second fade, and recovery. The effectiveness test was performed at three different braking speeds (50, 80, and 100 km/h) and eight different decelerations (0.1–0.8 g). The fade and recovery steps were a series of tests for evaluating the effects of brake systems changed by continuous braking. In the first step of fade and recovery, 0.3 g deceleration was performed 10 times for 35 s at 100 km/h, and the same conditions were repeated 15 times in the second step of fade and recovery. As shown in Figure 9b, the temperature, frictional force effectiveness, and pressure were analyzed via brake dynamometer experiments of the topology-optimized and reference models.

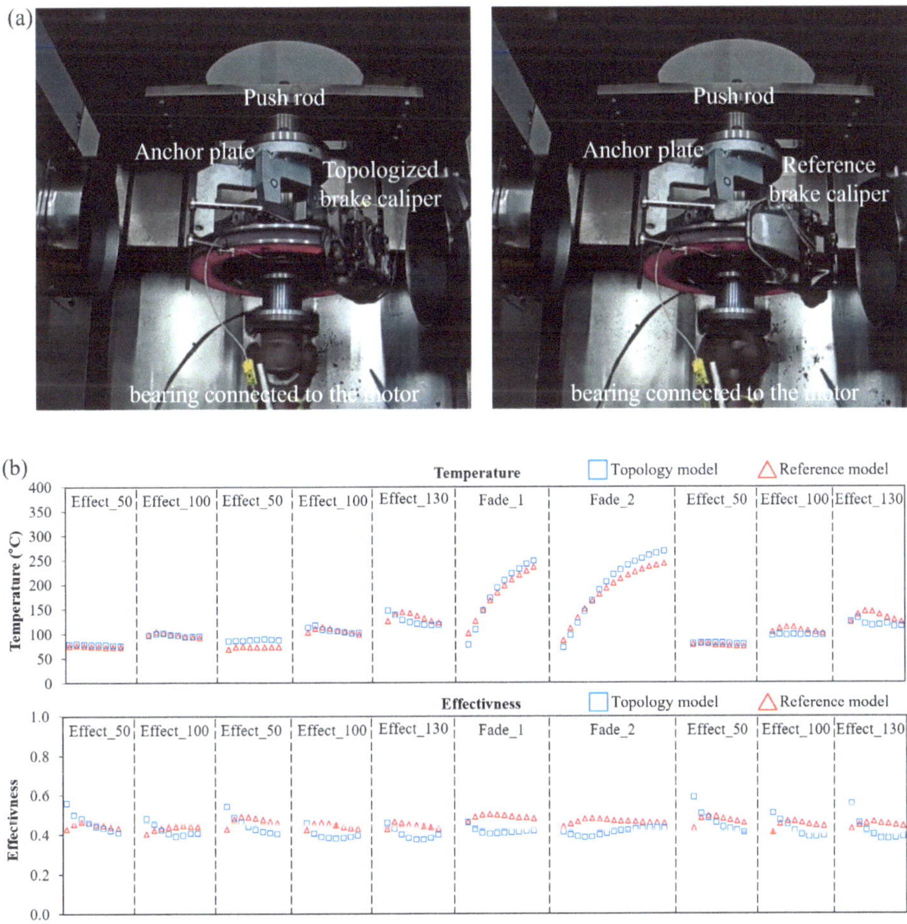

Figure 9. Dynamometer experiments (JASO C406), (**a**) topology-optimized brake caliper and commercial product (casting product); and (**b**) test results of dynamometer test.

In terms of temperature, both models showed values within 15 °C. However, a temperature gradient appeared because the thermal conductivity of titanium is lower than that of aluminum, therefore, heat accumulation occurred in the fade experiment. In terms of the frictional force effectiveness, the friction effect of topology-optimized model was as high as 0.01~0.03 in the initial brake dynamometer of each step. However, continuous experimentation with a topological brake caliper can result in performance degradation due to heat accumulation in the Ti-6Al-4V material, attributed to its low thermal conductivity. Nevertheless, it maintains frictional performance similar to that of the commercial product. Based on the results of JASO C 406, the performance of the additively manufactured brake caliper using the effective PBF process is comparable to that that of the commercial product.

4. Summary and Conclusions

The topology optimization brake caliper model with lightweight and high-strength characteristics is precisely manufactured. Based on the numerical analysis and the experimental results of topological model, the following conclusions can be inferred:

- The residual stress was reduced by 5.47–8.41% and the thermal deformation of the cantilever by 8.33% compared to the strip process by applying an island pattern with a hatching length of 5 mm.
- The topology-optimized brake caliper is built with a vertical orientation that minimizes the melting area that causes thermal deformation and residual stress. With an effective PBF process, a topology model and a precise brake caliper with an error of up to 97 μm were fabricated.
- The brake performance of the topology-optimized brake caliper was evaluated based on JASO C406. The brake performance of the topology-optimized model equals or surpasses that of the commercial product.

As a result of this study, it is possible that additive manufacturing parts applicable to the automotive industry can be manufactured through an effective PBF process. In addition, through a new topology-optimized brake caliper, fuel efficiency can be improved and showed the possibility of using it as a commercial automobile brake system.

Author Contributions: J.Y.: conceptualization, methodology, visualization, experiments, investigation, formal analysis, software, writing—original draft, review and editing. Y.J.: conceptualization, methodology, visualization, experiments, investigation, formal analysis, software, writing—original draft, review and editing. J.J.: visualization, experiments, formal analysis, investigation. J.D.O.: experiments, formal analysis, investigation. S.C.: methodology, investigation, formal analysis. S.H.P.: methodology, investigation, formal analysis. T.H.L.: conceptualization, supervision, funding acquisition, project administration, writing—original draft, review and editing. J.P.: conceptualization, supervision, funding acquisition, project administration, writing—original draft, review and editing. All authors have read and agreed to the published version of the manuscript.

Funding: This study was supported by the Technology Innovation Program funded by the Ministry of Trade, Industry & Energy (MOTIE) of Korea (1415175160), and the Korea Institute of Industrial Technology (KITECH) internal project (JE240029) and by the Incheon Metropolitan City (UR240077/IZ240077).

Data Availability Statement: The original contributions presented in the study are included in the article, further inquiries can be directed to the corresponding authors.

Conflicts of Interest: Authors Youngsuk Jung and Shinhu Cho were employed by Hyundai Motor Group. Author Jae Dong Ock was employed by Woosin Industries Corporation. The remaining authors declare that the research was conducted in the absence of any commercial or financial relationships that could be construed as a potential conflict of interest.

References

1. Bartlett, J.L.; Li, X. An overview of residual stresses in metal powder bed fusion. *Addit. Manuf.* **2019**, *27*, 131–149. [CrossRef]
2. Seifi, M.; Salem, A.; Beuth, J.; Harrysson, O.; Lewandowski, J.J. Overview of materials qualification needs for metal additive manufacturing. *JOM* **2016**, *68*, 747–764. [CrossRef]

3. Murr, L.E.; Gaytan, S.M.; Ramirez, D.A.; Martinez, E.; Hernandez, J.; Amato, K.N.; Wicker, R.B. Metal fabrication by additive manufacturing using laser and electron beam melting technologies. *J. Mater. Sci. Technol.* **2012**, *28*, 1–14. [CrossRef]
4. Liu, R.; Wang, Z.; Sparks, T.; Liou, F.; Newkirk, J. Aerospace applications of laser additive manufacturing. In *Laser Additive Manufacturing*; Woodhead Publishing: Cambridge, UK, 2017; pp. 351–371.
5. Juechter, V.; Franke, M.M.; Merenda, T.; Stich, A.; Körner, C.; Singer, R.F. Additive manufacturing of Ti-45Al-4Nb-C by selective electron beam melting for automotive applications. *Addit. Manuf.* **2018**, *22*, 118–126. [CrossRef]
6. Li, J.; Hu, J.; Zhu, Y.; Yu, X.; Yu, M.; Yang, H. Surface roughness control of root analogue dental implants fabricated using selective laser melting. *Addit. Manuf.* **2020**, *34*, 101283. [CrossRef]
7. Harzheim, L.; Graf, G.; Klug, S.; Liebers, J. Topologieoptimierung im praktischen Einsatz. *ATZ-Automob. Z.* **1999**, *101*, 530–539. [CrossRef]
8. Kranz, J. *Methodik und Richtlinien für Die Konstruktion von Laseradditiv Gefertigten Leichtbaustrukturen*; Springer: Berlin/Heidelberg, Germany, 2017.
9. Dagkolu, A.; Gokdag, I.; Yilmaz, O. Design and additive manufacturing of a fatigue-critical aerospace part using topology optimization and L-PBF process. *Procedia Manuf.* **2021**, *54*, 238–243. [CrossRef]
10. Yıldız, A.R.; Kılıçarpa, U.A.; Demirci, E.; Doğan, M. Topography and topology optimization of diesel engine components for light-weight design in the automotive industry. *Mater. Test.* **2019**, *61*, 27–34. [CrossRef]
11. Bendsoe, M.P.; Sigmund, O. *Topology Optimization: Theory, Methods, and Applications*; Springer Science & Business Media: Berlin/Heidelberg, Germany, 2003.
12. Großmann, A.; Weis, P.; Clemen, C.; Mittelstedt, C. Optimization and re-design of a metallic riveting tool for additive manufacturing—A case study. *Addit. Manuf.* **2020**, *31*, 100892. [CrossRef]
13. Kumaran, M.; Senthilkumar, V. Generative design and topology optimization of analysis and repair work of industrial robot arm manufactured using additive manufacturing technology. *IOP Conf. Ser. Mater. Sci. Eng.* **2021**, *1012*, 012036. [CrossRef]
14. Mantovani, S.; Barbieri, S.G.; Giacopini, M.; Croce, A.; Sola, A.; Bassoli, E. Synergy between topology optimization and additive manufacturing in the automotive field. *Proc. Inst. Mech. Eng. Part B J. Eng. Manuf.* **2021**, *235*, 555–567. [CrossRef]
15. Chen, X.; Yin, J.; Wang, W.; Wu, L.; Tang, F. Approaches to diminish large unsprung mass negative effects of wheel side drive electric vehicles. *J. Adv. Mech. Des. Syst. Manuf.* **2016**, *10*, JAMDSM0064. [CrossRef]
16. Kandukuri, S.Y.; Pai, A.; Manikandan, M. Scope of Carbon Fibre-Reinforced Polymer Wheel Rims for Formula Student Racecars: A Finite Element Analytical approach. *J. Inst. Eng. (India) Ser. C* **2022**, *103*, 939–948. [CrossRef]
17. Natarajan, H.K. Study of silicon carbide-reinforced aluminum matrix composite brake rotor for motorcycle application. *Int. J. Adv. Manuf. Technol.* **2018**, *94*, 1461–1475.
18. Jamal, M.H.M.; Saleh, M.K.M. Design and Analysis an Efficient Lightweight Brake Caliper for Kuim Electric Vehicle. *J. Eng. Health Sci.* **2019**, *2*, 67–78.
19. Maijer, D.M.; Gao, Y.X.; Lee, P.D.; Lindley, T.C.; Fukui, T. A through-process model of an A356 brake caliper for fatigue life prediction. *Metall. Mater. Trans. A* **2004**, *35*, 3275–3288. [CrossRef]
20. Merulla, A.; Gatto, A.; Bassoli, E.; Munteanu, S.I.; Gheorghiu, B.; Pop, M.A.; Munteanu, D. Weight reduction by topology optimization of an engine subframe mount, designed for additive manufacturing production. *Mater. Today Proc.* **2019**, *19*, 1014–1018. [CrossRef]
21. Sanaei, N.; Fatemi, A.; Phan, N. Defect characteristics and analysis of their variability in metal L-PBF additive manufacturing. *Mater. Des.* **2019**, *182*, 108091. [CrossRef]
22. Roehling, J.D.; Smith, W.L.; Roehling, T.T.; Vrancken, B.; Guss, G.M.; McKeown, J.T.; Matthews, M.J. Reducing residual stress by selective large-area diode surface heating during laser powder bed fusion additive manufacturing. *Addit. Manuf.* **2019**, *28*, 228–235. [CrossRef]
23. Wang, Y.; Guo, W.; Xie, Y.; Li, H.; Zeng, C.; Xu, M.; Zhang, H. In-situ monitoring plume, spattering behavior and revealing their relationship with melt flow in laser powder bed fusion of nickel-based superalloy. *J. Mater. Sci. Technol.* **2024**, *177*, 44–58. [CrossRef]
24. Guo, Q.; Zhao, C.; Qu, M.; Xiong, L.; Hojjatzadeh, S.M.H.; Escano, L.I.; Parab, N.D.; Fezzaa, K.; Sun, T.; Chen, L. In-situ full-field mapping of melt flow dynamics in laser metal additive manufacturing. *Addit. Manuf.* **2020**, *31*, 100939. [CrossRef]
25. Srivastava, S.; Garg, R.K.; Sharma, V.S.; Alba-Baena, N.G.; Sachdeva, A.; Chand, R.; Singh, S. Multi-physics continuum modelling approaches for metal powder additive manufacturing: A review. *Rapid Prototyp. J.* **2020**, *26*, 737–764. [CrossRef]
26. Bian, P.; Shi, J.; Liu, Y.; Xie, Y. Influence of laser power and scanning strategy on residual stress distribution in additively manufactured 316L steel. *Opt. Laser Technol.* **2020**, *132*, 106477. [CrossRef]
27. Nadammal, N.; Mishurova, T.; Fritsch, T.; Serrano-Munoz, I.; Kromm, A.; Haberland, C.; Bruno, G. Critical role of scan strategies on the development of microstructure, texture, and residual stresses during laser powder bed fusion additive manufacturing. *Addit. Manuf.* **2021**, *38*, 101792. [CrossRef]
28. Liu, J.; Li, G.; Sun, Q.; Li, H.; Sun, J.; Wang, X. Understanding the effect of scanning strategies on the microstructure and crystallographic texture of Ti-6Al-4V alloy manufactured by laser powder bed fusion. *J. Mater. Process. Technol.* **2022**, *299*, 117366. [CrossRef]
29. Thompson, S.M.; Bian, L.; Shamsaei, N.; Yadollahi, A. An overview of Direct Laser Deposition for additive manufacturing; Part I: Transport phenomena, modeling and diagnostics. *Addit. Manuf.* **2015**, *8*, 36–62. [CrossRef]

30. Zhang, H.; Li, C.; Xu, M.; Dai, W.; Kumar, P.; Liu, Z.; Li, Z.; Zhang, Y. The fatigue performance evaluation of additively manufactured 304L austenitic stainless steels. *Mater. Sci. Eng. A* **2021**, *802*, 140640. [CrossRef]
31. Liu, S.; Shin, Y.C. Additive manufacturing of Ti6Al4V alloy: A review. *Mater. Des.* **2019**, *164*, 107552. [CrossRef]
32. Ganeriwala, R.K.; Strantza, M.; King, W.E.; Clausen, B.; Phan, T.Q.; Levine, L.E.; Hodge, N.E. Evaluation of a thermomechanical model for prediction of residual stress during laser powder bed fusion of Ti-6Al-4V. *Addit. Manuf.* **2019**, *27*, 489–502. [CrossRef]
33. Nguyen, H.D.; Pramanik, A.; Basak, A.K.; Dong, Y.; Prakash, C.; Debnath, S.; Buddhi, D. A critical review on additive manufacturing of Ti-6Al-4V alloy: Microstructure and mechanical properties. *J. Mater. Res. Technol.* **2022**, *18*, 4641–4661. [CrossRef]
34. Kardak, A.A.; Sinclair, G.B. Stress concentration factors for ASTM E8/E8M-16a standard round specimens for tension testing. *J. Test. Eval.* **2020**, *48*, 711–719. [CrossRef]
35. *Altair OptiStruct 2016: Users Guide*; Altair Engineering, Inc.: Troy, MI, USA, 2016.
36. *ABAQUS/Standard User's Manual for Version 6.5.1*; Hibbitt, Karlsson and Sorensen Inc.: Pawtucket, RI, USA, 2004.
37. Yang, Y.; Li, M.; Li, K.R. Comparison and analysis of main effect elements of machining distortion for aluminum alloy and titanium alloy aircraft monolithic component. *Int. J. Adv. Manuf. Technol.* **2014**, *70*, 1803–1811. [CrossRef]
38. Mugwagwa, L.; Yadroitsava, I.; Makoana, N.W.; Yadroitsev, I. Residual stress in laser powder bed fusion. In *Fundamentals of Laser Powder Bed Fusion of Metals*; Elsevier: Amsterdam, The Netherlands, 2021; pp. 245–276.
39. 3Dsystems 3Dxpert. User's Manual. Available online: https://www.3dsystems.com/software/3dxpert (accessed on 5 November 2024).
40. Li, C.; Liu, Z.Y.; Fang, X.Y.; Guo, Y.B. Residual stress in metal additive manufacturing. *Procedia Cirp* **2018**, *71*, 348–353. [CrossRef]
41. Aranganathan, N.; Mahale, V.; Bijwe, J. Effects of aramid fiber concentration on the friction and wear characteristics of non-asbestos organic friction composites using standardized braking tests. *Wear* **2016**, *354*, 69–77. [CrossRef]

metals

MDPI

Article

Preliminary Evaluation of Nickel Silicide (NiSi12-wt%) Laser Cladding for Enhancing Microhardness and Corrosion Resistance of S355 Steel

Mohammad Ibrahim [1,*], Christopher Hulme [2], Geir Grasmo [1] and Ragnhild E. Aune [1,3]

1 Department of Engineering Sciences, University of Agder (UiA), 4879 Grimstad, Norway; geir.grasmo@uia.no (G.G.); ragnhild.aune@ntnu.no (R.E.A.)
2 Department of Materials Science and Engineering, KTH Royal Institute of Technology, 100 44 Stockholm, Sweden; chrihs@kth.se
3 Department of Materials Science and Engineering, Norwegian University of Science and Technology (NTNU), 7491 Trondheim, Norway
* Correspondence: mohammad.ibrahim@uia.no

Abstract: S355 construction steel, a commonly used mild steel due to its exceptional strength, is prone to environmental degradation, especially pitting corrosion in highly corrosive marine environments. To address this vulnerability, applying a surface layer of nickel silicide (NiSi) cladding on such components could offer a solution, given that NiSi-based alloys are known for their high corrosion resistance and exceptional mechanical properties. Thus, the present study has investigated the corrosion resistance and microhardness of the NiSi12-wt% cladding deposited onto substrates of S355 steel using laser metal deposition. An accelerated ASTM G48 corrosion test and a Vickers microhardness test were conducted in a solution of 6% ferric chloride ($FeCl_3$) solution at room and elevated temperatures to represent marine environments, with uncladded sheet substrates exposed to the same test environments as a reference. All exposed S355 steel samples, with and without cladding, underwent microhardness testing and were characterized using light optical microscopy (LOM) and low-voltage field emission scanning electron microscopy (LVFESEM). The findings indicate that the NiSi12-wt% cladding significantly enhances the corrosion resistance and mechanical properties of the S355 steel samples, showcasing its potential for use in marine and industrial environments where corrosion and mechanical wear are expected.

Keywords: nickel silicide; surface engineering; corrosion behavior; ASTM G48; laser surface modification

1. Introduction

Steel is a widely used material in the construction of offshore installations and equipment, onshore infrastructure, and buildings, both commercial and residential, due to its strength and stability for load-bearing components in elements such as beams, columns, and shells at a low cost. Its strength-to-weight ratio is particularly suitable for spanning long distances while maintaining structural integrity, making it a fundamental material in construction projects [1,2]. Among structural steels, S355, defined under the European standards [3], is particularly valued for its yield strength of 355 MPa, which makes it suitable for high load-bearing structures. However, steel's susceptibility to corrosion, accelerated under harsh environmental conditions, presents significant challenges [4]. Factors such as industrial pollution, coastal atmospheres, and acid rain accelerate corrosion through electrochemical mechanisms involving moisture and oxygen [5].

Moreover, corrosion becomes a concern in industrial areas with high levels of pollutants like sulfur dioxide (SO_2) and nitrogen oxides (NO_x). These pollutants, upon reacting with atmospheric moisture, form acidic compounds that accelerate the deterioration of nearby structures [6]. This makes the development of protective measures, such as effective cladding techniques, crucial for extending the lifespan of steel structures. Enhancing corrosion resistance and mechanical properties of steel through cladding can provide robust protective layers against aggressive environments.

Microhardness, a fundamental mechanical property, directly influences a material's ability to withstand wear, deformation, and damage under various loads and environmental conditions. It is approximately proportional to the material's yield stress [7,8].

1.1. Cladding Techniques

Cladding enhances steel's mechanical and corrosion-resistant properties by forming a robust protective layer. It involves adding a thick layer of material, often several millimeters thick, to protect the metal from corrosion and improve surface strength [9]. Laser cladding, a precision surface engineering technique, creates a metallurgically bonded protective layer with excellent adherence and tailored properties, enhancing resistance to wear, corrosion, and other forms of degradation [10]. It involves the focused application of a high-energy laser beam to melt and fuse a powdered or wire-form cladding material onto a substrate surface, particularly in challenging environments [9].

Additive manufacturing (AM) techniques offer a promising solution to enhancing the low-temperature ductility of different materials by providing precise control of the microstructure and porosity of the final product. Through AM, materials can be deposited layer by layer, optimizing grain size, shape, and orientation, thereby improving the ductility of the substrate material. Furthermore, integrating AM with advanced modeling and simulation techniques enables the optimization of manufacturing processes, identification of potential failure modes, and enhancement of material properties.

Laser-based direct energy deposition (DED), also known as laser metal deposition (LMD), is an AM technique for high-speed laser cladding [11]. LMD has gained popularity due to its ability to deposit brittle materials using the gradient laser power deposition method, in which the laser power is reduced layer by layer to obtain a dendritic microstructure [12,13].

1.2. Nickel Silicide (NiSi)

Various nickel silicide (NiSi) based alloys have demonstrated superior corrosion resistance compared to low-alloyed construction steels [14]. This enhanced resistance is due to silicon in key phases such as Ni_3Si, which promotes the formation of a protective, thin oxide layer on the surface of the nickel silicide when exposed to corrosive conditions [15]. However, nickel silicide-based alloys often have inadequate room-temperature ductility [16], limiting their practical use. This challenge can be overcome by applying them as protective thin films or thicker claddings.

Nickel silicide-based alloys can be deposited onto construction steel surfaces through various techniques, each with advantages and drawbacks depending on desired thickness, substrate adherence, and specific properties needed for corrosion protection. In processes like sputter deposition, precursor gases containing nickel-based compounds will react with, for example, a silicon surface to form a nickel silicide-based coating [17]. This technique allows for precise control of the film thickness and can produce uniform claddings. Additionally, nickel silicide-based alloys can be deposited on surfaces using plasma or thermal spraying techniques [18], as well as different AM techniques [19]. AM or laser cladding could be considered an effective technique for applying a nickel silicide-based coating to S355 steel due to its precise control of cladding parameters. By leveraging the small laser spot diameter, the thickness, surface roughness, and layer interface of the cladding layer can be finely optimized, ensuring superior coating performance and adhesion [20].

The present study's novelty lies in its application of a nickel silicide-based alloy as a cladding material for S355 steel, an approach not widely explored. This approach leverages the exceptional corrosion resistance and mechanical strength of NiSi12-wt%, offering a significant improvement over traditional coatings and cladding materials.

1.3. Objective

As S355 steel is prone to localized pitting corrosion in harsh environments, its use in construction can be challenging. To address this, the present study investigates the effectiveness of NiSi12-wt% cladding in enhancing the corrosion resistance and mechanical properties of S355 steel. Specifically, the pitting corrosion behavior and microhardness of both unprotected and NiSi12-wt% cladded S355 steel samples are examined. All samples were exposed to an accelerated corrosion test (ASTM G48) at room and elevated temperatures for different durations to assess pitting corrosion rates. Vickers microhardness tests were also performed to evaluate the cladding's influence on the steel's mechanical properties. This initial assessment will serve as a foundation for further, more comprehensive investigations into the long-term corrosion behavior and protective mechanisms offered by the NiSi12-wt% cladding.

2. Experimental Procedure and Analyses

2.1. Raw Material, Gas-Atomization, and Analysis

The raw materials used were nickel (sourced as crowns, 99.8% purity) and silicon (99.995% purity). The NiSi12-wt% powder was produced using a gas atomizer and used for cladding S355 steel samples through LMD. The raw materials were provided for the trials by ELKEM Silicon AS (Kristiansand, Norway).

The amount of nickel and silicon powders needed to produce NiSi12-wt% was introduced into a 75 kW induction furnace, and the resulting melt was cast into sheets with a thickness of 10 mm. Subsequently, the sheets were cut into smaller pieces suitable for loading into an atomization crucible. LECO analysis was performed on the metal pieces produced using a combustion testing unit (Model: RC612), revealing 0.2 wt% carbon contamination. The metal pieces were later used in a VIGA-type gas atomizer with a batch capacity of 10 kg of steel equivalents at the INSTM research unit (National Interuniversity Consortium of Materials Science and Technology) of the Polytechnic University of Turin (Alessandria, Italy) to produce the desired NiSi12-wt% powder. Low-voltage field emission scanning electron microscopy (LV-FESEM, Supra 55VP, Zeiss, Oberkochen, Germany) was used to investigate the morphology of the atomized powders using secondary electron imaging. The accelerating voltage was 15 keV, and a working distance of 11 mm was maintained for all samples.

2.2. Laser Metal Deposition (LMD) and Sample Preparation

The gas-atomized powder, falling within the size range of 60–120 μm, was utilized for cladding sheet samples of S355 steel. The LMD technique (LT 65 3D hybrid, DMG MORI AG, Pfronten, Germany), housed at the Mechatronics Innovation Lab (MIL) situated at the University of Agder (UiA, Grimstad, Norway), was used for this purpose. The unit, equipped with a 2500 W fiber laser operating at a wavelength of 1020 nm, deposited five layers of NiSi12-wt% powder particles onto the sheet samples. The dimensions of the samples were 28 mm × 28 mm × 6.5 mm ($L \times W \times H$). The LMD laser power, scanning speed, and deposition rate were systematically adjusted and analyzed to identify the optimal parameters for achieving crack-free multi-layer claddings at a single-layer height of 0.3 mm. To prevent excessive heat accumulation and subsequent warping of the clads, a 30 s delay time was introduced between successive layers, as well as a laser power reduction of 100 W. The deposition speed and feeding rate were kept constant at 1 m/min and 10.5 g/min, respectively. The laser powers used are summarized in Table 1.

Table 1. Laser power for the 5-layer cladding of the NiSi12-wt% alloy onto a sheet sample of S355 steel using the LMD technique.

Layers	Laser Power (W)
1 layer	2350
2 layers	2250

Table 1. *Cont.*

Layers	Laser Power (W)
3 layers	2150
4 layers	2050
5 layers	1950

The same SEM apparatus, detector, accelerating voltage, and working distance were used to analyze the atomized powder and check the quality of the claddings.

2.3. Corrosion Sample Preparation

To prepare the produced claddings for the ASTM G48 test, a grinding machine (SAPHIR 200, ATM, Mammelzen, Germany) and silicon carbide grinding paper with progressively finer grain sizes, ranging from 200 μm to 15 μm, were used. Following grinding, the samples were polished to achieve a uniform surface using a DiaPro Dac 3 μm diamond suspension solution on a standardized polishing machine (Tegramin-30, Struers, Copenhagen, Denmark). After grinding and polishing, the cladding thickness was 1.96 mm.

2.4. ASTM G48 Test

Samples were prepared for preliminary pitting corrosion testing using the ASTM G48 test in ferric chloride ($FeCl_3$) solution at room and elevated temperatures. After testing, ethanol was sprayed on the sample surfaces to prevent further corrosion. Weight loss analysis was performed to facilitate mass loss measurements.

Four samples, each measuring 2.8 cm × 1.5 cm ($L \times W$), were prepared, i.e., two consisting of the unprotected S355 steel sheet sample and two NiSi12-wt% cladded sheet samples. The solution for immersing the samples was prepared from 100 g of ferric chloride hexahydrate ($FeCl_3 \cdot 6H_2O$) dissolved in 900 mL of deionized water, resulting in a ferric chloride concentration of 6% by mass. The temperature of the solution was adjusted according to the requirements of each test. Specifically, two samples were placed in individual glass beakers and immersed in solutions maintained at 25 ± 2 °C for 24 and 72 h, respectively. The other two samples, also placed in separate glass beakers, were immersed in solutions heated to 50 ± 2 °C for 24 and 72 h, respectively. A thermometer was used to verify and monitor the temperature of the solutions during the tests. Figure 1 illustrates the ASTM G48 test setup [21].

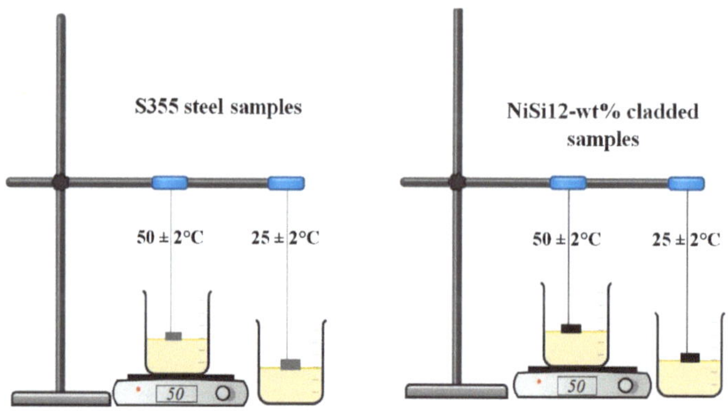

Figure 1. A schematic of the ASTM G48 corrosion test setup.

At the termination of each test, ethanol was sprayed onto the sample surfaces to prevent further corrosion. Additionally, each sample was weighed to the nearest 0.001 g to facilitate weight loss analysis.

2.5. Analysis of the Sheet Samples of the S355 Steel

Light optical microscopy (LOM) was used to characterize the localized corrosion pits on the surface of the NiSi-12 wt% cladded samples. Multiple images of the sample faces were secured using an inverted optical microscope (Model: Axiovert A1, Zeiss, Oberkochen, Germany) unit equipped with the Zeiss Axiovision Image Analysis Software (Release 4.1). For LOM analysis, the polished samples were exposed to 30% iron (III) chloride ($FeCl_3$) solution for 10 s as a final step of the metallographic preparation. Stereo microscopy (Stemi 508, Zeiss, Oberkochen, Germany) was used to image the complete deposited clad.

Microhardness testing was performed using a Viker's microhardness tester (VMH MOT Vickers, Leica, Wetzlar, Germany) under a load of 2.94 N for 15 s. For each sample, five separate measurements were taken to ensure reliability and accuracy, with the indentations being sufficiently large to represent the microhardness of the samples. The average value of these five measurements was calculated and reported for each condition.

3. Results and Discussion

3.1. Microstructural Analysis

The surface of unprotected S355 steel samples and the NiSi12-wt% cladded samples were analyzed using LOM and SEM. The NiSi12-wt% cladding exhibited directional grains and dendritic structures resulting from cooling during solidification. As seen in Figure 2a, the microstructure of the base material reveals a ferritic matrix and pearlite made of alternating lamellas of ferrite and cementite. The ferrite layers in the pearlite appear darker in the LOM images due to their faster reaction with the $FeCl_3$ solution during etching. Figure 2b presents the more complex structure of the NiSi12-wt% cladding, revealing directional grains formed by the LMD process, aligning parallel to the direction of the moving heat source.

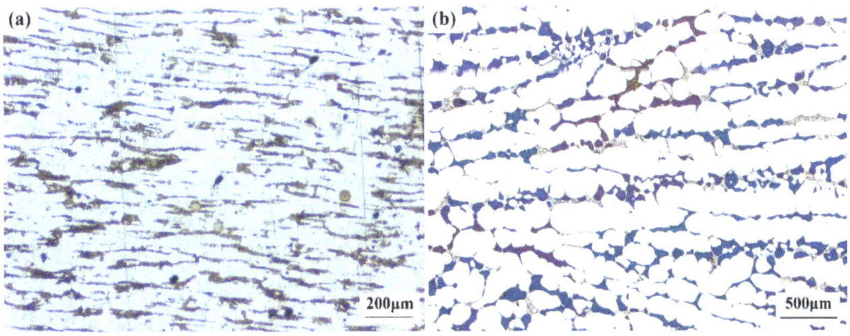

Figure 2. LOM images of the microstructure of the (**a**) unprotected S355 steel sample and (**b**) NiSi12-wt% cladding surface comprising multiple layers of NiSi12-wt% deposited onto an S355 steel sample.

SEM images were obtained for further clarity, see Figure 3. As seen in Figure 3b, the structure comprises dendritic grains resulting from rapid cooling during the LMD process [10,22]. Additionally, carbon contamination is evident in the microstructure, manifesting as carbon-rich regions along the grain boundaries. This contamination potentially influences the corrosion behavior by creating preferential sites for corrosion initiation.

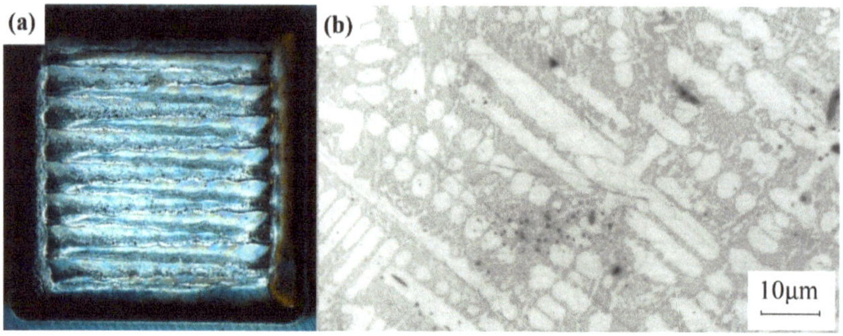

Figure 3. (**a**) Five layers of the NiSi12-wt% cladding deposited onto an S355 steel sample using the LMD technique. (**b**) The microstructure of the NiSi12-wt% cladding secured using SEM.

3.2. Corrosion Behavior

The ASTM G48 test showed superior corrosion resistance of the NiSi12-wt% cladded samples compared to the unprotected S355 steel samples. The test evaluated the resistance of both sample types against corrosion in a 6% $FeCl_3$ solution. Figure 4 shows the polished samples before exposure to the corrosive environment. The LOM and SEM micrographs in Figures 5 and 6 illustrate the corrosion patterns after exposure at room temperature (25 °C). As can be seen from the figures, the base material exhibits extensive pitting corrosion (highlighted by arrows), characterized by pits or holes from the dissolution of iron ions [23]. In contrast, the NiSi12-wt% cladded sample demonstrates minimal signs of corrosion, indicating effective protection.

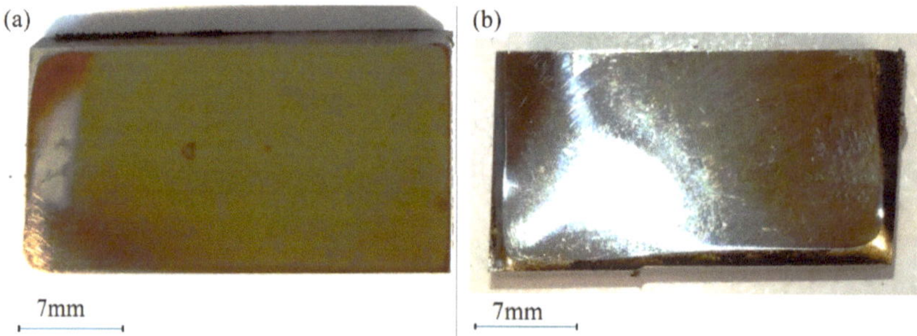

Figure 4. (**a**) Polished unprotected S355 steel sample and (**b**) polished NiSi12-wt% cladded sample.

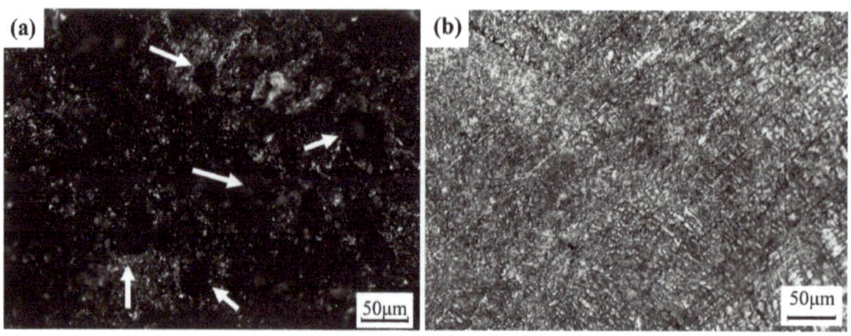

Figure 5. LOM images of the samples after the ASTM G48 test at 25 °C for 72 h. (**a**) Unprotected S355 steel sample, and (**b**) NiSi12-wt% cladded sample.

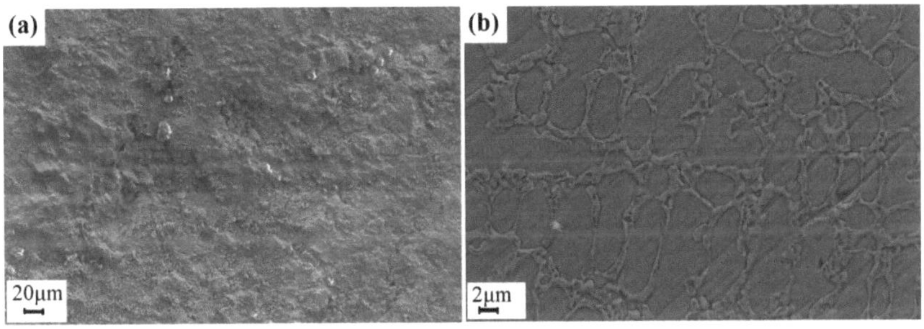

Figure 6. LV-FESEM images of the samples after the ASTM G48 test were performed at 25 °C for 72 h. (**a**) Unprotected S355 steel sample, and (**b**) NiSi12-wt% cladded sample.

Further investigation into the impact of temperature on corrosion behavior involved subjecting the samples to the 6% $FeCl_3$ solution at elevated temperatures. The LOM and SEM micrographs in Figures 7 and 8 illustrate the corrosion patterns at 50 °C. As can be seen from the figures, the corrosion damage to the base material intensifies with increasing temperature, leading to a more pronounced pitting corrosion. Although some corrosion is observed on the NiSi12-wt% cladded sample, it remains localized, indicating continued protection.

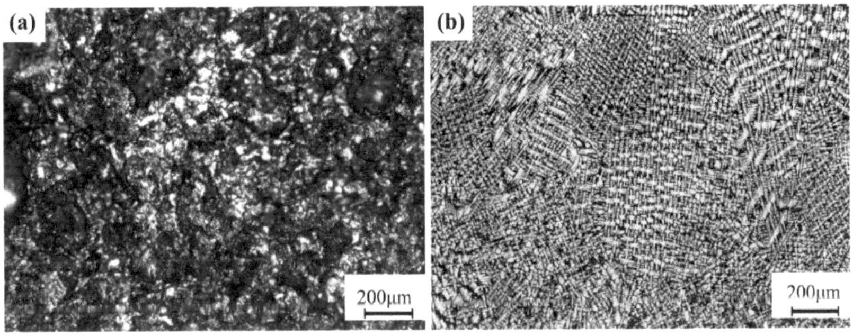

Figure 7. LOM images of the samples after the ASTM G48 test were performed at 50 °C for 72 h. (**a**) Unprotected S355 steel sample, and (**b**) NiSi12-wt% cladded sample.

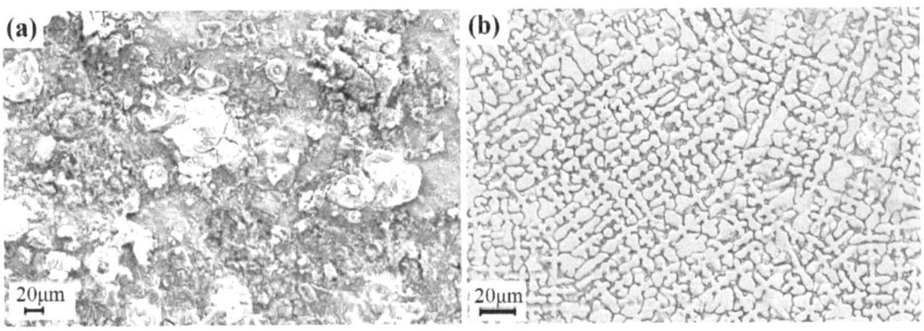

Figure 8. LV-FESEM images of the samples after the ASTM G48 test at 50 °C for 72 h. (**a**) Unprotected S355 steel sample and (**b**) NiSi12-wt% cladded sample.

The reaction of Ni_3Si and Ni_2Si in the presence of $FeCl_3$ solution would depend on the specific conditions and temperatures involved. When exposed to an oxidizing agent like $FeCl_3$, they can undergo oxidation reactions. The chloride ions (Cl^-) present in the $FeCl_3$

solution may interact with the metal ions (Ni^{2+} and Fe^{2+}) produced during the oxidation of Ni_3Si and Ni_2Si. In the presence of excess chloride ions (Cl^-) from the $FeCl_3$ solution, Ni^{2+} and Fe^{2+} ions can form soluble chlorides like $NiCl_2$ and $FeCl_2$. These chlorides would remain in solution due to their solubility in water [24], increasing the ionic strength and conductivity of the solution. This is also the mechanism for using $FeCl_3$ as an etchant of nickel alloys [25]. Nickel usually exhibits thermodynamic reactivity and theoretically should dissolve readily in dilute mineral acids while releasing hydrogen gas (H_2 (g)). Still, it often demonstrates a surprising degree of inertness towards the dilute acids. This is mainly due to nickel readily undergoing "passivation", resulting in a highly adherent oxide layer that presents significant kinetic resistance to further chemical reactions [26].

3.3. Corrosion Rate Measurements

At ambient temperature (25 °C), the unprotected S355 steel exhibited significantly higher mass loss rates compared to the NiSi12-wt% cladded samples, particularly during the first 24 h, as shown in Table 2. This observation highlights the S355 steel's lack of inherent resistance to corrosion, which is exacerbated by its initial exposure to the corrosive $FeCl_3$ environment. The NiSi12-wt% cladded samples, in contrast, demonstrated consistently lower mass loss rates, attributed to the formation of protective nickel- and silicon-based oxide layers that effectively mitigate corrosion.

Table 2. Mass loss rates for the unprotected and NiSi12-wt% cladded S355 steel samples after the ASTM G48 test in a 6% $FeCl_3$ solution at 25 °C and 50 °C for 1 and 3 days.

Sample Name	Time (Days)	Mass Loss (mg/mm ^{2}/Day)
Unprotected S355 steel sample (25 °C)	1	4.41 ± 0.25
	3	1.78 ± 0.15
NiSi12-wt% cladded sample (25 °C)	1	4.12 ± 0.11
	3	1.60 ± 0.06
Unprotected S355 steel sample (50 °C)	1	4.36 ± 0.37
	3	3.98 ± 0.13
NiSi12-wt% cladded sample (50 °C)	1	4.30 ± 0.16
	3	3.10 ± 0.17

Over the 72-h test period, both materials exhibited a reduction in corrosion rates. This decrease is likely due to the buildup of corrosion products on the surface, which acted as partial and chemical barriers to further attack. The protective effect was more pronounced for the NiSi12-wt% cladded samples, emphasizing their superior performance in corrosion-prone environments.

At higher temperatures (50 °C), corrosion rates increased for both materials, highlighting the role of temperature in accelerating electrochemical and diffusion-controlled reactions. The unprotected S355 steel showed a pronounced rise in mass loss rate, particularly during the first 24 h, as seen in Figure 9. This rapid degradation reflects its susceptibility to temperature-induced material breakdown. In contrast, the NiSi12-wt% cladded samples maintained significantly lower corrosion rates at 50 °C, demonstrating their enhanced resistance to high-temperature conditions. Notably, the corrosion rate of the cladded samples after 72 h at 50 °C was comparable to that observed at 25 °C, underscoring their thermal stability and effectiveness in prolonged exposure scenarios.

The temperature-dependent trends observed also shed light on the dynamic nature of corrosion processes. Initial rates were higher due to active surface reactions and rapid electrochemical activity. However, the gradual decline in rates over the 72-h period indicates stabilization mechanisms, such as the formation of denser oxide layers or the reduced availability of active corrosive species. This behavior underscores the material's ability to self-passivate, particularly in the case of the cladded samples.

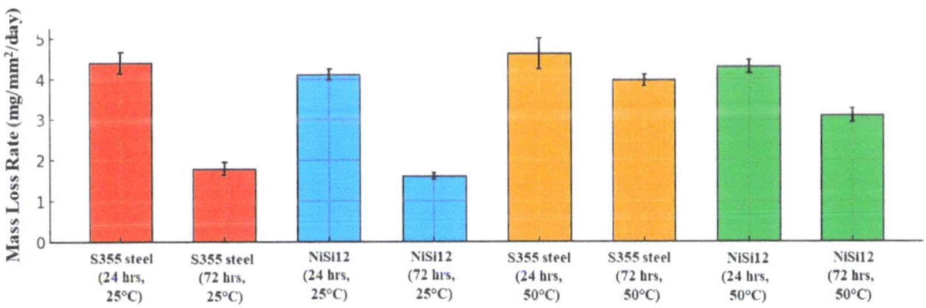

Figure 9. Graphical visualization of the mass loss rates for the unprotected and NiSi12-wt% cladded S355 steel samples after the ASTM G48 test in a 6% $FeCl_3$ solution at 25 °C and 50 °C for 72 h.

The LV-FESEM images in Figure 8 further validate these observations. The unprotected S355 steel sample (Figure 8a) shows extensive surface damage after the ASTM G48 test at 50 °C for 72 h, with clear evidence of pit formation and widespread material loss. In contrast, the NiSi12-wt% cladded sample (Figure 8b) retains a relatively intact surface morphology, indicating the effectiveness of the cladding in mitigating both localized and uniform corrosion.

3.4. Mechanical Properties

Microhardness measurements were performed both before (control samples) and after performing the ASTM G48 test at 25 °C and 50 °C for 72 h to evaluate the impact of the cladding process on the mechanical properties. Microhardness measurements showed a significant increase for the NiSi12-wt% cladded sample compared to the base material, indicating enhanced mechanical properties. Table 3 summarizes the measured microhardness values of the control sample, as well as for the unprotected and NiSi12-wt% cladded S355 steel samples. The results are further visualized in Figure 10. As seen in the figure, a significant increase in microhardness is demonstrated for the NiSi12-wt% cladded sample compared to the base material. This result indicates that a 1.5 mm NiSi12-wt% cladding enhances the mechanical properties of the base material. The high solidification rates associated with the LMD technique contribute to the observed increase in microhardness [21]. The enhancement offers several advantages, including improved resistance to wear, impact, and deformation, making the NiSi12-wt% cladding well-suited for applications requiring high mechanical strength and durability.

Table 3. Microhardness, measurements of the unprotected and NiSi12-wt%, cladded S355 steel samples before and after the ASTM G48 test in a 6% $FeCl_3$ solution at 25 °C and 50 °C for 72 h.

Sample	Microhardness (HV)
Unprotected S355 steel sample—Control	842 ± 25.3
NiSi12-wt% cladded sample—Control	1258 ± 18.9
Unprotected S355 steel sample (25 °C)	385 ± 11.5
NiSi12-wt% cladded sample (25 °C)	1116 ± 16.7
Unprotected S355 steel sample (50 °C)	353 ± 10.6
NiSi12-wt% cladded sample (50 °C)	913 ± 13.7

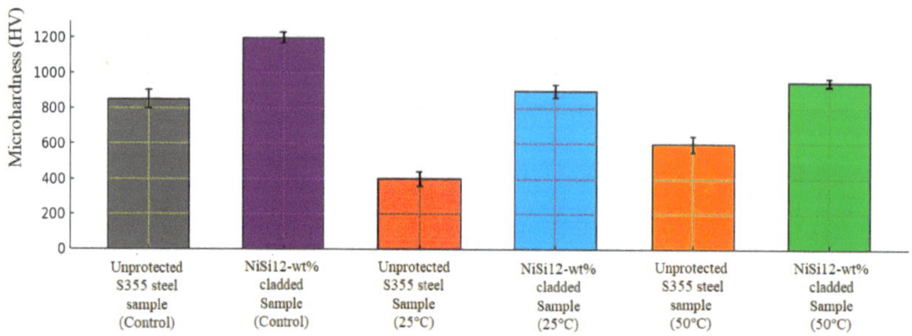

Figure 10. Graphical visualization of microhardness variation of the unprotected and NiSi12-wt% cladded S355 steel samples before and after the ASTM G48 test in a 6% $FeCl_3$ solution at 25 °C and 50 °C for 72 h.

3.5. Key Considerations

Understanding the kinetics and thermodynamics of oxidation reactions involving nickel silicide phases under different temperatures and solution conditions is essential for evaluating the long-term performance of the NiSi12-wt% cladding. The unique microstructure of the NiSi12-wt% cladding, characterized by dendritic grains, influences its mechanical strength [27]. Furthermore, the interaction between the clad material and the corrosive environment, particularly the oxidation reactions involving the nickel silicide phases Ni_3Si and Ni_2Si, requires further investigation. Gaining insights into these reactions under different temperatures and solution conditions can provide valuable insights into the long-term performance of the NiSi12-wt% cladding material in practical applications.

4. Conclusions

The NiSi12-wt% alloy was successfully deposited onto S355 steel samples using LMD, enhancing both corrosion resistance and mechanical properties. Preliminary results from the ASTM G48 pitting corrosion test and the Vickers microhardness test indicate that the NiSi12-wt% cladded samples exhibit superior corrosion resistance and mechanical properties compared to the base material. Initial SEM and LOM analysis showed minimal signs of corrosion on the NiSi12-wt% cladded samples, with only minor localized effects, while widespread surface pitting was observed on the base material.

Furthermore, the unprotected S355 steel samples experienced greater mass loss than the NiSi12-wt% cladded samples, highlighting the protective nature of the NiSi12-wt% cladding. Moreover, corrosion rates were more pronounced at higher temperatures, likely due to the increased kinetics and the highly acidic nature of the 6% $FeCl_3$ solution.

Overall, the preliminary results from the present study highlight the potential of the NiSi12-wt% cladding technology to specifically enhance the corrosion resistance and mechanical properties of structural materials such as S355 steel. By mitigating corrosion and improving mechanical strength, this cladding material offers a viable solution for extending the service life and durability of components subjected to harsh operating conditions. These findings provide a foundation for further, more detailed investigations and underscore the potential of the NiSi12-wt% cladding for improving the longevity of construction steel components in challenging environments.

5. Future Work

In future studies, MP-AES will be utilized to analyze both soluble and insoluble corrosion products to provide a more comprehensive understanding of the corrosion mechanisms. This approach will allow for the quantification of soluble ions in the solution and the characterization of insoluble corrosion products released during the corrosion process. Investigating the kinetics of corrosion, the formation of corrosion products, and the influence of different environmental factors can provide important insights into the corrosion process. Exploring the adhesion strength and integrity of the NiSi12-wt% cladding under mechanical stresses, including cyclic loading and thermal cycling, can offer valuable information about its mechanical robustness. Additionally, examining the long-term durability and performance of the NiSi12-wt% cladding in real-world environments, considering factors such as exposure to varying temperatures, humidity levels, and corrosive agents over extended periods, will be beneficial.

Furthermore, it is crucial for practical implementation to explore potential methods for optimizing the cladding's application process, including deposition parameters and improving the uniformity and adhesion of the NiSi12-wt% cladding to the substrate material. Investigating methods to enhance the cladding's mechanical robustness and long-term durability under real-world conditions will further improve its effectiveness.

Author Contributions: Conceptualization, M.I., C.H. and R.E.A.; Methodology, M.I. and G.G.; Formal analysis, R.E.A.; Investigation, M.I.; Resources, G.G. and R.E.A.; Writing—original draft, M.I.; Writing—review & editing, C.H., G.G. and R.E.A.; Visualization, C.H.; Supervision, C.H., G.G. and R.E.A. All authors have read and agreed to the published version of the manuscript.

Funding: The authors wish to express their sincere gratitude to the Norwegian Research Council (NFR) for supporting the project *"Development and Production of Metal Alloys for Powder-Based Additive Manufacturing (MADAM)"*, under Ref. no. 309856, which enabled the completion of the present research.

Data Availability Statement: The original contributions presented in this study are included in the article. Further inquiries can be directed to the corresponding author.

Acknowledgments: The authors extend their appreciation to Future Materials Norwegian Catapult Centre in Grimstad, Norway, and the Department of Material Science and Engineering at the Norwegian University of Science and Technology (NTNU) in Trondheim, Norway, for providing access to essential analytical equipment. Special thanks are also due to Product Manager Jan Ove Odden from Elkem ASA, Vågsbygd, Norway, and Marco Actis Grande at the University of Turin in Turin, Italy. Their hospitality, guidance, and expertise have significantly contributed to the success of the present research activities.

Conflicts of Interest: The authors declare no conflicts of interest.

References

1. Skoglund, O.; Leander, J.; Karoumi, R. Overview of Steel Bridges Containing High Strength Steel. *Int. J. Steel Struct.* **2020**, *20*, 1294–1301. [CrossRef]
2. Mohd Salim, M.N. (Ed.) Economies in Structural Steel Design Using Different Steel Grade. In *Secondary Economies in Structural Steel Design Using Different Steel Grade*; U.T. PETRONAS: Seri Iskandar, Perak, Malaysia, 2013. Available online: https://utpedia.utp.edu.my/id/eprint/13420/1/21.pdf (accessed on 15 January 2024).
3. *EN 10025-2*; Hot Rolled Products of Structural Steels—Part 2: Technical Delivery Conditions for Non-Alloy Structural Steels. CEN: Brussels, Belgium, 2019.
4. He, X.; Song, R.; Kong, D. Microstructure and Corrosion Behaviours of Composite Coatings on S355 Offshore Steel prepared by Laser Cladding Combined with Micro-Arc Oxidation. *Appl. Surf. Sci.* **2019**, *497*, 143703. [CrossRef]
5. Refait, P.; Grolleau, A.-M.; Jeannin, M.; Rémazeilles, C.; Sabot, R. Corrosion of Carbon Steel in Marine Environments: Role of the Corrosion Product Layer. *Corros. Mater. Degrad.* **2020**, *1*, 198–218. [CrossRef]
6. Betancourt, N.; Corvo, F.; Cuesta, O.; Labrador, R.; González, M.L.; Pérez, J.; León, Y.; Cruz, P. Influence of SO_2 and NO_x on Atmospheric Corrosion of Steel. *Rev. CENIC Cienc. Químicas* **2002**, *33*, 71–75.
7. Taylor, W. The Hardness Test as a Means of Estimating the Tensile Strength of Metals. *Aeronaut. J.* **1942**, *46*, 198–209. [CrossRef]
8. Kuhn, H. *ASM Handbook Volume 8: Mechanical Testing and Evaluation*; ASM International: Materials Park, OH, USA, 2000; Volume 8, pp. 232–243.
9. Pawlowski, L. Thick Laser Coatings: A Review. *J. Therm. Spray Technol.* **1999**, *8*, 279–295. [CrossRef]
10. Clemeña, G.G. (Ed.) Testing of selected metallic reinforcing bars for extending the service life of future concrete bridges: Summary of conclusions and recommendations. In *Secondary Testing of Selected Metallic Reinforcing Bars for Extending the Service Life of Future Concrete Bridges: Summary of Conclusions and Recommendations*; U.S. Department of Transportation, Bureau of Transportation Statistics: Washington, DC, USA, 2002.
11. Szala, M.; Szafran, M.; Macek, W.; Marchenko, S.V.; Hejwowski, T. Abrasion Resistance of S235, S355, C45, AISI 304 and Hardox 500 Steels with Usage of Garnet, Corundum and Carborundum Abrasives. *Adv. Sci. Technol. Res. J.* **2019**, *13*, 151–161. [CrossRef] [PubMed]

12. Liu, X.; Yu, L.; Wang, H. Synthesis of a Nickel Silicide-base Composite Coating on Austenitic Steel by Laser Cladding. *J. Mater. Sci. Lett.* **2001**, *20*, 1489–1492. [CrossRef]
13. Wilms, M.B.; Pirch, N.; Gökce, B. Manufacturing Oxide-Dispersion-Strengthened Steels Using the Advanced Directed Energy Deposition Process of High-Speed Laser Cladding. *Prog. Addit. Manuf.* **2023**, *8*, 159–167. [CrossRef]
14. Xu, L.; Chai, Z.; Zhang, X.; Peng, B.; Zhou, W.; Chen, X. A New Approach to Improve Strength and Ductility of Laser Powder Deposited Inconel 718 Thin-Wall Structure. *Mater. Sci. Eng. A* **2022**, *855*, 143871. [CrossRef]
15. Xu, L.; Chai, Z.; Chen, H.; Zhang, X.; Xie, J.; Chen, X. Tailoring Laves Phase and Mechanical Properties of Directed Energy Deposited Inconel 718 Thin-Wall via a Gradient Laser Power Method. *Mater. Sci. Eng. A* **2021**, *824*, 141822. [CrossRef]
16. Wagle, S.; Priyotomo, G.; Kaneno, Y.; Iwase, A.; Takasugi, T.; Nishimura, R. Pitting Corrosion of Intermetallic Compound Ni_3 (Si, Ti) in Sodium Chloride Solutions. *Corros. Sci.* **2011**, *53*, 2514–2517. [CrossRef]
17. Cai, L.; Wang, H.; Wang, C. Corrosion Resistance of Laser Clad Cr-alloyed Ni_2Si/NiSi Intermetallic Coatings. *Surf. Coat. Technol.* **2004**, *182*, 294–299. [CrossRef]
18. Zheng, Y.F.; Zhang, J.B.; Sun, Y.N.; Fan, D.; Zheng, M. Effect of Cr Additives on the Microstructures and Properties of Laser Cladded Ni_3Si. *Key Eng. Mater.* **2008**, *368–372*, 1319–1321. [CrossRef]
19. Ishikawa, M.; Kada, T.; Machida, H.; Ohshita, Y.; Ogura, A. Ni Precursor for Chemical Vapor Deposition of NiSi. *Jpn. J. Appl. Phys.* **2004**, *43* (Suppl. S4), 1833. [CrossRef]
20. Zhao, H.; Zhao, C.; Xie, W.; Wu, D.; Du, B.; Zhang, X.; Wen, M.; Ma, R.; Li, R.; Jiao, J. Research Progress of Laser Cladding on the Surface of Titanium and its Alloys. *Materials* **2023**, *16*, 3250. [CrossRef] [PubMed]
21. Van Dyck, S.; Delaey, L.; Froyen, L.; Buekenhout, L. Microstructural Evolution and its Influence on the Mechanical Properties of a Nickel Silicide based Intermetallic Alloy. *Intermetallics* **1997**, *5*, 137–145. [CrossRef]
22. Ibrahim, M.; Sætre, T.O.; Aune, R.E. Laser Metal Deposition of Nickel Silicide on S355 Structural Steel. In Proceedings of the TMS 2022 151st Annual Meeting & Exhibition Supplemental Proceedings, Anaheim, CA, USA, 27 February–3 March 2022; Springer: Berlin/Heidelberg, Germany, 2022; pp. 249–258. [CrossRef]
23. *ASTM G48-11(2020)e1*; Standard Test Methods for Pitting and Crevice Corrosion Resistance of Stainless Steels and Related Alloys by Use of Ferric Chloride Solution. ASTM International: West Conshohocken, PA, USA, 2009. [CrossRef]
24. Wexler, P.; Anderson, B.D. *Encyclopedia of Toxicology*; Academic Press: Cambridge, MA, USA, 2005; Volume 1, pp. 223–225.
25. Walker, P.; Tarn, W.H. *CRC Handbook of Metal Etchants*; CRC Press: Boca Raton, FL, USA, 1990; p. 862.
26. Nelson, N.J. *Nickel Etching Process and Solution*; Google Patents: San Francisco, CA, USA, 1985.
27. Cui, C.; Zhang, J.; Wu, K.; Ma, Y.; Liu, L.; Fu, H. Microstructure and Properties of Ni–Ni_3Si Composites by Directional Solidification. *Phys. B Condens. Matter* **2012**, *407*, 3566–3569. [CrossRef]

metals

MDPI

Article

Integration of Arc and Microstructural Analysis for Anomaly Detection in Walls Manufactured by GMA-Based WAAM

Lucas J. E. B. Santos [1], Joyce I. V. Souto [2], Igo J. S. Azevedo [1], Walman B. Castro [1], Jefferson S. Lima [1,3,4], João M. P. Q. Delgado [5,*], Renato A. C. Santana [1], Ricardo S. Gomez [1], André L. D. Bezerra [6] and Antonio G. B. Lima [1]

1 Department of Mechanical Engineering, Federal University of Campina Grande, Campina Grande 58429-900, Brazil; lucas.elias@estudante.ufcg.edu.br (L.J.E.B.S.); igo.jose@estudante.ufcg.edu.br (I.J.S.A.); walman.castro@ufcg.edu.br (W.B.C.); jefferson.segundo@ufpe.br (J.S.L.); renato.alexandre@professor.ufcg.edu.br (R.A.C.S.); ricardo.soares@professor.ufcg.edu.br (R.S.G.); antonio.gilson@ufcg.edu.br (A.G.B.L.)

2 Mechanical Engineering Faculty, Federal University of Uberlândia, Uberlândia 38410-337, Brazil; joycesouto@ufu.br

3 Department of Mechanical Engineering, Federal University of Pernambuco, Recife 50670-901, Brazil

4 Federal Institute of Education, Science and Technology of Rio Grande do Norte, Santa Cruz 59200-000, Brazil

5 Institute of R&D in Structures and Construction (CONSTRUCT-LFC), Civil Engineering Department, Faculty of Engineering, University of Porto, 4200-465 Porto, Portugal

6 Post-Graduate in Process Engineering, Federal University of Campina Grande, Campina Grande 58429-900, Brazil; dr.andreldb@gmail.com

* Correspondence: jdelgado@fe.up.pt; Tel.: +351-225-081-404

Abstract: Wire Arc Additive Manufacturing (WAAM) is a process for fabricating metal parts known for its high productivity and material flexibility. However, defects such as overheating, residual stresses, distortions, porosity, and a non-homogeneous microstructure limit its commercial applications. Therefore, the present study aims to analyze the correlation between electrical sensing anomalies in the Gas Metal Arc (GMA) during WAAM and the occurrence of microscopic defects caused by external contamination. To achieve this, experiments were conducted to fabricate walls using WAAM with controlled contaminant introduction. Simultaneously, electrical arc data, specifically voltage and current, were segmented and acquired during the wall deposition process. Metallographic analysis confirmed the presence of microscopic defects or changes in the solidification patterns in regions with contaminant inclusion, distinguishing them from other areas of the analyzed samples. Similarly, the contaminations were proven to cause anomalies in attributes associated with the electrical arc. Therefore, this approach confirms the criticality of electrical arc monitoring in WAAM, as it demonstrates that anomalies in the electrical arc could lead to microstructural consequences.

Keywords: wire arc additive manufacturing; gas metal arc; electrical signals; contamination; features; defects

1. Introduction

In recent years, the industry has presented itself in an innovative way, encompassing a broad system of advanced technologies that drive progress in production methods and business models on a global scale [1,2]. As a result, there has been significant development in the field of Additive Manufacturing (AM), as the use of this technology is linked to the reduction of production steps, as well as contributing to shorter product delivery times, thus adding value to distribution, storage, and production processes [3–5].

https://doi.org/10.3390/met15020110

The AM process is defined as a material joining process for manufacturing components by depositing layer by layer from a 3D model, differing from conventional subtractive manufacturing (machining) [6,7]. Additionally, this process can be applied to various materials, such as polymers, ceramics, composites, and even metals [6,7]. In Metal Additive Manufacturing (MAM) processes, material deposition can be performed by melting a metal wire onto a substrate [8], using energy sources like a laser, electron beam, or even an electric arc, as well as by the fusion/sintering of metals in powder form [9–11].

Laser and electron-beam-based additive manufacturing processes are common, but their high cost and low efficiency greatly limit their application in the production of large metallic components [12]. On the other hand, Wire Arc Additive Manufacturing (WAAM) utilizes the heat generated by an electric arc, which melts the metal material (wire) that is continuously deposited, overcoming the described obstacles [13]. Regarding its terminology, according to ISO/ASTM 52900:2021 [14], the WAAM process is classified as a direct energy deposition of metals with an electric arc (DED-Arc) process. The electric arc used in WAAM can be a Gas Metal Arc (GMA), Gas Tungsten Arc (GTA), or plasma arc (PA) [15]. Arc-based techniques, such as GMA and GTA, offer energy efficiencies of up to 90% [16] and allow nearly all of the wire material to be used in the component. These methods are more affordable and suitable for wire-fed additive manufacturing processes compared to laser and electron beam techniques [17].

However, one of the major limitations associated with WAAM is the defects resulting from improper process parameter settings, environmental disturbances, and process instability [18]. These can lead to serious issues, affecting product performance, durability, and even causing safety incidents [19]. Similarly, connections have been identified between data collected during the manufacturing process and the emergence of defects in the produced parts, enabling the application of methodologies for monitoring control parameters [20,21] or other types of responses [22–24].

Recent studies have explored different monitoring methods for defect analysis in Wire and Arc Additive Manufacturing (WAAM) processes. Ramalho et al. [20] investigated the use of acoustic emissions to detect contaminations during the additive manufacturing of 316L stainless steel, highlighting its application in fault identification. Similarly, Chen et al. [24] developed an optical spectral-physics-informed attention network for condition monitoring in WAAM, demonstrating the potential of optical signal-based techniques for predictive defect analysis. These works emphasize the variety of approaches employed to enhance defect detection and quality control in the WAAM process.

Among the monitoring techniques in WAAM, the electrical sensing method is based on the perspective that electric arc signals are closely related to droplet transfer and arc ignition [25]. While these methods have advanced the understanding of WAAM monitoring, their focus often remains on macroscopic defect detection or signal-based assessments, without direct exploration of the microstructural implications of anomalies. In this context, Lupo et al. [26] analyzed microstructural images captured between powder layers to assess how the quality of the powder layer in the Selective Laser Sintering (SLS) process affects defects in the final product.

Furthermore, this study bridges this critical gap by integrating electric arc signal analysis with microstructural investigation for anomaly detection in WAAM. Unlike prior works, which primarily concentrate on process monitoring or defect prediction, this research experimentally validates the correlation between electric arc characteristics and microstructural changes in austenitic stainless steel 316L-Si walls. Through a controlled experimental design, electric arc data (voltage and current) are acquired during the deposition process, with conditions tailored to induce specific defects. The subsequent microstructural analysis evaluates how these defects manifest at the microscopic level, confirming the legitimacy of using electric arc signals for anomaly detection in WAAM. This novel approach ensures a deeper understanding of the relationship between process parameters, arc instability, and material integrity, offering critical insights for improving defect detection and ensuring the industrial reliability of WAAM components.

2. Materials and Methods

In this work, C-Mn plates (Aço Brazil, Campina Grande, Brazil) with dimensions of 18 × 160 × 170 mm were used as the substrate. For the filler metal, austenitic stainless steel wire ER 316LSi (Avesta ER316Si/316L Si®, from Avesta, Sweden) with a nominal diameter of 1.2 mm was used, and commercially pure argon (99.98% Ar) was used as the shielding gas. Table 1 presents the nominal chemical composition of the wire used, as specified by the AWS A5.9 classification and measured by X-ray fluorescence (XRF).

Table 1. Nominal chemical composition of filler metal AISI 316L-Si.

	C%	Cr%	Ni%	Mo%	Mn%	Si%	Cr_{eq}/Ni_{eq} *	FN
AWS	0.03 Max.	18.0–20.0	11.0–14.0	2.0–3.0	1.0–2.5	0.65–1.0	Varied	5–8 **
XRF	0.02	18.04	12.11	2.67	2.20	0.96	1.62	5

* Calculated based on the WRC 1992. ** Estimated FN interval provided by the filler wire manufacturer (Avesta ER316Si/316L Si®).

The walls produced by WAAM were made by depositing 5 weld beads, each 100 mm in length, using a zig-zag strategy in the manufacturing direction to achieve a more uniform wall height [27,28]. The process used was GMAW with controlled short-circuit transfer mode (GMAW-CCC) [29]. For this, an IMC Soldagem (Palhoça, Brazil) power source, model Digiplus A7, was used along with the STA 20-D wire feeder model, from IMC Soldagem. The torch displacement system was a CNC robotic mechanism with three-axis movement (X, Y, and Z), controlled by MACH3 CNC software version 3.043, from the ArtSoft company (Hood River, OR, USA), with G-code language. The deposition parameters were chosen to ensure minimal dimensional stability and mechanical properties of the deposits, according to the literature related to the material [30], as presented in Table 2.

Table 2. Deposition parameters for the WAAM manufacturing of 316L-Si steel walls.

Parameter	Unity	Value
Peak current	A	250
Base current	A	50
Torch speed	$mm \cdot min^{-1}$	300
Interpass temperature	°C	80
CTWD	mm	10
Wire feed rate	$m \cdot min^{-1}$	5
Gas flow rate	$l \cdot min^{-1}$	18
Shielding gas	-	Commercial Argon (99.98%)

To investigate defects and arc instability due to the influence of contaminants during layer deposition, three types of contaminants were used: sand, oil, and chalk [20,31]. Thus, two walls were manufactured for each contaminant, with intercalation in the deposition of the contaminants, represented by strategies P1 (Figure 1b) and P2 (Figure 1b). This pattern was applied to all three contaminants, deposited in the layers and locations indicated, as shown in Figure 1a.

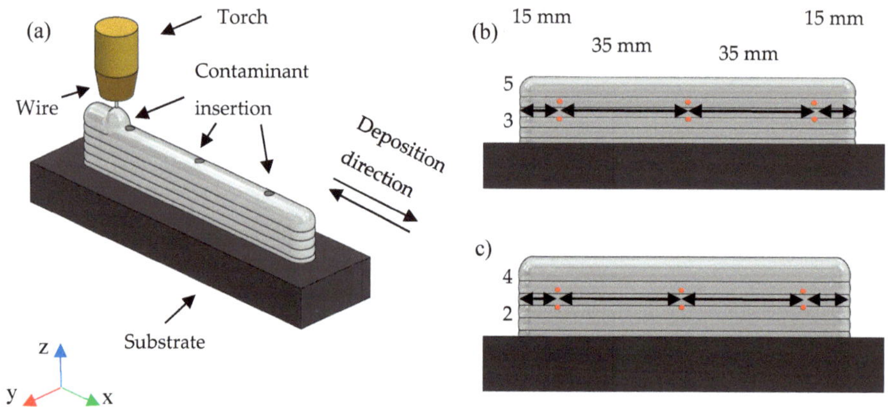

Figure 1. (**a**) Schematic diagram of contaminant insertion during layer deposition by MADA, according to strategies (**b**) P1 and (**c**) P2. The red dots indicate the exact locations of contaminant introduction.

The fundamental parameters determining the CCC waveform shape included the peak current (1), peak current time (2), current drop rate (3), current rise rate (4), short-circuit wait current (5), and background current (6), as depicted in Figure 1.

For better visualization and future reference for results and discussions, acronyms were adopted to identify the walls according to the contamination strategy and the introduced contaminant. This standardization is presented in Table 3.

Table 3. Nomenclature of the fabricated walls.

Strategy/Contaminant	Chalk	Oil	Sand
P1	W2	W4	W6
P2	W3	W5	W7
-	W1 (Reference)		

For metallographic preparation, samples were taken from the walls where contaminants were introduced into the deposits. After cutting, 5 mm-thick samples were embedded, ground to a 2400 grit, polished with 1 μm and 0.5 μm alumina, and subjected to electrolytic etching using 10% oxalic acid for 50 s at 2 V. Digital photographs were obtained using an Olympus (Tokyo, Japan) optical microscope, SC30 model, with 5× magnification. These images were taken in all interpass regions of the three samples removed transversely from the contamination areas for each manufactured wall. This was done to compare the appearance of microscopic defects in the contamination regions with the others, as well as to distinguish the microstructural morphology of an interpass region with and without the presence of external contaminants.

2.1. Data Labeling

Before analyzing the electric arc data to investigate the effect of contaminant introduction on defect occurrence in the walls, it is necessary to perform a preprocessing step to remove irrelevant data. Some data are considered useless for analysis; for example, when both current and voltage levels are zero before and after the start and end of the electric arc due to delays in communication between the welding and robotic systems, or when only the current levels are zero due to delays between the start and end of data acquisition and the start and end of the electric arc. After this data cleaning, the welding current and voltage datasets are normalized using the Min-Max technique.

All preprocessed files were labeled to enable the correlation between raw or feature-extracted electric arc data and electric arc anomalies, which were categorically assessed. This categorical label was assigned according to the type of electric arc disturbance and could assume the following values: 0 for no disturbance; 1 for disturbance associated with the start of the arc; 2 for disturbance associated with contamination; and 3 for disturbance associated with the end of the arc.

2.2. Feature Selection and Extraction Process

Next, a rolling window is used to divide the normalized data into sections containing 100 samples each. There is overlap between neighboring windows in this work. If the last data window contains fewer than 100 samples, the script will re-sample this last window to 100 samples using data from the previous window. This windowing process aims to calculate features related to the electric arc (voltage and current).

Based on a complete cycle of the events occurring during the CCC metal transfer, eleven statistical features can be extracted for both current and voltage data, similarly to what was proposed by Shi et al. [12] and Li et al. [13]. Another selected feature was the IVcc (Vilarinho Index of Regularity) for CCC metal transfer, based on the short-circuit and arc burn time intervals. This feature was chosen considering domain knowledge that reflects the understanding that electric arc anomalies caused by contaminants can effectively lead to irregularities in the electric arc during metal transfer. For calculating IVcc, the electrical voltage and current data were acquired at a rate of 5 kHz and 22 V for short-circuit voltage. The acquisition and storage of the electric arc data were performed using the data acquisition system SAP V4 from IMC Soldagem.

To determine the IVcc, as previously done in other works [27,32,33], a dedicated script developed in Python 3.12 was used, through which the short-circuit time values and their respective standard deviations are calculated numerically [27,28], as shown in Equation (1).

$$IVsc = \frac{\sigma_{tsc}}{t_{sc}} + \frac{\sigma_{tarcing}}{t_{arcing}}, \quad (1)$$

where σ_{tsc} is the standard deviation of the mean short-circuit time, $\sigma_{tarcing}$ is the standard deviation of the mean arc burn time, t_{sc} is the mean short-circuit time, and t_{arcing} is the mean arc burn time. In this context, the regularity index IVcc takes into account both the mean values and standard deviations. Thus, a lower IVcc value indicates greater stability of the arc for short-circuit transfer [21].

Thus, knowing that all selected features are calculated for both voltage and current data, except for IVcc, there are a total of 23 electric arc features calculated for each data window. These features are listed in Table 4 below.

Table 4. Features related to acquired electric arc data.

Time Domain	Frequency Domain
Average	Average peak width
Standard deviation	Number of peaks
Maximum/Minimum	Standard deviation of peak width
Curtose/Skewness	Average distance between neighboring peaks
IVcc	Standard deviation of the distance between neighboring peaks

To concatenate the extracted features with the labels, which did not undergo the data windowing process, the labels were also windowed, with the most frequent value (mode) calculated for each window.

3. Results

3.1. Visual Inspection of the Walls

The walls were manufactured according to the manufacturing parameters of the WAAM process and the contaminants—chalk, sand, and oil—indicated in Table 3. The visual characteristics of the walls are shown in Figure 2.

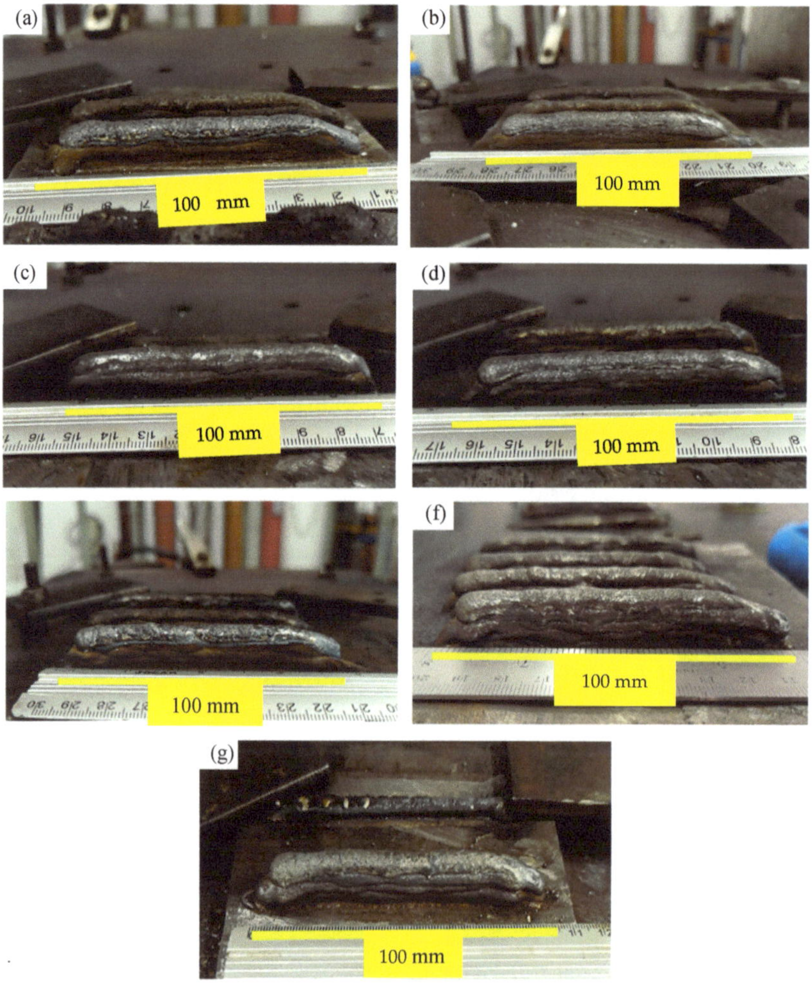

Figure 2. Fabricated walls: (**a**) W2, (**b**) W3, (**c**) W4, (**d**) W5, (**e**) W6, (**f**) W7, and (**g**) W1.

Visual inspection revealed distinct characteristics on the surface of wall W1 (Figure 2g) compared to those where contaminants were introduced (Figure 2a–f). It was observed that the walls contaminated with sand (W6 and W7) were the most geometrically inconsistent, showing lateral deformations along the entire length of the layers, as indicated in Figure 2e,f. This can be justified by the physical interference with arc stability, as sand particles may create points of low adhesion between the deposited layers, compromising material cohesion and leading to geometric deformations. This is also described by Meier et al. [34], who noted that the presence of sand particles can interfere with deposition continuity in the metal powder bed fusion process, resulting in irregular surfaces and inadequate adhesion. Roy et al. [35] highlighted that interparticle cohesion can result in layer structures with variable density, creating low-density areas susceptible to adhesion failures between subsequent layers. Additionally, it's possible to infer chemical contamination due to undesirable reactions. Sand, primarily composed of silicon dioxide (SiO^2) [36,37], might react with elements present in the stainless steel 316L-Si and these reactions can form compounds that alter the material properties of the deposited layers and compromise their geometric integrity. This is also discussed by Mouayd et al. [38], who concluded that the presence of silica or silicon dioxide delays scale growth by creating a barrier at the metal interface due to the low diffusivity of iron in silica.

Another perspective for justifying the greater deformations in the walls contaminated with sand (Figure 2e,f) is the thermal effects of this inclusion. Sand may alter thermal distribution during the deposition process, causing differences in thermal contraction and expansion along the wall, which leads to deformations in the piece. This is also discussed by Meier et al. [34]. Additionally, sand particles can act as nucleation points for solidification, accelerating the cooling of molten metal in specific areas and creating thermal stresses, which also cause deformations. This effect can be compared to the behavior observed in metal welding processes, where contaminants can result in microcracks and structural failures due to accumulated residual stresses, as described by Roy et al. [35].

Additionally, the walls W4 and W5, contaminated with oil, also exhibited the same discontinuities as those contaminated with sand, but to a lesser extent, along with the presence of lateral spatter in the same direction as the layer deposition. This is due to the interference with the electric arc caused by the inclusion of oil. The presence of oil on the surface of the molten metal can alter the surface tension of the metal, causing spatter and irregularities in the deposition. As the oil vaporizes, it can create small explosions that disperse the molten metal, resulting in lateral spatter, as discussed by Li et al. [39], who investigated the effects of oxides on the surface tension of molten metal and arc behavior in the GTAW process.

In addition, oil can create a thin layer between deposits, reducing cohesion between the deposited layers and resulting in smaller, yet still present, discontinuities. This is similar to the discussion by Yoo, Lee, and Kim [40], who indicated that oil contamination can lead to fusion and adhesion problems, creating weak points in the WAAM-produced piece. It should also be noted that the introduction of oil can reduce the wettability of the deposited metal, hindering proper fusion of successive layers and resulting in discontinuities. This is consistent with discussions by other authors, who emphasize the critical nature of oil contamination during additive manufacturing, creating physical barriers that prevent adequate fusion between layers, resulting in structural defects [41,42]. This problem is particularly significant in DED-AM processes, where layer adhesion is directly related to the quality of the final product.

The walls contaminated with oil (Figure 2c,d) exhibited reductions in width towards the end of their length. Additionally, the walls contaminated with chalk (W2 and W3) showed excessive surface oxidation in the layers where the contaminant was introduced, significantly compromising arc stability when evaluated layer by layer, as also discussed by Bevans et al. [31]. As a preliminary assessment in the context of investigating defects and arc stability in walls manufactured by WAAM from 316L-Si stainless steel, visual inspection was complemented by quantitative and experimental indicators, as shown in the following sections.

3.2. Metallographic Analysis

The metallographic characterization of the samples, strategically taken from the locations where contaminants were introduced into the walls manufactured according to the contamination strategies P1 and P2, can be seen in Figures 3–8. To demonstrate the effects of contaminant insertion on the microscopic aspect of the material, images from the contamination regions (lower and upper zones) and other interpass regions of the sample (middle zone) were selected for each of the three samples taken per wall for comparative purposes. Beyond the scope of this study, Figures 3–8 display the typical microstructure of austenitic stainless steel manufactured by WAAM, consisting of austenite dendrites, visualized in yellowish regions, with ferrite at the boundaries, visible in darker regions of the figures [43–45]. The results for the walls with contaminants, W2 to W7, are shown in Figures 3–8, in sequence.

From Figure 3, the microstructure of the W2 sample can be observed. In the analysis of Figure 3, regarding the upper zone, the start/contamination sample showed a relatively homogeneous microstructure with a darker region prominently highlighted, possibly related to the accumulation of chalk during deposition. In this region, the middle/contamination sample also showed a relatively homogeneous microstructure with elongated grains in the direction of solidification. The end/contamination sample exhibited a more evenly distributed microstructure, but with a highlighted area of material accumulation resulting from the buildup of chalk in that location.

For the lower zone, the start/contamination sample displays a more irregular solidification pattern, characteristic of the interpass regions in the WAAM process, and a dark region prominently highlighted, possibly associated with residues from the chalk accumulation. The middle/contamination sample shows the characteristic solidification structure, revealing a different solidification pattern at the edges of the sample, related to the natural cooling gradient of the deposit to the environment. Similarly, the end/contamination sample displayed the characteristic solidification pattern of the WAAM process but with a region of solidification typical of the sample edges, where there is greater contact with the environment. This is featured in the physical characteristics of the chalk, which can act as a thermal insulator, affecting heat and mass transfer in the contamination regions and their surroundings, leading to a heterogeneous distribution of microconstituents.

In the samples shown in Figure 4, there is no significant differentiation compared to what was observed in Figure 6, except for a drastic reduction in the identification of contaminant accumulation, which is only present in the highlighted section of the upper zone of the start/contamination sample. It is noted that the middle zone of all samples, as illustrated in Figures 3 and 4, shows a more uniform and fine structure compared to the upper and lower zones, possibly due to a lesser influence from the contaminants introduced in the other zones during the process. The microstructural and solidification patterns of the other zones in the samples are characteristic of the WAAM process, indicating that there was no significant influence of chalk insertion for this wall. Alternatively, the imperfections in the deposit, as shown in Figure 2, especially at the edges, may have affected the precision of sample extraction. This could justify why a sample does not necessarily reflect the intended manufacturing/contamination conditions.

The presence of inclusions, highlighted in Figures 3 and 4, located in the interpass region and exhibiting irregular morphology, is featured in the contamination by chalk and the reduction in the activity of the electric arc caused by this type of contamination during the metal deposition. This finding aligns with the discussions by Wang et al. [46], who developed a strategy for detecting porosity in WAAM, and Li et al. [47], who established an internal system for tracking defects in WAAM.

Figures 5 and 6 illustrate the microstructures of the interpass regions for the walls W4 and W5, respectively, when contaminated with oil. From the analysis of Figure 5, which displays the microstructure of sample W4, it is observed that in the upper zone, the initial/contaminated sample has a heterogeneous microstructure with elongated dendrites, originating from a prominent dark region featuring the effect of oil vaporizing during deposition. In the same region, the medium/contaminated sample exhibits a more homogeneous microstructure with fewer visible disturbances, but still with some indications of solidification anomalies compared to other regions of the sample. Similarly, the end/contaminated sample also shows a more homogeneous distribution of microconstituents, with signs of solidification anomalies in specific locations.

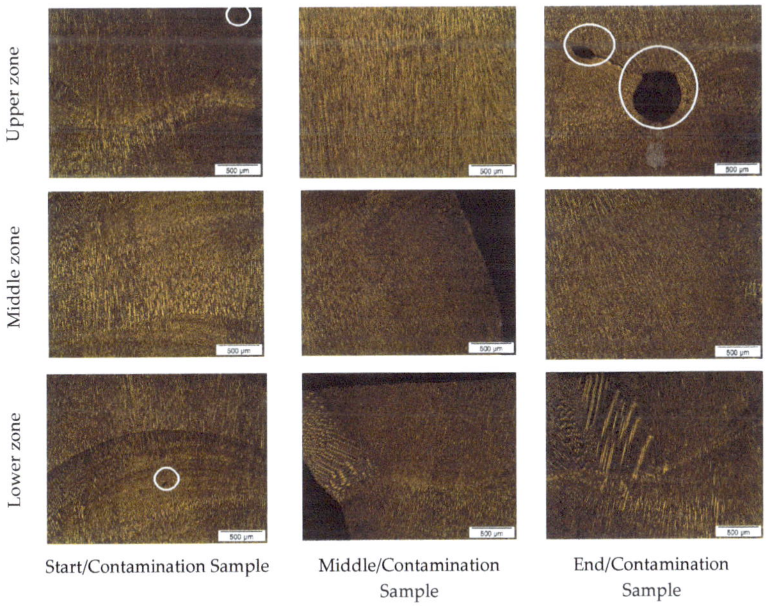

Figure 3. Microstructure of wall W2 at 50× magnification. The white circles indicate noticeable microstructural defects.

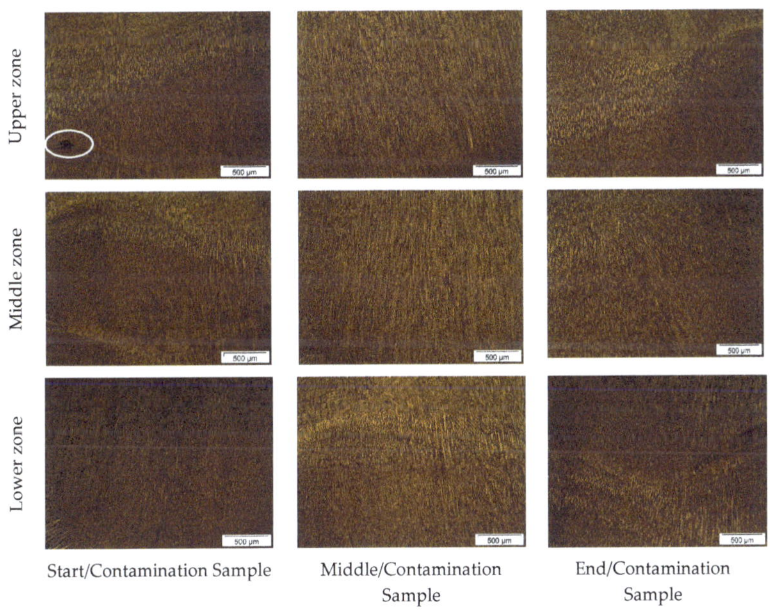

Figure 4. Microstructure of wall W3 at 50× magnification. The white circle indicates noticeable microstructural defects.

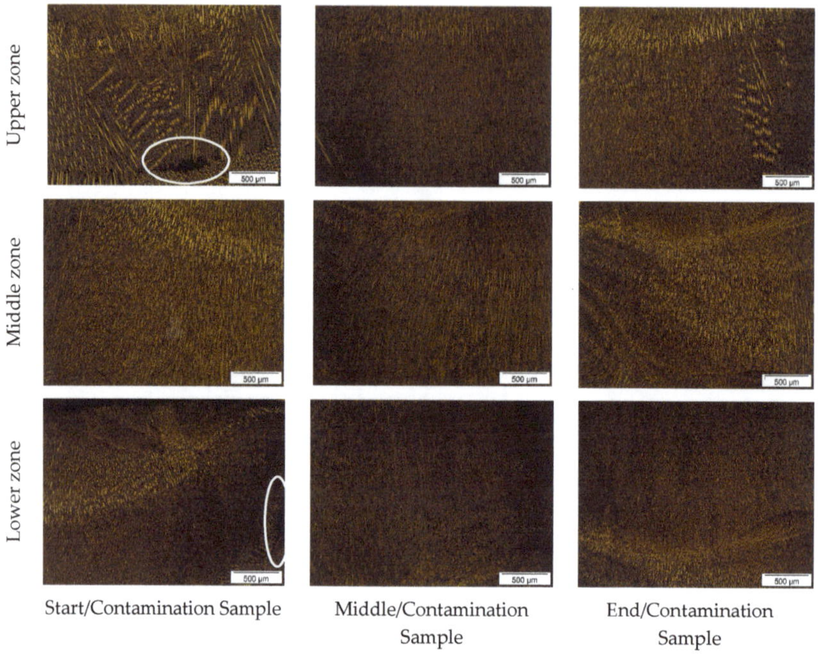

Figure 5. Microstructure of wall W4 at 50× magnification. The white circles indicate noticeable microstructural defects.

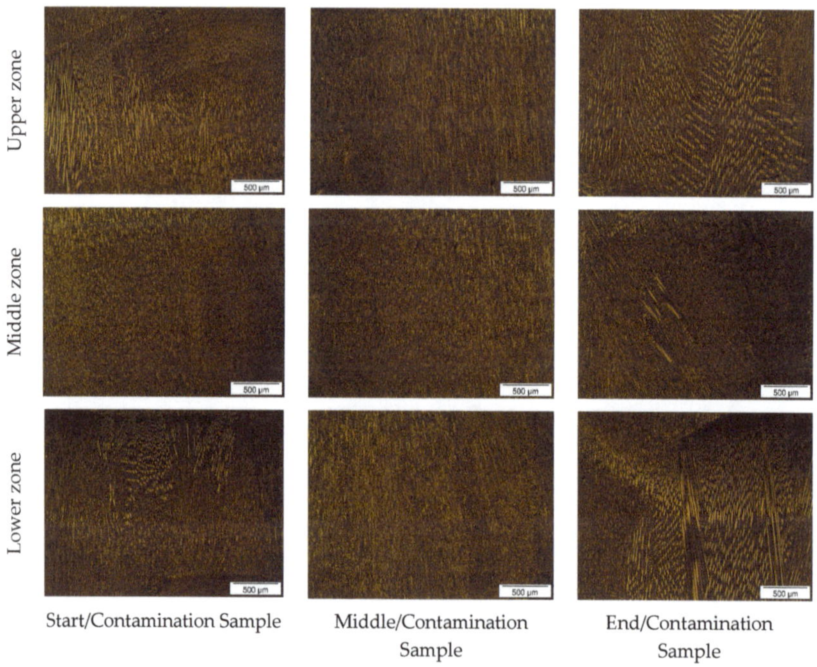

Figure 6. Microstructure of wall W5 at 50× magnification.

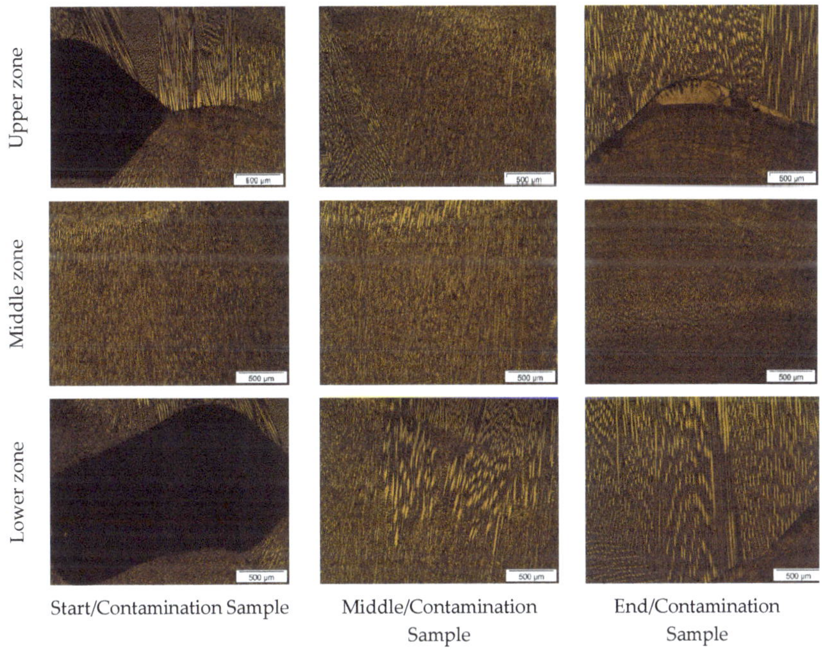

Figure 7. Microstructure of wall W6 at 50× magnification.

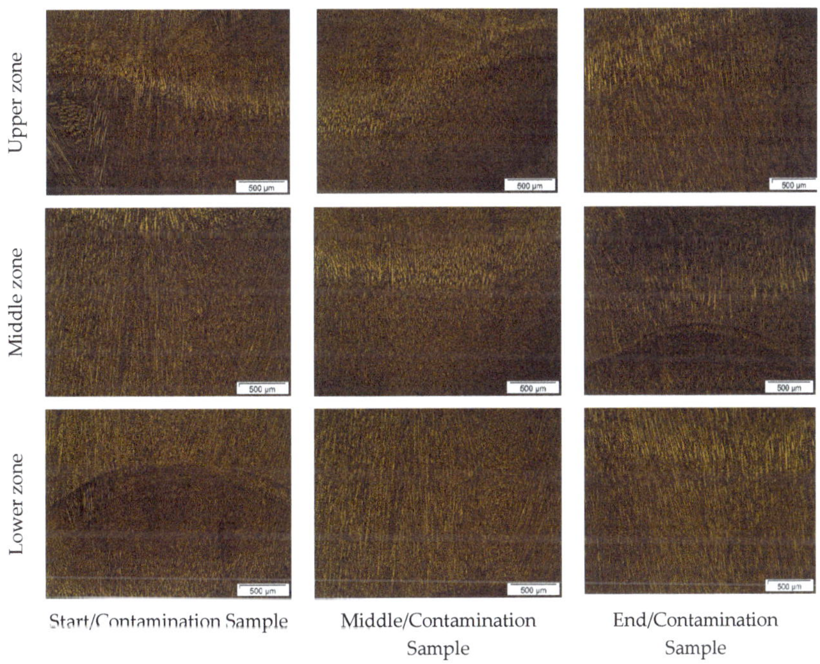

Figure 8. Microstructure of wall W7 at 50× magnification.

Regarding the lower zone, the initial/contaminated sample shows a marked segregation in the microstructure, characterized by faster solidification, with elongated dendrites originating from a prominent fissured region. In contrast, the medium/contaminated and end/contaminated samples display a microstructure and morphology characteristic of the WAAM process.

Figure 6 shows similar behavior in the microstructural characteristics of the oil-contaminated samples, as previously analyzed in Figure 5. In both the upper and lower zones, the initial/contaminated and end/contaminated samples exhibited irregular solidification patterns, deviating from the typical WAAM process characteristics. This suggests that the presence of oil during deposition caused irregular metal solidification due to gas generation. In contrast, the upper and lower zones of the medium/contaminated sample reveal a more homogeneous microstructure with a finer grain size, which could be attributed to the lower influence of oil in this part of the deposit or to the imprecision in sampling from the wall.

Similar to Figures 3 and 4, it is noteworthy that the middle zone of the samples illustrated in Figures 5 and 6 shows finer and more uniform granulation. This is featured to the proper selection of deposition parameters and the reduced influence of oil in the subsequent layers after its introduction, considering that it vaporizes shortly after deposition.

The presence of voids and cracks in the samples extracted from walls W4 and W5 can be featured to the evaporation mechanism of this contaminant, which is absorbed by the molten pool and subsequently generates pores upon solidification. This phenomenon was discussed when examining the influence of different types of contamination on the acoustic spectrum in WAAM, as demonstrated by Huang et al. [48] and Li et al. [47].

Figures 7 and 8 illustrate the microstructure of the interpass regions for walls W6 and W7, respectively, when contaminated with sand. As observed in Figure 7, in the upper zone, the initial/contamination and end/contamination samples show pronounced dark areas and unusual solidification patterns in the interpass region. This indicates that the sand interacted with the material during solidification, compromising its fusion and the typical annealing process in the interpass region of WAAM. The middle/contamination sample shows a more uniform microstructure but still has some areas of irregular solidification, suggesting a lesser effect of the sand in this region. In the lower zone, the initial/contamination sample displays a large dark region associated with rapid solidification patterns in its vicinity, indicating that the concentration of the contaminant in this region may have acted as an anchor for internal metal solidification. The middle/contamination and end/contamination samples in the lower zone showed solidification patterns similar to those observed in the contaminant accumulation area in the initial/contamination sample, indicating that sand caused internal metal solidification differences, although not explicitly identified in the images, as illustrated in the initial/contamination sample of the lower zone.

Based on Figure 8, both the upper and lower zones exhibit microstructural characteristics and solidification conditions considered suitable for the process. When compared to other samples from walls contaminated with sand in these zones, this would be unusual as there are no evident effects of the contaminant. This can be featured as inaccuracies in sample removal from wall W7, given the geometric imperfections at the wall's edges. The middle zones of the samples presented in Figures 7 and 8, similar to previous ones in this region, show fine and homogeneous grain structure, reflecting the lesser direct influence of contaminants in this area. The irregularities in the microstructure observed in sand-contaminated samples, as presented in Figures 7 and 8, were also noted by Wang et al. [46] and Li et al. [47].

3.3. Feature Extraction

The routines for extracting features related to the electrical arc, outlined in Section 2.2, were applied to the original datasets collected during the manufacturing of the walls. In this study, only anomalies related to contamination of the deposit in the middle region of the wall (AC), or at the start (ONC) or end (OFC) of the electrical arc are targeted. Some of these anomalies identified in the electrical arc data are represented in Figure 9. It is preliminarily estimated that the average values of the electrical arc signals (Figure 9d), along with their respective standard deviation values (Figure 9a), minimum values (Figure 9b), and the number of peaks (Figure 9c), provide good indicators for distinguishing between normal and anomalous behavior of the electrical arc under the current set of experimental conditions related to manufacturing anomalies.

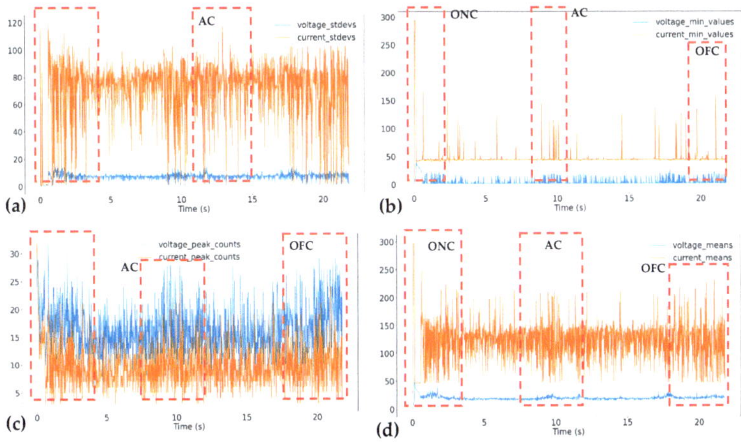

Figure 9. Examples of the features extracted from the electrical arc data: (**a**) standard deviation values, (**b**) minimum values, (**c**) number of peaks, and (**d**) average values. The dotted red squares indicate regions of the oscillogram where some type of electric arc anomaly occurred, marked by the acronyms ONC, AC, and OFC.

3.4. Correlation Between Features and Target Parameter

The feature extraction routines were aggregated into the datasets with labeled responses, along with the electric arc data collected during the fabrication of the walls, through the application of the rolling window technique. The same parameters were applied to the labeled data to adjust the dimensionality of the data matrices resulting from both scripts. By combining the feature dataset with the original labeled dataset, it was possible to observe the correlation matrix between the features and the categorical responses of the label, represented by the feature 'Anomaly', as shown in Figure 10.

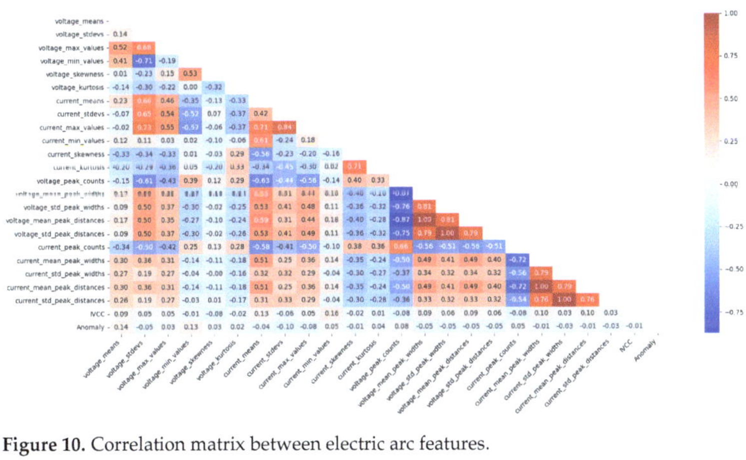

Figure 10. Correlation matrix between electric arc features.

Based on the results for electric arc features presented in Figure 9a, which displays the correlation matrix between the features and the response ('Anomaly') as a heatmap, it can be observed that the features 'voltage_max_values' (0.52) and 'current_means' (0.66) showed moderately strong correlations with the response. This indicates that higher voltage values and current averages may be associated with the occurrence of electric arc anomalies. Additionally, the features 'voltage_peak_counts' (−0.34) and 'current_peak_counts' (−0.34) exhibited moderate negative correlations, suggesting that a lower frequency of voltage and current peaks may be related to anomalous characteristics in the electric arc, considering CCC metal transfer.

In addition, the feature IVCC has a low correlation with 'Anomaly' (0.01), suggesting that this feature does not have a significant influence on the target parameter, or at least does not establish a linear relationship with the target parameter, given the multi-complexity of the metal transfer process in WAAM, as the correlation matrix uses Pearson's coefficient. Furthermore, there are strong positive correlations between variables related to electrical current, such as 'current_max_values' and 'current_std_peak_widths' (0.81), which is expected since both parameters are related to peak analysis in current waveforms. A similar behavior is observed between voltage variables, such as 'voltage_max_values' and 'voltage_std_peak_distances' (0.87), suggesting an intrinsic relationship between voltage peak characteristics and their statistical variations.

4. Conclusions

The presented study revealed significant discussions regarding arc analysis techniques for anomaly detection in walls fabricated by Wire Arc Additive Manufacturing (WAAM) using the GMA process. The following conclusions were highlighted based on the visual inspection of the walls, microstructural analysis of cross-sectioned samples, and analysis of arc-related features:

(1) The controlled introduction of contaminants, such as chalk, oil, and sand, led to observable geometric and microstructural changes due to their effect on the molten pool's solidification and the appearance of microscopic defects;
(2) The analysis of arc-related features, selected based on the controlled short-circuit transfer mode, identified strong correlations between variations in arc features—such as mean, minimum, standard deviation values, as well as the number of peaks—and the occurrence of anomalies in the WAAM process, as confirmed by metallographic analyses;
(3) The study of features indicated that applying machine learning techniques for defect prediction could be enabled through the collected data. This expands the resources for controlling and mitigating anomalies in metal additive manufacturing processes;
(4) The study confirmed that arc analysis can effectively detect anomalies in additive manufacturing processes, suggesting that such techniques could be refined to improve the reliability and quality of fabricated components.

All the evidence presented in this work highlights the significant potential of integrated methods to advance monitoring and quality control in additive manufacturing processes.

Author Contributions: Conceptualization, L.J.E.B.S., J.S.L., R.A.C.S., R.S.G., I.J.S.A. and A.L.D.B.; investigation, L.J.E.B.S. and I.J.S.A.; formal analysis, L.J.E.B.S., J.I.V.S., I.J.S.A., W.B.C. and J.S.L.; validation, J.I.V.S., J.S.L., I.J.S.A. and W.B.C.; writing—original draft preparation, L.J.E.B.S., J.I.V.S., J.M.P.Q.D., R.S.G., A.G.B.L. and A.L.D.B.; writing—review and editing, J.I.V.S., J.S.L., W.B.C., J.M.P.Q.D., R.A.C.S., A.L.D.B. and A.G.B.L.; supervision, J.I.V.S. and J.S.L.; funding acquisition, J.S.L., R.A.C.S., R.S.G., J.M.P.Q.D. and A.G.B.L. All authors have read and agreed to the published version of the manuscript.

Funding: This research was funded by the Brazilian National Council for Scientific and Technological Development (CNPq): grant numbers 409585/2022-0 and 408253/2022-3. Also, this work was financially supported by Base Funding—UIDB/04708/2020, with DOI: 10.54499/UIDB/04708/2020; Programmatic Funding—UIDP/04708/2020, with DOI: 10.54499/UIDP/04708/2020 of the CONSTRUCT funded by national funds through the FCT/MCTES (PIDDAC); and FCT through the individual Scientific Employment Stimulus 2020.00828.CEECIND/CP1590/CT0004, DOI: 10.54499/2020.00828.CEECIND/CP1590/CT0004.

Data Availability Statement: The raw data supporting the conclusions of this article will be made available by the authors upon request due to privacy concerns related to the confidentiality of proprietary processes.

Acknowledgments: The authors thank the Welding Laboratory (LABSOL) at the Federal University of Campina Grande (UFCG) and the startup WIRE—Technology and Smart Manufacturing for the loan of machines and equipment, and to the Tutorial Education Program of Mechanical Engineering (PETMEC-CG) at UFCG.

Conflicts of Interest: The authors declare no conflicts of interest.

References

1. Javaid, M.; Haleem, A.; Singh, R.P.; Suman, R.; Gonzalez, E.S. Understanding the adoption of Industry 4.0 technologies in improving environmental sustainability. *Sustain. Oper. Comput.* **2022**, *3*, 203–217. [CrossRef]
2. Sehnem, S.; Lara, A.C.; Benetti, K.; Schneider, K.; Marcon, M.L.; da Silva, T.H.H. Improving startups through excellence initiatives: Addressing circular economy and innovation. *Environ. Dev. Sustain.* **2023**, *26*, 15237–15283. [CrossRef]
3. Sireesha, M.; Lee, J.; Kranthi Kiran, A.S.; Babu, V.J.; Kee, B.B.T.; Ramakrishna, S. A review on additive manufacturing and its way into the oil and gas industry. *RSC Adv.* **2018**, *8*, 22460–22468. [CrossRef]
4. Minguella-Canela, J.; Morales Planas, S.; Gomà Ayats, J.R.; De los Santos López, M.A. Assessment of the Potential Economic Impact of the Use of AM Technologies in the Cost Levels of Manufacturing and Stocking of Spare Part Products. *Materials* **2018**, *11*, 1429. [CrossRef] [PubMed]
5. Fernandes, V.; Matos, F.; Oliveira, J.P.; Neves, A.; Godina, R. Identifying strategic opportunities through the development of a roadmap for additive manufacturing: The example of Portugal. *Heliyon* **2023**, *9*, e19672. [CrossRef] [PubMed]
6. Gonzalez-Gutierrez, J.; Cano, S.; Schuschnigg, S.; Kukla, C.; Sapkota, J.; Holzer, C. Additive Manufacturing of Metallic and Ceramic Components by the Material Extrusion of Highly-Filled Polymers: A Review and Future Perspectives. *Materials* **2018**, *11*, 840. [CrossRef] [PubMed]
7. Ferretti, P.; Santi, G.M.; Leon-Cardenas, C.; Freddi, M.; Donnici, G.; Frizziero, L.; Liverani, A. Molds with Advanced Materials for Carbon Fiber Manufacturing with 3D Printing Technology. *Polymers* **2021**, *13*, 3700. [CrossRef] [PubMed]
8. Kumar, N.; Bhavsar, H.; Mahesh, P.V.S.; Srivastava, A.K.; Bora, B.J.; Saxena, A.; Dixit, A.R. Wire Arc Additive Manufacturing—A revolutionary method in additive manufacturing. *Mater. Chem. Phys.* **2022**, *285*, 126144. [CrossRef]
9. Ron, T.; Shirizly, A.; Aghion, E. Additive Manufacturing Technologies of High Entropy Alloys (HEA): Review and Prospects. *Materials* **2023**, *16*, 2454. [CrossRef]
10. Hassan, A.; Alnaser, I.A. A Review of Different Manufacturing Methods of Metallic Foams. *ACS Omega* **2024**, *9*, 6280–6295. [CrossRef]
11. Liu, D.; Lee, B.; Babkin, A.; Chang, Y. Research Progress of Arc Additive Manufacture Technology. *Materials* **2021**, *14*, 1415. [CrossRef] [PubMed]
12. Shi, J.; Li, F.; Chen, S.; Zhao, Y.; Tian, H. Effect of in-process active cooling on forming quality and efficiency of tandem GMAW–based additive manufacturing. *Int. J. Adv. Manuf. Technol.* **2019**, *101*, 1349–1356. [CrossRef]
13. Li, Y.; Polden, J.; Pan, Z.; Cui, J.; Xia, C.; He, F.; Mu, H.; Li, H.; Wang, L. A defect detection system for wire arc additive manufacturing using incremental learning. *J. Ind. Inf. Integr.* **2022**, *27*, 100291. [CrossRef]
14. *ISO/ASTM 52900:2021(en)*; Additive Manufacturing—General Principles—Fundamentals and Vocabulary. ASTM: West Conshohocken, PA, USA, 2021. Available online: https://www.iso.org/obp/ui/#iso:std:iso-astm:52900:ed-2:v1:en (accessed on 3 December 2024).
15. Wang, J.F.; Sun, Q.J.; Wang, H.; Liu, J.P.; Feng, J.C. Effect of location on microstructure and mechanical properties of additive layer manufactured Inconel 625 using gas tungsten arc welding. *Mater. Sci. Eng. A* **2016**, *676*, 395–405. [CrossRef]
16. Ding, D.; Pan, Z.; Cuiuri, D.; Li, H. A multi-bead overlapping model for robotic wire and arc additive manufacturing (WAAM). *Robot. Comput. Integr. Manuf.* **2015**, *31*, 101–110. [CrossRef]

17. Zhu, L.; Luo, Y.; Han, J.; Zhang, C.; Xu, J.; Chen, D. Energy characteristics of droplet transfer in wire-arc additive manufacturing based on the analysis of arc signals. *Measurement* **2019**, *134*, 804–813. [CrossRef]
18. Tomar, B.; Shiva, S.; Nath, T. A review on wire arc additive manufacturing: Processing parameters, defects, quality improvement and recent advances. *Mater. Today Commun.* **2022**, *31*, 103739. [CrossRef]
19. He, X.; Wang, T.; Wu, K.; Liu, H. Automatic defects detection and classification of low carbon steel WAAM products using improved remanence/magneto-optical imaging and cost-sensitive convolutional neural network. *Measurement* **2021**, *173*, 108633. [CrossRef]
20. Ramalho, A.; Santos, T.G.; Bevans, B.; Smoqi, Z.; Rao, P.; Oliveira, J.P. Effect of contaminations on the acoustic emissions during wire and arc additive manufacturing of 316L stainless steel. *Addit. Manuf.* **2022**, *51*, 102585. [CrossRef]
21. de Souto, J.I.V.; de Lima, J.S.; de Castro, W.B.; de Santana, R.A.C.; Silva, A.A.; de Abreu Santos, T.F.; Tavares, J.M.R.S. Effects of Contaminations on Electric Arc Behavior and Occurrence of Defects in Wire Arc Additive Manufacturing of 316L-Si Stainless Steel. *Metals* **2024**, *14*, 286. [CrossRef]
22. Chabot, A.; Laroche, N.; Carcreff, E.; Rauch, M.; Hascoët, J.Y. Towards defect monitoring for metallic additive manufacturing components using phased array ultrasonic testing. *J. Intell. Manuf.* **2020**, *31*, 1191–1201. [CrossRef]
23. Lopez, A.; Bacelar, R.; Pires, I.; Santos, T.G.; Sousa, J.P.; Quintino, L. Non-destructive testing application of radiography and ultrasound for wire and arc additive manufacturing. *Addit. Manuf.* **2018**, *21*, 298–306. [CrossRef]
24. Chen, L.; Yang, F.; Wang, R.; Zhang, Y.; Diao, Z.; Rong, M. Optical Spectral Physics-Informed Attention Network for Condition Monitoring in WAAM. *IEEE Trans. Ind. Electron.* **2023**, *71*, 9708–9718. [CrossRef]
25. Li, Y.; Su, C.; Zhu, J. Comprehensive review of wire arc additive manufacturing: Hardware system, physical process, monitoring, property characterization, application and future prospects. *Results Eng.* **2022**, *13*, 100330. [CrossRef]
26. Lupo, M.; Ajabshir, S.Z.; Sofia, D.; Barletta, D.; Poletto, M. Experimental metrics of the powder layer quality in the selective laser sintering process. *Powder Technol.* **2023**, *419*, 118346. [CrossRef]
27. Rodideal, N.; Machado, C.M.; Infante, V.; Braga, D.F.O.; Santos, T.G.; Vidal, C. Mechanical characterization and fatigue assessment of wire and arc additively manufactured HSLA steel parts. *Int. J. Fatigue* **2022**, *164*, 107146. [CrossRef]
28. Zhang, T.; Li, H.; Gong, H.; Wu, Y.; Chen, X.; Zhang, X. Study on location-related thermal cycles and microstructure variation of additively manufactured inconel 718. *J. Mater. Res. Technol.* **2022**, *18*, 3056–3072. [CrossRef]
29. de Lima, J.S.; da Silva Neto, J.F.; Maciel, T.M.; López, E.A.T.; de Santana, R.A.C.; de Abreu Santos, T.F. Effect of wire arc additive manufacturing parameters on geometric, hardness, and microstructure of 316LSi stainless steel preforms. *Int. J. Adv. Manuf. Technol.* **2024**, *131*, 5981–5996. [CrossRef]
30. De Souto, J.I.V.; Ferreira, S.D.; de Lima, J.S.; de Castro, W.B.; Grassi, E.N.D.; Santos, T.F. de A. Effect of GMAW Process Parameters and Heat Input on Weld Overlay of Austenitic Stainless Steel 316L-Si. *Soldag. Inspeção* **2023**, *28*, e2809. [CrossRef]
31. Bevans, B.; Ramalho, A.; Smoqi, Z.; Gaikwad, A.; Santos, T.G.; Rao, P.; Oliveira, J.P. Monitoring and flaw detection during wire-based directed energy deposition using in-situ acoustic sensing and wavelet graph signal analysis. *Mater. Des.* **2023**, *225*, 111480. [CrossRef]
32. Surovi, N.A.; Soh, G.S. Acoustic feature based geometric defect identification in wire arc additive manufacturing. *Virtual Phys. Prototyp.* **2023**, *18*, e2210553. [CrossRef]
33. Alcaraz, J.Y.; Foqué, W.; Sharma, A.; Tjahjowidodo, T. Indirect porosity detection and root-cause identification in WAAM. *J. Intell. Manuf.* **2023**, *35*, 1607–1628. [CrossRef]
34. Meier, C.; Weissbach, R.; Weinberg, J.; Wall, W.A.; Hart, A.J. Critical influences of particle size and adhesion on the powder layer uniformity in metal additive manufacturing. *J. Mater. Process. Technol.* **2019**, *266*, 484–501. [CrossRef]
35. Roy, S.; Xiao, H.; Angelidakis, V.; Pöschel, T. Structural fluctuations in thin cohesive particle layers in powder-based additive manufacturing. *Granul. Matter* **2024**, *26*, 43. [CrossRef]
36. Jeanne Rampe, M.; Zeth Lombok, J.; Arini Tiwow, V.; Milian Tompunu Tengker, S.; Bua, J. Characterization of silica (SiO2) based on beach sand from Sulawesi and Sumatra as silicon carbide (SiC) base material. *J. Chem. Technol. Metall.* **2023**, *58*, 467–476. [CrossRef]
37. Akl, M.A.; Aly, H.F.; Soliman, H.M.A.; Abd ElRahman, A.M.E.; Abd-Elhamid, A.I. Preparation and characterization of silica nanoparticles by wet mechanical attrition of white and yellow sand. *J. Nanomedicine Nanotechnol.* **2013**, *4*, 2. [CrossRef]
38. Alaoui Mouayd, A.; Koltsov, A.; Sutter, E.; Tribollet, B. Effect of silicon content in steel and oxidation temperature on scale growth and morphology. *Mater. Chem. Phys.* **2014**, *143*, 996–1004. [CrossRef]
39. Li, C.; Shi, Y.; Gu, Y.; Yang, F. Effect of oxide on surface tension of molten metal. *RSC Adv.* **2017**, *7*, 53941–53950. [CrossRef]
40. Yoo, S.-W.; Lee, C.-M.; Kim, D.-H. Effect of Functionally Graded Material (FGM) Interlayer in Metal Additive Manufacturing of Inconel-Stainless Bimetallic Structure by Laser Melting Deposition (LMD) and Wire Arc Additive Manufacturing (WAAM). *Materials* **2023**, *16*, 535. [CrossRef]
41. Huang, W.; Nelson, B.; Ding, H. Surface wettability patterning of metal additive manufactured parts via laser-assisted functionalization. *J. Laser Appl.* **2023**, *35*, 42067. [CrossRef]

42. Zhai, W.; Zhou, W.; Nai, S.M. Effect of Interface Wettability on Additively Manufactured Metal Matrix Composites: A Case Study of 316L-Y2O3 Oxide Dispersion-Strengthened Steel. *Metals* **2024**, *14*, 170. [CrossRef]
43. Zhao, D.S.; Long, D.F.; Niu, T.R.; Liu, Y.J. Effect of Current Mode on Anisotropy of 316L Stainless Steel Wire Arc Additive Manufacturing. *J. Mater. Eng. Perform.* **2023**, *33*, 8728–8732. [CrossRef]
44. Wu, W.; Xue, J.; Wang, L.; Zhang, Z.; Hu, Y.; Dong, C. Forming process, microstructure, and mechanical properties of thin-walled 316L stainless steel using speed-cold-welding additive manufacturing. *Metals* **2019**, *9*, 109. [CrossRef]
45. Xie, B.; Xue, J.; Ren, X.; Wu, W.; Lin, Z. A comparative study of the CMT+P process on 316L stainless steel additive manufacturing. *Appl. Sci.* **2020**, *10*, 3284. [CrossRef]
46. Wang, Z.; Zimmer-Chevret, S.; Léonard, F.; Bourlet, C.; Abba, G. In Situ Monitoring of Internal Defects by a Laser Sensor for CMT Based Wire-Arc Additive Manufacturing Parts. *Defect Diffus. Forum* **2022**, *417*, 67–72. [CrossRef]
47. Li, Y.P.; Wang, C.R.; Du, X.D.; Tian, W.; Zhang, T.; Hu, J.S.; Li, B.; Li, P.C.; Liao, W.H. Research status and quality improvement of wire arc additive manufacturing of metals. *Trans. Nonferrous Met. Soc. China (Engl. Ed.)* **2023**, *33*, 969–996. [CrossRef]
48. Huang, Y.; Yue, C.; Tan, X.; Zhou, Z.; Li, X.; Zhang, X.; Zhou, C.; Peng, Y.; Wang, K. Quality Prediction for Wire Arc Additive Manufacturing Based on Multi-source Signals, Whale Optimization Algorithm–Variational Modal Decomposition, and One-Dimensional Convolutional Neural Network. *J. Mater. Eng. Perform.* **2023**, *33*, 11351–11364. [CrossRef]

metals

MDPI

Article

The Microstructure and Wear Resistance of Laser Cladding Ni60/60%WC Composite Coatings

Junxiao Liu [1,2], Chen Liu [1,2], Xiaoyu Zhang [1,2], Xiuyuan Yin [1,2], Fanyu Meng [1,2] and Changsheng Liu [1,2,*]

1 School of Materials Science and Engineering, Northeastern University, Shenyang 110819, China; 2300632@stu.neu.edu.cn (J.L.); 1810156@stu.neu.edu.cn (C.L.); 2410250@stu.neu.edu.cn (X.Z.); 2110176@stu.neu.edu.cn (X.Y.); 2470502@stu.neu.edu.cn (F.M.)
2 Key Laboratory for Anisotropy and Texture of Materials, Ministry of Education, Northeastern University, Shenyang 110819, China
* Correspondence: csliu@mail.neu.edu.cn

Abstract: In this study, Ni60/60%WC composite coatings were fabricated on 45 steel by laser cladding. The optimum process was selected through high throughput optimization experiments which had a laser power of 2400 W, scanning speed of 8 mm/s, and powder feeding rate of 20 g/min. The single-layer multilayer coatings were prepared without any cracks and pores, and the thickness of the coatings was 1.52 mm. The coating and the substrate were found to have an effective metallurgical connection. WC was distributed relatively uniformly throughout the coating, which involved the γ-(Fe, Ni), WC, W_2C, $Cr_{23}C_6$, and $Fe_{3.57}W_{9.43}C_{3.54}$ phases. The average microhardness of the coating was 1416.14 $HV_{0.2}$, approximately 5.47 times that of the substrate, and the average coefficient of friction of the coating was 0.5144, which was 43.5% lower than that of the substrate. The wear rate was reduced by 79.13%.

Keywords: laser cladding; WC; microstructure; wear resistance

1. Introduction

An exemplary specimen of carbon structural steel is 45 steel. It is distinguished by a multitude of advantageous qualities, including its low cost, high strength, good plasticity, and excellent wear resistance, among others [1–4]. These properties render it a material of choice in numerous industrial applications. Its applications encompass a diverse range of mechanical components, such as rolls, bearings, gears, and bushings [5,6]. However, when subjected to harsh environments characterized by high loads and extended periods of operation, the wear resistance of 45 steel falls short of meeting requirements. To address this limitation, scholars have resorted to laser cladding technology, a process which involves the modification of 45 steel's surface to enhance its wear resistance.

Laser cladding, a surface modification technology, has attracted significant attention from researchers due to its low dilution rate, small heat-affected zone, and good metallurgical bonding, among other advantages [7,8]. Among the materials commonly used for laser cladding, Ni-based alloys are notable for their excellent wettability, self-melting, and good mechanical properties [9,10]. An exemplary Ni-based alloy of note is that of Ni60, which exhibits exceptional hardness and wear resistance [11]. Lu et al. [12] have demonstrated the preparation of Ni60 coatings through the utilization of laser cladding techniques, yielding coatings that possess an average microhardness of 499.4 $HV_{0.5}$, which is approximately 2.4 times that of 45 steel. This process has also been shown to lead to a substantial reduction in wear, measuring at 68.3% less than that of 45 steel. Furthermore, Shen et al. [13] utilized high-speed laser cladding to prepare Ni60 coatings on the substrate of 45 steel, and the maximum microhardness of the coatings was measured at 800 HV, representing a significant enhancement compared to the substrate. In a similar study, Zhang et al. [14] employed laser cladding of Ni60A coatings, resulting in an average microhardness of 580 HV, which was a substantial enhancement compared with that of 45 steel. The mean microhardness of the Ni60A coating was found to be 580 HV, which is approximately 2.6 times higher than that of the 45 steel. The wear rate of the coating under load (2.94 N) was determined to be approximately 1/6.2 of that for the 45 steel. However, the Ni60 coating is inadequate in resisting wear under high loads of up to 20 N. To enhance its wear resistance, the coating is typically strengthened by adding ceramic reinforcing phases, such as WC [15], TiC [16], and SiC [17], among others.

WC has the advantage of high hardness, a high melting point, and good wettability with Ni-based materials [18], so it is used in large quantities in Ni60 coatings to enhance the wear-resistant properties. The type, size, and amount of WC added are all significant factors in Ni60/WC composite coatings. Zhang et al. [8] investigated the impact of three distinct categories of WC, including spherical, profiled, and flocculent WC, on the microstructure and wear-resistant characteristics of Ni60/WC composite coatings. In these coatings, spherical WC exhibited the highest microhardness of $2000HV_{0.05}$ and the lowest coefficient of friction of 0.2. In addition, flocculent WC demonstrated the most homogeneous distribution in these coatings. In the study by Chen et al. [19], the wear resistance of Ni60 coatings with added micron, nanometer, and micron/nanometer WC, respectively, was compared. The results demonstrated that nanometer WC had the most significant effect in refining the coating structure and exhibited the highest microhardness of 890.8 HV. Conversely, micron/nanometer WC exhibited the optimal wear resistance, exhibiting an average friction coefficient of 0.41. Hu et al. [20] investigated the wear resistance of Ni60/WC composite coatings with variable WC contents and determined that an increase in WC content results in a corresponding enhancement of the hardness and wear resistance of the coating. Shen et al. [21] established that when the WC addition exceeded50 wt.%, the coating microhardness exhibited a substantial increase, while the coating cracking behavior underwent a marked deterioration. In their study, Meng et al. [22] examined the forming characteristics and wear resistance of Ni60/WC composite coatings with compositions of (65/35), (50/50), and (35/65). The composite coatings exhibited extremely high levels of wear resistance as the WC content increased. However, it was also observed that when the WC content was increased, the thickness of the coating layers was found to be limited, and the composite coating exhibited a more pronounced tendency to crack. The maximum thickness of the single layer of the Ni35/WC65 coating was found to be 0.26 mm, and the thickness of eight layers was 2 mm. Liu et al. [23] prepared single-layer Ni/60% WC coatings at various powder feeding rates. The coatings had an average microhardness of 2066 $HV_{0.5}$ and a thickness of 600–900 μm, with just a few holes. Additionally, Wang et al. [24] created Ni60/50% WC coatings with a coating thickness of 0.12 mm and an average microhardness of 1053.5 $HV_{0.2}$, which is around 4.04 times greater than that of the substrate Q345.

Therefore, in-depth research on the cracking issues commonly encountered during the preparation of coatings with high WC is of significant academic value and practical importance. In this paper, the focus is on the preparation of a crack-free high-WC coating with a single-layer thickness exceeding 1 mm on a 45 steel substrate by laser cladding, with a subsequent investigation of its microstructure and wear resistance.

2. Materials and Methods

In this experiment, the substrate material chosen was 45 steel, measuring 150 mm × 100 mm × 10 mm. The substrate surface was polished to an average roughness (Ra) of 1.2 μm using 240-grit SiC sandpaper (Yuying Abrasives Co., Ltd., Foshan, China) to ensure good metallurgical bonding and subsequently cleaned with anhydrous ethanol. The cladding materials were 53–106 μm Ni60 powder (Liaoning Guanda New Materials Technology Co., Ltd., Anshan, China) and 53–150 μm WC powder (WEL Trading Co., Ltd., Shanghai, China), which were mixed according to the weight ratio of 4:6, and the powder was mixed by a mechanical ball mill with a rotational speed of 100 rpm and a mixing time of 3 h. Table 1 showed the chemical composition of Ni60 and WC. The mixed powder was dried at 120 °C for 2 h to remove the moisture before the laser cladding experiments. Figure 1 shows the particle size distribution of the mixed Ni60/60% WC powder after mixing. The results of the mixed powder morphology and EDS surface scanning are shown in Figure 2.

Table 1. Chemical composition of Ni60 and WC powder (wt.%).

Material \ Element	C	Si	B	Cr	Fe	W	Ni
Ni60	0.8	3.4	3.4	17	15.2	-	Bal.
WC	6					Bal.	

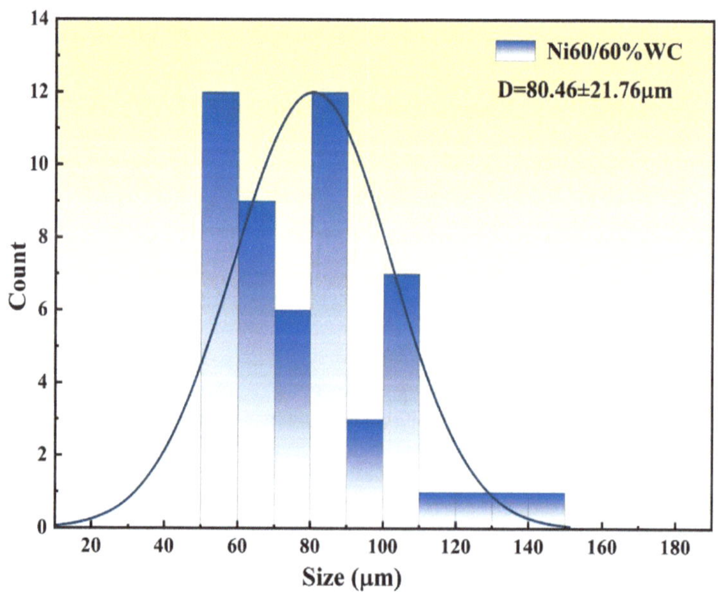

Figure 1. Distribution of Ni60/60% WC powder particle size.

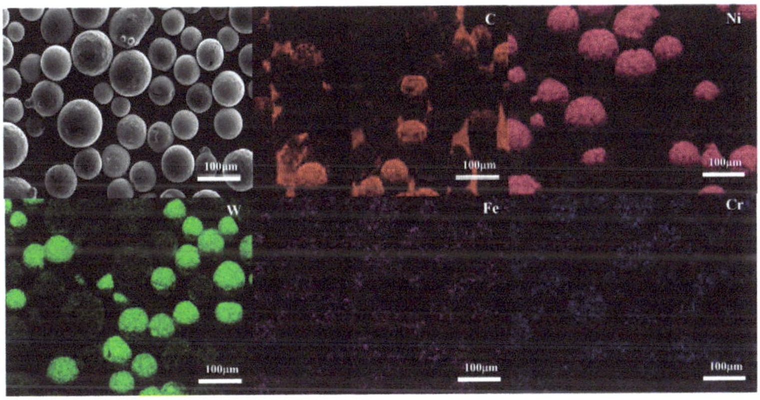

Figure 2. SEM morphology and EDS elemental distribution of powder after combining.

The laser cladding tests utilized a GWLASER 6 kW fiber laser cladding system (GW Laser Tech, YLIS-6000-W, Wuhan, China), employing a flow rate of 20 L/min of 99.9% pure argon as the shielding gas, alongside a coaxial powder feeder for delivery of the cladding powder. Initially, laser power, scanning speed, and powder feeding rate were identified as the three variables for the design of the single-pass laser cladding experiment. The single-pass organization was observed using a metallurgical microscope (Olympus Corporation, GX71, Tokyo, Japan) and the height of the single-pass cladding layer was measured using ImageJ 1.45 The optimum parameters were selected on the basis of unidirectional experiments with the aim of maximizing the coating thickness. The laser cladding process was optimized with a laser power of P = 2400 W, a scanning speed of v = 8 mm/s, a powder feeding rate of 20 g/min, and a 50% overlap rate during the multi-pass trials. As shown in Figure 3, the multi-pass laser coating experiment was shown as a simplified diagram. The coating was inspected for cracks using liquid penetrant defect detection after the cladding layer was cooled to 25 °C.

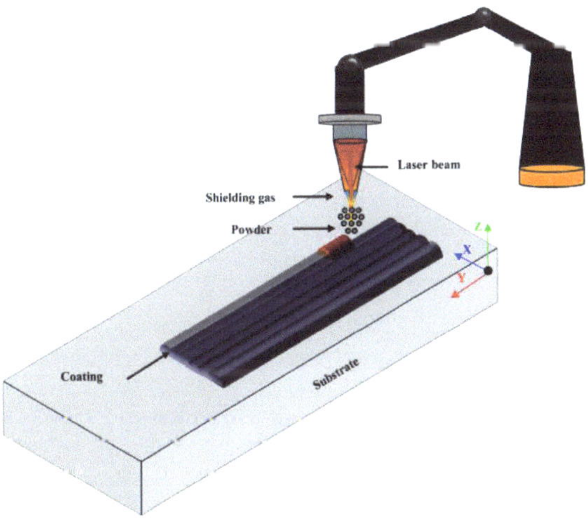

Figure 3. Multi-pass laser cladding experiment schematic diagram.

Using electrical spark wire cutting, the worn specimen and the Ni60/60%WC coated metallographic specimen were machined into blocks sized 10 mm × 10 mm × 12 mm. For the purpose of better observing the microstructure, the specimens were polished, ground, and etched. Etching was carried out with a 10% oxalic acid solution and a direct current of 0.02 A at a voltage of 5 V. It took 15 to 20 s to electrolytically etch. An X-ray diffractometer (Shimadzu XRD, Kyoto, Japan, 20 keV, 20 mA, scanning range 20–90°, 4°/min) was used to determine the phase composition. The microstructure was analyzed using a field emission scanning electron microscope (Thermo Fisher Scientific Apreo2C, Waltham, MA, USA), while the composition was analyzed using an energy dispersive spectrometer (Oxford Instruments, Ultim Max, Oxford, UK). Under the test conditions of a 1.96 N load and a 15 s loading period, the microhardness of the coating cross-section was assessed using a micro Vickers hardness tester (Shanghai Hengyi Precision Instrument, MH-500, Shanghai, China). The wear resistance of coatings was tested using a friction and wear testing machine (Bruker Corporation, UMT-02, Billerica, MA, USA). A 5 mm diameter Si_3N_4 ball was used as the counter material, and the wear pattern was linear reciprocating wear at room temperature with a load of 20 N, an amplitude of 8 mm, a frequency of 2 Hz, and a duration of 60 min. The wear trace was measured using a 3D laser confocal microscope (Olympus Corporation, LEXT OLS4100, Tokyo, Japan) and the wear rate W ($mm^3 \cdot N^{-1} \cdot m^{-1}$) was calculated using Equation (1).

$$w = \frac{v}{L \times F} \quad (1)$$

where V represents the wear volume in mm^3, L represents the sliding distance in m, and F represents the applied load in N.

3. Results

3.1. Ni60/60%WC Coating Process Optimization

In Figure 4, the synopsis of the single-pass cross-section with varying process parameters is illustrated, and the thickness of each single-pass fused cladding layer was determined using ImageJ-Pro. The shedding of spherical WC during the grinding and polishing procedure may have resulted in the holes that are noted in Figure 4g,h. The single-pass layer thickness of (m) in Figure 4 was 0.961 mm, which is the maximum thickness in Figure 4. The process parameters of (m) were laser power 2400 w, scanning speed 8 mm/s, powder feeding rate 20 g/min. Consequently, the optimal parameter was selected for the subsequent experiments. Figure 5a showed that the surface quality of Ni60/60%WC coatings created using the optimal parameters was excellent, with no holes or cracks, as evidenced by the results of penetration flaw detection in Figure 5b.

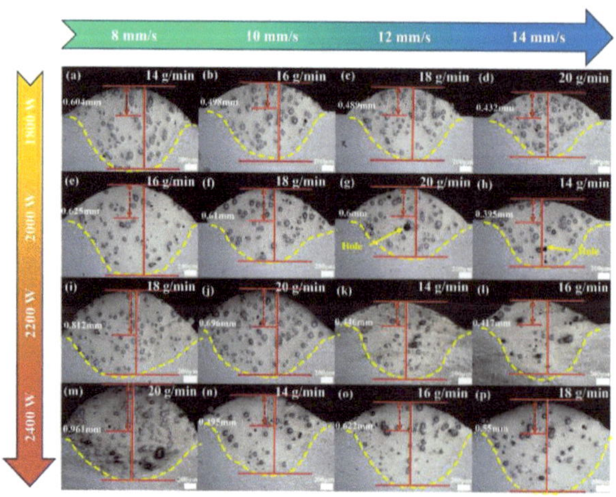

Figure 4. Metallographic summary of single-pass laser cladding cross-section.

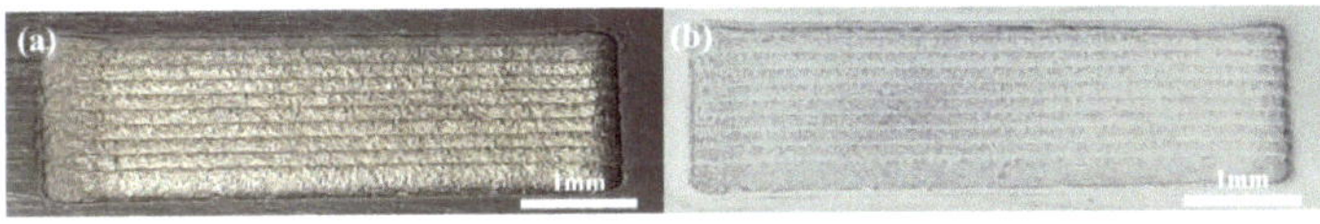

Figure 5. (**a**) Multi-pass Ni60/60%WC coating surface morphology; (**b**) results of penetration flaws.

3.2. Phase Analysis

Figure 6a shows the XRD results of the laser clad Ni60/60%WC coating, which consisted mainly of γ-(Fe, Ni), WC, W_2C, $Fe_{3.57}W_{9.43}C_{3.54}$, and $Cr_{23}C_6$ phases. The γ-(Fe, Ni) phase was the matrix phase of the coating, with the appearance of the hard phases of WC, W_2C, and $Cr_{23}C_6$. Figure 6b shows the local zoomed-in region in Figure 6a. It could be found that the diffraction peaks of the two phases of γ-(Fe, Ni) and $Cr_{23}C_6$ were not overlapped. In the laser cladding process, some of the WC was melted, which let elements diffuse into the melting pool and mix with other alloying elements., resulting in the W_2C, $Cr_{23}C_6$ hard phase.

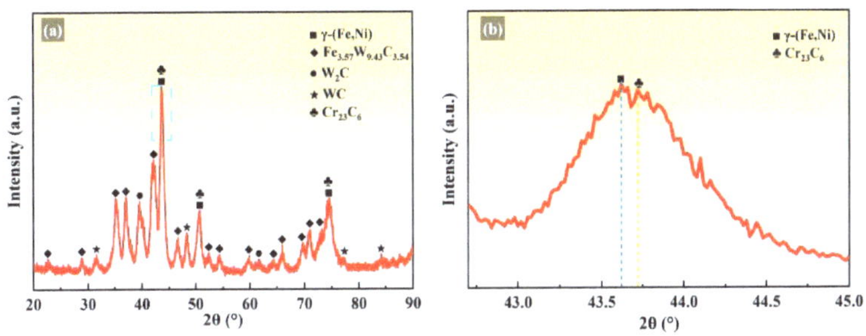

Figure 6. (**a**) XRD spectrum of the Ni60/60%WC coating, (**b**) magnified area of the dashed part.

3.3. Microstructure Analysis

The coating in this study's metallographic image is displayed in Figure 7. The height of the Ni60/60%WC coating was determined to be 1.52 mm. There were no pores or cracks seen in the covering, which had WC dispersed equally throughout. The middle and lower strata exhibited a propensity for WC accumulation, which is in accordance with the research conducted by Ortiz et al. [25].

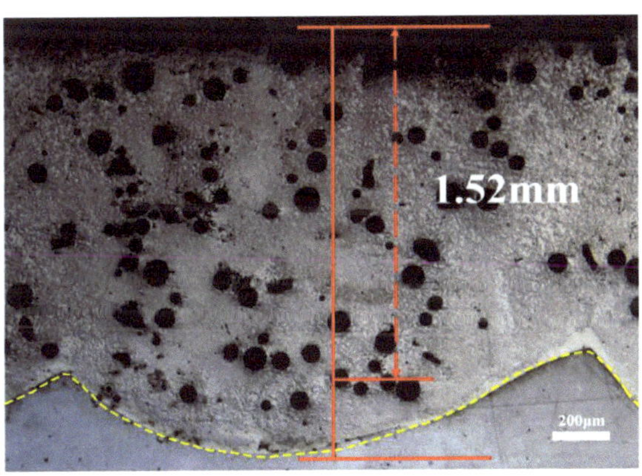

Figure 7. Metallographic diagram of Ni60/60%WC single layer coating.

In Figure 8, the microstructures of various regions of Ni60/60%WC coatings are illustrated. The microstructures and partial magnification pictures of the coatings' top part are displayed in Figure 8a,b. The top part of the coatings contained smaller equiaxial crystals, but there were still some WC particles present, and melting also took place around the WC particles. According to measurements, the diffusion zone's thickness was between 1 and 2 μm. Figure 8c shows the microstructure of the middle portion of the coating, dendrites were found. In Figure 8c, partially melted WC and W_2C are depicted. W_2C was generated as a result of the decarburization of WC during the laser cladding process [26]. The middle portion of Figure 8d was partially magnified, revealing the thermal collapse and partial decomposition of the WC spheres. Elemental analyses of the two points of Figure 8d were performed, and the elemental compositions are shown in Table 2. Combined with the data in Table 2, it was observed that Point 1 was primarily composed of the γ-(Fe, Ni), while the main composition of the dendrite at Point 2 was likely W_2C. The microstructure of the bottom of the coating was depicted in Figure 8e, where a melting line between the coating and the substrate was obviously apparent. The growth of columnar crystals at the bottom of the coating, perpendicular to the melting line, is consistent with the rapid solidification properties of laser cladding [27] as the addition of WC effectively organized the grain growth, reduced the grain size, and refined the grain. Figure 8f showed a magnified section of the microstructure at the bottom of the coating, where a diffusion zone of approximately 3–5 μm thickness was observed around the WC particles, with melting occurring in the surrounding area.

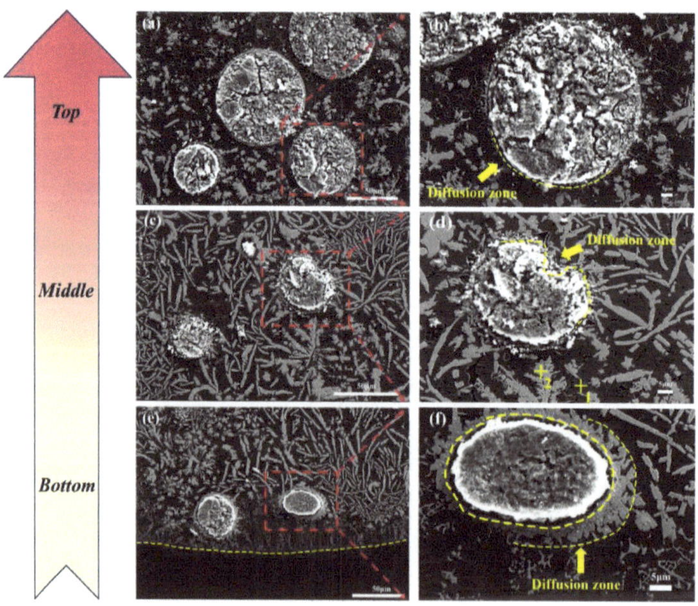

Figure 8. The cross-sectional microstructure of the Ni60/60%WC composite coating by laser cladding. (**a**) Top microstructure, (**b**) enlarged area of top microstructure, (**c**) middle microstructure, (**d**) enlarged area of middle microstructureand point EDS position, (**e**) bottom microstructure, (**f**) enlarged area of bottom microstructure.

Table 2. EDS point-scan results of microstructure in the middle of the coatings (wt.%).

Element / Position	C	Si	Cr	Fe	Ni	W
Point 1	7.84	1.11	0.79	37.67	46.53	6.06
Point 2	8.70	-	0.88	9.12	8.41	72.90

Comparing the melting of WC particles in the three sections reveals that the middle of the coating has the most severe melting, followed by the bottom and the top. With a melting temperature of roughly 2870 °C, WC was more difficult to dissolve during the laser cladding process and absorbed more heat, which lowered the melt pool's mobility. The density of WC (15.63 g/cm^3) differed from that of the γ-(Fe, Ni) substrate (7.53 g/cm^3), and the heavier WC particles were more likely to be dispersed by gravity in the middle and lower coating layers [28]. Consequently, there was a greater chance that WC in the coating's bottom portion melted and created a secondary phase. The top portion of the coating exhibited the highest cooling rate, resulting in less WC melting in that area. In Figure 9, the EDS line scan of the bottom part of the coating reveal the melting of WC particles and elemental diffusion; the surrounding C and W elemental content increased, and the elemental content of Fe increased at the bottom of the coating and the elemental diffusion substrate. Figure 10 depicts the surface scanning results of the WC on top of the coating. The results indicated that there was a minor amount of melting and elemental diffusion in the vicinity of the deconstructed WC. The distribution of W elements was clearly visible in the spectra, and C elements were enriched in their periphery. The appearance of carbides accelerated nucleation and refined the surrounding equiaxial crystal grains.

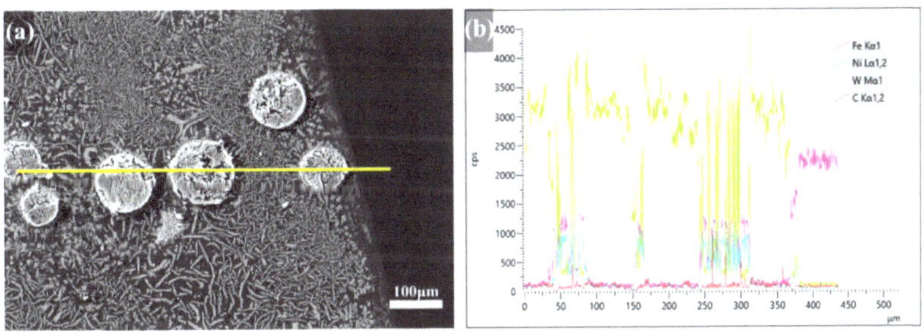

Figure 9. (**a**) Section morphology and line scan region (yellow line), (**b**) EDS line scan result.

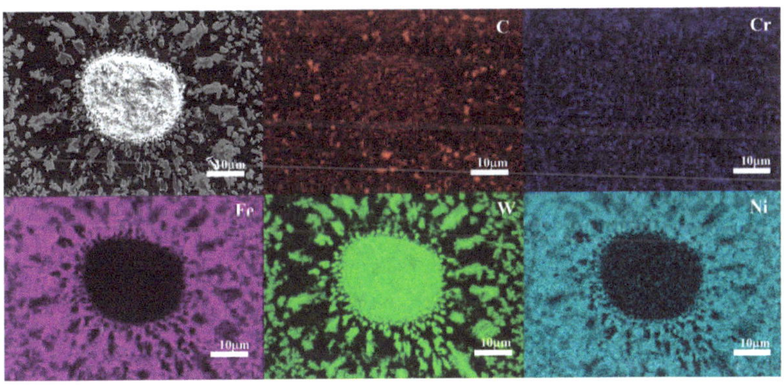

Figure 10. SEM morphology and EDS elemental distribution of the top portion of the coatings.

3.4. Microhardness Analysis

Figure 11 illustrates the cross-sectional hardness distribution of the Ni60/60%WC coating utilized in this investigation. The 45 steel substrate had an average microhardness of 259 $HV_{0.2}$, while the laser-clad Ni60/60%WC coating had an average microhardness of 1416.14 $HV_{0.2}$, which was approximately 5.47 times that of the substrate. The addition of WC significantly increased the number of WC, W_2C, and $Cr_{23}C_6$ hard phases in the coating, which significantly contributed to increasing the coating's hardness. The uniformly distributed WC particles in the coating allowed the coating to reach a maximum hardness of 2697.5 $HV_{0.2}$. Furthermore, WC encouraged grain refinement to serve as a fine grain strengthening agent, the Hall–Petch connection stated [29]. Combined with the indentation plot in Figure 11, the microhardness fluctuations in the curve were due to the indenter coming into contact with the WC in the coating. The coating and substrate had different levels of hardness in the fusion zone, and the microhardness gradually dropped. The microhardness of the coating near the substrate decreased in comparison to the middle and upper portions due to the mutual dilution of the coating and the substrate, which was exacerbated by the absence of WC reinforcement in this area. This is further supported by the line scan results in Figure 9. In comparison to the middle and top portions, the coating's microhardness diminished in the vicinity of the substrate.

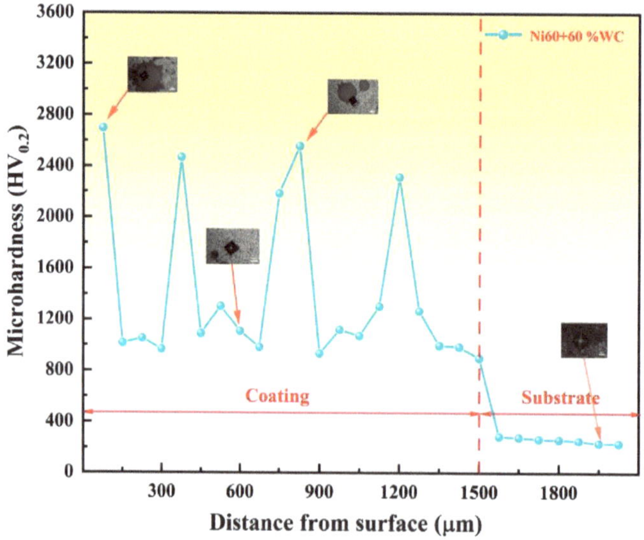

Figure 11. Microhardness diagram of Ni60/60%WC coating in cross-sectional direction.

3.5. Wear Behavior

Typically, the coefficient of friction and wear rate served as the primary reference indices for assessing the coating's wear resistance, which was a crucial criterion for assessing the coating's dependability and real service life. Figure 12a displays the friction coefficients of Ni60/60%WC coating and substrate 45 steel over time. The coating's average friction coefficient was 0.5144, while the substrate's was 0.9103. The coating's friction coefficient was 43.5% lower than the substrate's. The presence of a significant amount of WC also improved the wear resistance by lowering the coating's coefficient of friction. According to Figure 12a, the friction coefficient fluctuated a lot during the break-in period and tended to steadily stabilize after 600 s. When the friction coefficient between the coating and the substrate was compared during the stabilization stage, it was evident that the coating's fluctuation was significantly greater due to the presence of WC and other hard phases on its surface. This resulted in intense friction when the coating came into contact with the wear material Si_3N_4. In Figure 12b, the substrate and coating are compared in terms of their wear rates. The mean wear rate of the substrate was 3.848×10^{-6} $mm^3 \cdot N^{-1} \cdot m^{-1}$, while the wear rate of the coating was 8.031×10^{-7} $mm^3 \cdot N^{-1} \cdot m^{-1}$. The wear rate of the coating was reduced by 79.13% compared to the substrate. Figure 13 illustrates the abrasion morphology and depth map. According to Figure 13a, the substrate's abrasion width was approximately 1200 μm with a depth of 12 μm, and the coating's abrasion width was approximately 600 μm with a depth of 4 μm. In comparison to the substrate, the coating's depth was reduced by 66.7%, and its width was significantly diminished by 50%. When viewed in conjunction with the laser confocal images in Figure 13b,c, it is evident that the substrate had noticeable wear markings, while the coating's wear marks were extremely shallow.

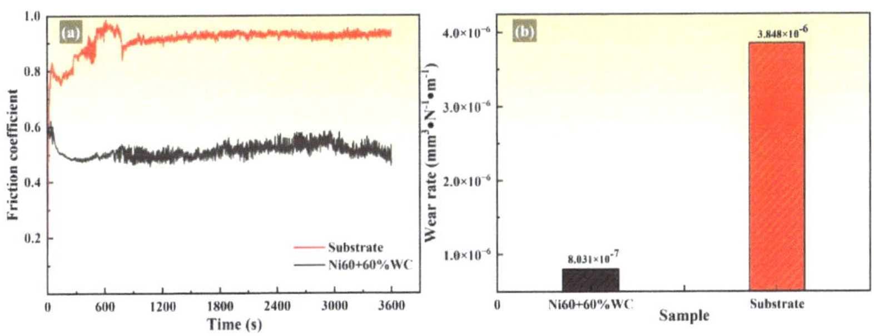

Figure 12. (**a**) Coating and substrate coefficient of friction with time, (**b**) comparison graphs of coating and substrate wear rates.

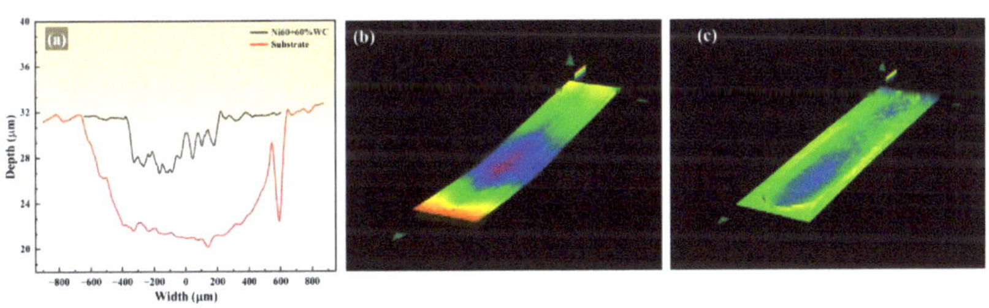

Figure 13. (**a**) Abrasion profile curves of coating and substrate, (**b**) 3D morphology of coating abrasion, (**c**) 3D morphology of substrate abrasion.

3.6. Wear Mechanisms

The microscopic wear morphology of Ni60/60%WC is depicted in Figure 14a. At this point, the type of wear is abrasive wear, and the wear marks contain abrasive debris, scratches, and some plow grooves. Figure 14b,d shows the microscopic wear morphology and local magnification of a 45-steel substrate. After the surface was worn, a lot of delamination and plowing grooves appeared, as well as a tiny amount of abrasive debris. The substrate wear this time was characterized by abrasive particle wear and adhesive wear. A partial enlargement of the coating wear and elemental analysis is shown in Figure 14c. The findings of the local zoom region elemental analysis are shown in Table 3. It was found that Points 1 and 2 are abrasive debris. Point 1 has a Si and O content of 11.87% and 29.39%, respectively, and Point 2 has a Si and O content of 21.32% and 29.68%. Point 2 has a higher Si and O content than Point 1. This may be because Si_3N_4 friction vice produces a lot of heat during the abrasion process of the wear process for a short time. Due to poor thermal conductivity, oxidation reactions occur in the local area of contact, resulting in the formation of SiO_2 oxide film and broken WC particles. The softer substrate γ-(Ni, Fe) is the first to wear, exposing WC particles, which gradually break and fall out of the wear track. The WC particles decompose and detach in the wear track over time. The W_2C and $Cr_{23}C_6$ hard stages that are made in the coating also make chip cutting less effective. The friction subsurface's surface experiences some abrasion from the WC particles shed in the wear track, and the broken SiO_2 turns into abrasive debris that builds up on the coating's surface as it moves.

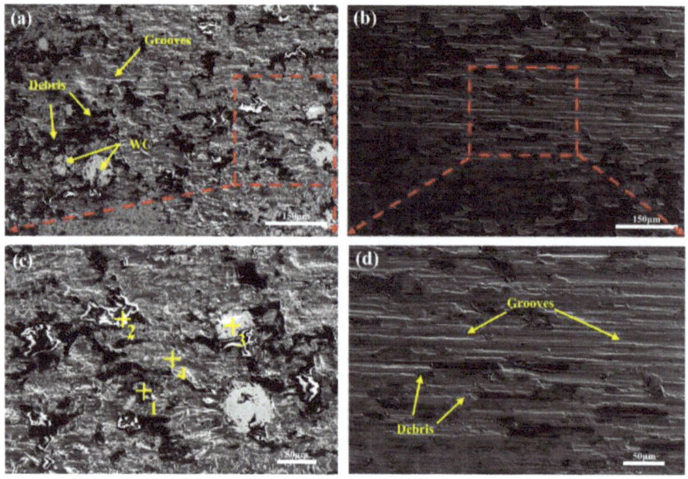

Figure 14. Microscopic morphology of wear marks, (**a**) micro-morphology of coating wear marks, (**b**) microscopic morphology of 45 steel wear marks, (**c**) localized zoomed-in area of coating and point EDS position, (**d**) localized zoomed-in area of 45 steel.

Table 3. EDS results for different locations on the coating surface after wear (wt.%).

Position \ Element	C	W	O	Si	Cr	Fe	Ni
1	6.21	31.13	29.39	11.87	0.93	4.39	16.08
2	15.06	15.52	29.68	21.32	0.51	3.66	14.25
3	15.41	82.96	1.64	-	-	-	-
4	3.75	2.81	0.37	1.39	1.55	18.29	71.83

Figure 15 illustrates the wear mechanism of Ni60/60%WC coatings. The WC was more evenly distributed in the coating, and there was more of it. Partially melted WC increases the strength of the γ-(Ni, Fe) matrix, and unmelted WC acts as both a solid solution reinforcement and a reinforcing phase to make the coating stronger and more resistant to wear. Figure 15a illustrates the wear scenario at the early stages of wear, when there was little contact area between the coating and the anti-wear material Si_3N_4, and the friction coefficient rose quickly. The process of reaching the stable stage of wear is depicted in Figure 15b. As the wear intensifies and the contact area was greater, the organization's coating became more involved. The WC particles eventually lost their ability to withstand wear, ruptured, shed, and transformed into abrasive particles. While the coating surface wore at the same rate on the wear material Si_3N_4, the wear process was influenced by two forces, the applied load and relative sliding. The WC particles grinding on the wear material Si_3N_4 caused the surface of the tiny bumps to overlap and deform until they fell off as abrasive debris. Figure 15c shows the abrasive debris that was created by the details of the schematic diagram. It was the abrasive material's movement that caused the coating surface to wear. As the wear went on, the contact area between the coating surface and the abrasive material gradually grew. Eventually, the friction coefficient reached a stable level.

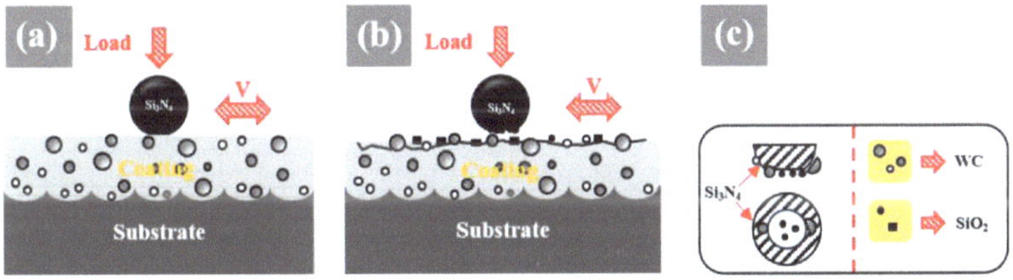

Figure 15. Coating wear mechanism (**a**) early wear stage, (**b**) wear stabilization stage, and (**c**) localized detail of debris generation.

4. Conclusions

The microstructure and wear resistance properties of Ni60/60%WC composite coatings were investigated, and the following conclusions were obtained:

1. Laser cladding effectively produced a Ni60/60%WC composite coating with a single layer thickness of up to 1.52 mm, exhibiting no pores or cracks on a 45 steel substrate.
2. The microstructure of the coating mostly consists of columnar and equiaxial crystals, the incorporation of WC contributes to grain refinement and solid solution strengthening. However, some elemental diffusion and WC particles appear in varying degrees of diffusion zones.
3. The composite coating contains γ-(Fe, Ni) as the matrix phase, as well as unmelted WC, which generates W_2C, $Cr_{23}C_6$, and $Fe_{3.57}W_{9.43}C_{3.54}$ hard phases through elemental diffusion. These hard phases significantly affect the coating's microhardness, with an average microhardness of 1416.14 $HV_{0.2}$, which is approximately 5.47 times that of the substrate.
4. The average friction coefficient of the Ni60/60%WC coating is 0.5144, which is 43.5% lower than that of the substrate, and the wear rate is decreased by 79.13%, indicating good wear resistance. Wear resistance of the Ni60/60%WC coating is improved by the combination of the hard phase produced during the laser cladding process and the WC in the coating.

Author Contributions: Conceptualization, J.L. and C.L. (Changsheng Liu); methodology, J.L., C.L. (Chen Liu) and C.L. (Changsheng Liu); software, J.L. and C.L. (Chen Liu); validation, J.L., X.Z. and C.L. (Changsheng Liu); formal analysis, J.L., X.Z. and X.Y.; investigation, J.L., X.Y. and F.M.; resources, C.L. (Changsheng Liu); data curation, J.L., X.Z., and F.M.; writing—original draft preparation, J.L., X.Y. and C.L. (Chen Liu); writing—review and editing, J.L., C.L. (Chen Liu) and C.L. (Changsheng Liu); visualization, J.L., X.Z., and C.L. (Chen Liu); supervision, C.L. (Chen Liu); project administration, J.L., X.Y. and C.L. (Changsheng Liu); funding acquisition, C.L. (Changsheng Liu). All authors have read and agreed to the published version of the manuscript.

Funding: This research received no external funding.

Data Availability Statement: The original contributions presented in this study are included in the article. Further inquiries can be directed to the corresponding author.

Conflicts of Interest: The authors declare no conflicts of interest.

References

1. Li, Q. Mechanism of Laser Carbonitriding Enhancing the Wear Resistance of 45# Steel. *Mater. Today Commun.* **2024**, *38*, 108327.
2. Huang, G.; Qu, L.; Lu, Y.; Wang, Y.; Li, H.; Qin, Z.; Lu, X. Corrosion Resistance Improvement of 45 Steel by Fe-Based Amorphous Coating. *Vacuum* **2018**, *153*, 39–42. [CrossRef]
3. Priymak, E.Y.; Stepanchukova, A.V.; Yakovleva, I.L.; Tereshchenko, N.A.; Chirkov, E.Y. Effect of Carbonitriding on the Susceptibility of Medium-Carbon Alloy Steels to Temper Brittleness. *Met. Sci. Heat Treat.* **2017**, *59*, 243–249. [CrossRef]
4. Zhao, L.; Yu, H.; Wang, Y.; Zhao, Z.; Song, W.; Chen, C. NiCrBSi Coatings Fabricated on 45 Steel Using Large Spot Laser Cladding. *Materials* **2022**, *15*, 6246. [CrossRef]
5. Cai, W.; Meng, F.; Gao, X.; Hu, J. Effect of QPQ Nitriding Time on Wear and Corrosion Behavior of 45 Carbon Steel. *Appl. Surf. Sci.* **2012**, *261*, 411–414. [CrossRef]
6. Zhu, L.; Wang, S.; Pan, H.; Yuan, C.; Chen, X. Research on Remanufacturing Strategy for 45 Steel Gear Using H13 Steel Powder Based on Laser Cladding Technology. *J. Manuf. Process.* **2020**, *49*, 344–354. [CrossRef]
7. Zhang, X.; Yin, X.; Liu, C.; Liu, C. Laser Melting of Prefabrication AlOOH-Activated Film on the Surface of Nodular Cast Iron and Its Associated Properties. *Materials* **2023**, *16*, 5486. [CrossRef]
8. Zhang, P.; Pang, Y.; Yu, M. Effects of WC Particle Types on the Microstructures and Properties of WC-Reinforced Ni60 Composite Coatings Produced by Laser Cladding. *Metals* **2019**, *9*, 583. [CrossRef]
9. Sun, S.; Wang, J.; Xu, J.; Cheng, X.; Jing, C.; Chen, Z.; Ru, H.; Liu, Y.; Jiao, J. Preparing WC-Ni Coatings with Laser Cladding Technology: A Review. *Mater. Today Commun.* **2023**, *37*, 106939. [CrossRef]
10. Devojno, O.G.; Feldshtein, E.; Kardapolava, M.A.; Lutsko, N.I. On the Formation Features, Microstructure and Microhardness of Single Laser Tracks Formed by Laser Cladding of a NiCrBSi Self-Fluxing Alloy. *Opt. Lasers Eng.* **2018**, *106*, 32–38. [CrossRef]
11. Shi, B.; Li, T.; Wang, D.; Zhang, X.; Zhang, H. Investigation on Crack Behavior of Ni60A Alloy Coating Produced by Coaxial Laser Cladding. *J. Mater. Sci.* **2021**, *56*, 13323–13336. [CrossRef]
12. Lu, P.; Jia, L.; Zhang, C.; Heng, X.; Xi, X.; Duan, M.; Lu, Z.; Zhou, Y. Optimization on Laser Cladding Parameters for Preparing Ni60 Coating along with Its Friction and Wear Properties. *Mater. Today Commun.* **2023**, *37*, 107162. [CrossRef]
13. Shen, X.; Chen, L.; Wu, Y.; Chen, C.; Zhang, X.; Zhou, W.; Ren, X. Effect of Substrate Material Properties on Microstructure and Mechanical Properties of Ni60 Coatings Prepared by Extreme High-Speed Laser Cladding. *Steel Res. Int.* **2024**, *95*, 2300531. [CrossRef]
14. Zhang, J.; Hu, Y.; Tan, X.; Guo, L.; Zhang, Q. Microstructure and High Temperature Tribological Behavior of Laser Cladding Ni60A Alloys Coatings on 45 Steel Substrate. *Trans. Nonferrous Met. Soc. China* **2015**, *25*, 1525–1532. [CrossRef]
15. Ma, C.; Li, H.; Xia, F.; Yan, P. Preparation and Characterization of Ni60-WC Composites Fabricated Using Laser Cladding Technique. *J. Mater. Eng. Perform.* **2024**. [CrossRef]
16. Cai, Q.; Li, G.; Wu, B.; Xu, S.; Wang, L.; Guo, Y. Effect of TiC Content on Microstructure and Properties of TiC/Ni60 Coatings on Ti6Al4V Alloy Deposited by Laser Cladding. *Opt. Laser Technol.* **2024**, *168*, 109854. [CrossRef]
17. Lou, B.; Chen, Z.; Bai, W.; Dong, G. Structure and Erosion Resistance of Ni60A/SiC Coatting by Laser Cladding. *Trans. Nonferrous Met. Soc. China* **2006**, *16*, 643–646. [CrossRef]
18. Leunda, J.; Sanz, C.; Soriano, C. Laser Cladding Strategies for Producing WC Reinforced NiCr Coatings inside Twin Barrels. *Surf. Coat. Technol.* **2016**, *307*, 720–727. [CrossRef]
19. Chen, C.; Feng, A.; Wei, Y.; Wang, Y.; Pan, X.; Song, X. Effects of WC Particles on Microstructure and Wear Behavior of Laser Cladding Ni60 Composite Coatings. *Opt. Laser Technol.* **2023**, *163*, 109425. [CrossRef]
20. Hu, Z.; Li, Y.; Lu, B.; Tan, N.; Cai, L.; Yong, Q. Effect of WC Content on Microstructure and Properties of High-Speed Laser Cladding Ni-Based Coating. *Opt. Laser Technol.* **2022**, *155*, 108449. [CrossRef]
21. Shen, X.; Peng, H.; Xue, Y.; Wang, B.; Su, G.; Zhu, J.; Li, A. Microstructure and Properties of WC/Ni-Based Laser-Clad Coatings with Different WC Content Values. *Materials* **2022**, *15*, 6309. [CrossRef] [PubMed]
22. Meng, L.; Zhu, B.; Liu, X.; Zeng, X. Investigation on the Ni60-WC Composite Coatings Deposited by Extreme-High-Speed Laser-Induction Hybrid Cladding Technology: Forming Characteristics, Microstructure and Wear Behaviors. *Surf. Coat. Technol.* **2023**, *473*, 130033. [CrossRef]
23. Liu, Z.; He, C.; Kong, D. Effect of Powder Feeding Speed on Microstructure and Corrosive–Wear Performance of Laser Cladded Ni–60%WC Coatings. *Opt. Laser Technol.* **2024**, *168*, 109801. [CrossRef]
24. Wang, Q.; Li, Q.; Zhang, L.; Chen, D.X.; Jin, H.; Li, J.D.; Zhang, J.W.; Ban, C.Y. Microstructure and Properties of Ni-WC Gradient Composite Coating Prepared by Laser Cladding. *Ceram. Int.* **2022**, *48*, 7905–7917. [CrossRef]
25. Ortiz, A.; García, A.; Cadenas, M.; Fernández, M.R.; Cuetos, J.M. WC Particles Distribution Model in the Cross-Section of Laser Cladded NiCrBSi + WC Coatings, for Different Wt% WC. *Surf. Coat. Technol.* **2017**, *324*, 298–306. [CrossRef]
26. Cao, Q.; Fan, L.; Chen, H.; Hou, Y.; Dong, L.; Ni, Z. Wear Behavior of Laser Cladded WC-Reinforced Ni-Based Coatings under Low Temperature. *Tribol. Int.* **2022**, *176*, 107939. [CrossRef]

27. Konstantinov, A.; Chizhikov, A.; Zhidovich, A.; Avdeeva, V.; Agasieva, S.; Bazhin, P. Effect of Argon and Nitrogen Environments on the Structure and Properties of Protective Cermet Coatings Based on Titanium Borides Obtained by Electric Arc Surfacing. *Ceram. Int.* **2024**, *50*, 51338–51345. [CrossRef]
28. Xiao, Q.; Sun, W.L.; Yang, K.X.; Xing, X.F.; Chen, Z.H.; Zhou, H.N.; Lu, J. Wear Mechanisms and Micro-Evaluation on WC Particles Investigation of WC-Fe Composite Coatings Fabricated by Laser Cladding. *Surf. Coat. Technol.* **2021**, *420*, 127341. [CrossRef]
29. Sriraman, K.R.; Ganesh Sundara Raman, S.; Seshadri, S.K. Synthesis and Evaluation of Hardness and Sliding Wear Resistance of Electrodeposited Nanocrystalline Ni–W Alloys. *Mater. Sci. Eng. A* **2006**, *418*, 303–311. [CrossRef]

metals

MDPI

Article

Experimental and Numerical Study of Behavior of Additively Manufactured 316L Steel Under Challenging Conditions

Lenka Kunčická [1,*], Radim Kocich [1] and Marek Pagáč [2]

[1] Department of Metallurgical Technologies, Faculty of Materials Science and Technology, VŠB—Technical University of Ostrava, 17. listopadu 2172-15, 708 00 Ostrava, Czech Republic
[2] Department of Machining, Assembly and Engineering Metrology, Faculty of Mechanical Engineering, VŠB—Technical University of Ostrava, 708 00 Ostrava, Czech Republic
* Correspondence: lenka.kuncicka@vsb.cz; Tel.: +420-596-991-265

Abstract: AISI 316L stainless steel, widely used in numerous industrial fields, can be fabricated by conventional methods, but also by additive manufacturing. As materials prepared by additive manufacturing typically feature various printing defects deteriorating their mechanical and utility properties, post-processing by plastic deformation is able to enhance their performance. The determination of optimized post-processing conditions can advantageously be performed by combining experimental work and numerical simulations using the finite element method. The presented research focuses on investigating the deformation behavior of AISI 316L stainless steel prepared by additive manufacturing under a variety of thermomechanical conditions (temperatures of 900–1250 °C, strain rates of 0.1–100 s^{-1}). Together with the deformation behavior of the steel, the kinetics of the occurring softening processes is also discussed. The experimentally acquired data are further used for numerical simulations to predict the expected magnitudes of force and imposed strains during prospective post-processing. Observing the microstructures and mechanical properties reveals that the prospective post-processing of AISI 316L stainless steel, prepared by additive manufacturing, via plastic deformation is the most favorable when performed at the temperature of 900 °C and using high strain rates. The flow stress/microhardness generally increase at lower temperatures and higher strain rates, as a result of the development of a substructure. On the contrary, higher temperatures support the recrystallization of grains and their coarsening, which consequently decreases the mechanical properties.

Keywords: additive manufacturing; 316L stainless steel; finite element method; microstructure; flow stress

1. Introduction

Additive manufacturing (AM) represents a manufacturing trend whose popularity is still increasing [1,2]. It has also found application in the (small-scale) production of metallic materials. The AM of metals involves two main technological solutions—direct energy deposition (DED), e.g., [3,4], and (laser) powder bed fusion (L-PBF), e.g., [5,6]. The L-PBF technologies include several methods such as Direct Metal Laser Sintering (DMLS) [7], Electron Beam Melting (EBM) [8], Selective Laser Sintering (SLS) [9], and Selective Laser Melting (SLM) [10,11]. SLM has been used to process a wide range of metallic materials, such as aluminum alloys [12,13], titanium alloys [14,15], stainless steels [16–19], high-entropy alloys [20,21], and even non-conventional materials and composites [22,23]. The primary advantage of AM techniques is their ability to create dimensionally complicated

 https://doi.org/10.3390/met15020169

products, the production of which would be difficult, or even impossible, when using conventional processing methods. Nevertheless, AM-prepared materials typically feature specific disadvantages, including surface roughness [24], internal porosity and voids [25], the inhomogeneous distribution of residual stress [26], and a limited product size [27]. The post-processing of AM-prepared materials typically includes heat treatments [28,29]. However, the sole application of heat is not sufficient to completely eliminate residual porosity, although it is rather effective for the homogenization of residual stress [30]. Regarding the elimination of residual porosity, combinations of high temperatures and high pressures are much more efficient. On the other hand, the Hot Isostatic Pressing (HIP) method, which enables the application of temperatures up to 2000 °C and pressures up to 200 MPa [31], is favorable for the overall densification of AM-prepared materials [32–34].

The properties and overall performance of AM-prepared materials can advantageously be improved by combining AM with post-processing using methods such as intensive plastic deformation (IPD, e.g., rotary swaging [35–37], incremental forming [38,39]), or severe plastic deformation (SPD). The greatest advantage of the IPD and SPD methods is that they predominantly feature compressive stress states, which contributes to the homogenization of residual stresses and the reduction/elimination of residual porosity [40,41]. For example, room temperature rotary swaging was previously shown to successfully eliminate residual porosity within stainless steel workpieces prepared by SLM [40]. The most commonly used SPD methods, which could also be used to process AM-prepared bulk metallic materials, involve Equal Channel Angular Pressing (ECAP) [42–44] and its modifications, such as ECAP with (Partial) Back Pressure [45,46] or Twist Channel Angular Pressing, TCAP [47,48]. Similar to SPD methods, the IPD method of rotary swaging features a predominantly compressive stress state and can thus provide various composite, powder-based, and even AM-prepared workpieces with all the aforementioned benefits [49–51]. The method is suitable for the processing of any metallic material. However, contrary to SPD methods, rotary swaging features very high strain rates. Examining the effects of high strain rates on AM-prepared materials, to be prospectively processed by methods like rotary swaging, is therefore of high interest.

The AISI 316L austenitic stainless steel features a variety of advantageous properties, such as favorable tensile and fatigue strengths, as well as excellent corrosion and oxidation resistances, even at elevated temperatures [52,53]. Its high stability makes the steel favorable for applications experiencing extreme temperature fluctuations, including those in corrosive environments, as the steel is also resistant to corrosion in salt water and numerous chemicals [54,55]. It is commonly used to fabricate durable parts and components within various industrial fields, including bioengineering [56,57], power engineering [58], construction [59], or transportation [60]. Given its austenitic structure, the 316L steel is hardly hardenable by heat treatments. On the other hand, its mechanical properties can be (substantially) improved by deformation processing, typically by cold working, or by implementing other (innovative) processing techniques, such as combinations of AM [61] and deformation or thermomechanical post-processing [62,63]. Maamoun et al. [64] applied the SPD method of Friction Stir Processing to refine grains, reduce the residual stress and increase the microhardness of a AlSi10Mg AM-prepared workpiece. Similar favorable effects of FSP were observed by Rubino et al. [65] and Zhao et al. [66] for AM-prepared Ti6Al4V workpieces. On the other hand, Hönnige et al. [67] used rolling to efficiently reduce the residual stress within an AM-prepared Al-based alloy wire, and Xu et al. [68] used rolling to thermomechanically post-process AM-prepared Inconel 718 to increase its ultimate strength to almost 400 MPa (compared to 220 MPa acquired for the AM-prepared state). The surfaces of AM-prepared materials can also be treated, for example, by shot peening, laser shock peening, or abrasive flow machining [69], and high-strain-rate processing [70], to improve the mechanical properties and wear behavior.

The primary motivation for this study was to find the most suitable conditions for the prospective post-process thermomechanical treatment of AM-prepared AISI 316L steel in order to enhance its performance and eliminate printing defects. The study was thus primarily focused on the characterization of the deformation behavior of AM-prepared AISI 316L steel samples by uniaxial hot compression tests performed under varying thermomechanical conditions. The first part of the study presents details of the experimental research and gives details on the performed numerical simulations, too, whereas the following part focuses on presenting the results of microstructure analyses and investigations of the mechanical properties, i.e., flow stress and Vickers microhardness measurements. The study concludes with the validation of the acquired experimental data with the numerically predicted results. Several studies investigating the high-temperature behavior of AM-prepared AISI 316L steel have been published (e.g., [71,72]); however, the works only discussed considerably lower strain rates (0.001 s^{-1}). As far as the authors' knowledge reaches, no study similar to the one presented herein has been published so far.

2. Materials and Methods

2.1. Experimental Works

The initial workpieces were an SLM-prepared stainless-steel powder, with a particle size varying between 15 and 45 μm using an AM400 3D printer (both powder and printer by Renishaw plc., Wootton-under-Edge, UK). The maximum possible dimensions of the components produced by the AM400 printer by Renishaw were 250 × 250 × 300 mm^3; the machine has the laser power of up to 400 W and enables the production of up to 20 cm^3/h of material volume, according to the actual setting. The original workpieces with circular cross-sections and a 10.5 mm diameter and 160 mm length were built from left to right, with the vertical direction parallel to the building substrate, using the meander printing strategy [40,73]. The printing was performed in an inert argon (purity of 99.998%) atmosphere with an applied laser power of 200 W. The AM-prepared 316L steel workpieces were machined into a total of sixteen individual cylindrical test samples that were 10 mm in diameter and 15 mm in length. These samples were then subjected to uniaxial hot compression tests using a Gleeble 3800 thermal–mechanical simulator equipped with a Hydrawedge II Mobile Conversion Unit (both machine and equipment by Dynamic System Inc., Poestenkill, NY, USA). The tests were performed under the combined conditions of four testing temperatures (900 °C, 1000 °C, 1100 °C, and 1250 °C), and four strain rates of 0.1 s^{-1}, 1 s^{-1}, 10 s^{-1}, and 100 s^{-1}. The temperatures and strain rates were selected to cover the widest range of possible forming conditions. Hot forming usually starts at relatively high temperatures, which can progressively decrease during the forming process, depending on the selected technology. The aim of this research was to find the lowest possible temperature enabling the successful forming of the AM-prepared steel without crossing the cohesion limit. Regarding the selection of strain rates, both low and high values were investigated. Low strain rates were investigated due to the prospective future processing of the AM-prepared steel on hydraulic presses, whilst high strain rates were investigated to characterize the development of flow stress during impact forming, such as hammer forging or rotary swaging. The test samples were heated by direct electric resistance heating up to the required deformation temperature with a heating rate of 10 $°C\cdot s^{-1}$, and then deformed by uniaxial compression up to a true strain of 1.0. After the sample had been heated to the required temperature, a time dwell of 300 s was applied to homogenize the temperature throughout the sample. A time of 300 s was selected based on our experience, as it is sufficient for the temperature distribution throughout the sample to homogenize, and, at the same time, not too long to negatively influence the microstructure. Figure 1a shows an example of a sample before and after testing, and Figure 1b depicts the scheme of the testing sample with dimensions. Temperature measurement was performed using a pair of thermocouple wires (R-type (Pt-13%Rh (+) and Pt (−)) for 1250 °C, and K-type (Ni-Cr (+) and Ni-Al (−)) for 900–1100 °C) welded onto the surface of a sample, along its middle length (thermocouples by OMEGA Engineering, Norwalk, CT, USA). The testing chamber was held under vacuum to reduce oxidation, and the sample–anvil interfaces were treated with a tantalum foil coated with a nickel-based high-temperature grease to protect the anvils and reduce friction.

Figure 1. Examples of samples before and after deformation (**a**); schematic depiction of testing sample with dimensions (**b**).

In order to assess the effects of different processing conditions on the microstructures and mechanical properties of the AM-prepared steel, scanning electron microscopy (SEM) and Vickers microhardness analyses were implemented. As for the SEM, Tescan Fera device equipped with an electron backscatter diffraction (EBSD) detector (equipment by Tescan Orsay Holding a.s., Brno, Czech Republic) was used. For the analysis and measurements, four samples subjected to the highest and lowest applied strain rates and temperatures were selected. To evaluate the microstructures on longitudinally (vertically) cut and processed samples, the EBSD method was used in particular. For the AM-prepared material state, a scan step of 2 μm was used, as the expected grain size was large. The processed material states, on the other hand, were scanned with a scan step of 0.1 μm, as they were expected to exhibit (highly) refined grains. For the evaluation of the data, the low- and high-angle grain boundaries (LAGBs and HAGBs) were defined using values of 5° and 15°. The microhardness was measured with a load of 1000 g (i.e., HV 1) on comparable longitudinal (vertical) sections of the deformed samples (as also further depicted graphically in Section 3.2), with an indent distance of 1 mm (FM ARS 900 equipment by Future Tech, Spectrographic Limited, Leeds, UK).

2.2. Numerical Simulation

In order to complete the mapping of the formability of the AM-prepared AISI 316L steel, numerical simulations using the finite element method (FEM) were performed. The simulated task consisted of a hot compression test performed at a strain rate of 0.1 s^{-1} and all the four deformation temperatures of 900 °C, 1000 °C, 1100 °C, and 1250 °C. The simulations were performed using the Forge® software (NxT version, by Transvalor S.A., Biot, France). The geometric assembly of the process consisted of three objects, as depicted in Figure 2: the upper die, billet, and lower die. The dimensions of all the components corresponded to those used in the real Gleeble experiment, i.e., the test specimen diameter and height were 10 mm and 15 mm, and the anvil diameter and height were 19 mm and 6.4 mm, respectively. The anvils were considered as rigid tools. The FEM volume mesh of the billet consisted of 6315 nodes, i.e., 29,656 tetrahedron elements in total, and the mesh size was 0.53 mm. The lower and upper anvils were surface-meshed, the mesh consisted of 1008 nodes (2012 elements) for each anvil. A finer mesh with a size of 1 mm was applied in the contact area with the test specimen, as shown in Figure 2, while the other areas of the anvils were meshed with a mesh size of 2 mm.

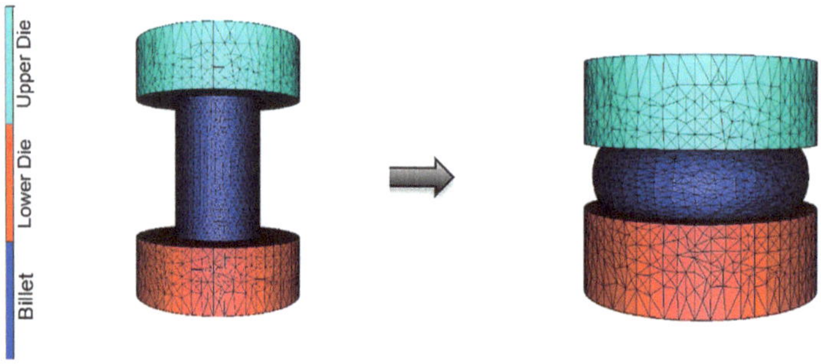

Figure 2. FEM assembly for simulation of the hot compression test.

The behavior of the AM-prepared steel necessary for the FEM simulations was characterized via the Hensel–Spittel rheology law (Equation (1), e.g., [74–76]),

$$\sigma = A \cdot e^{m_1 \cdot T} \cdot \varepsilon^{m_2} \cdot \dot{\varepsilon}^{m_3} \cdot e^{m_4/\varepsilon} \cdot (1+\varepsilon)^{m_5 \cdot T} \cdot e^{m_7 \cdot \varepsilon} \cdot \dot{\varepsilon}^{m_8 \cdot T} \cdot T^{m_9} \text{ (MPa)} \quad (1)$$

where ε (−) is the equivalent strain, $\dot{\varepsilon}$ (s^{-1}) is the equivalent strain rate, T (K) is the temperature, and A, m_1, m_2, m_3, m_4, m_5, m_7, m_8 and m_9 are regression coefficients (−), the values of which are the following: A = 3949.693 MPa, m_1 = −0.00292, m_2 = −0.01398, m_3 = 0.07648, m_4 = −0.00854 and $m_5 - m_9 = 0$.

The Hensel–Spittel rheology law mathematically describes the experimental data obtained using the above characterized set of uniaxial hot compression tests. In other words, it describes the flow stress response of a material formed in a uniaxial stress state, i.e., a material characteristic independent of the specific applied forming technology. During FEM simulations of real forming processes, this material flow stress response can thus be used as a starting point to predict the material behavior during a selected forming process.

As for the properties of the examined material, they are summarized in the material file containing the aforementioned constants of the Hensel–Spittel equation, as well as the coefficients necessary to define the thermal properties of the studied material. In the case of the herein examined AM-prepared steel, the following constant values were considered: density ρ (7350 $kg \cdot m^{-3}$), thermal conductivity λ (33.5 $W \cdot m^{-1} \cdot K^{-1}$) and specific heat capacity c (678 $J \cdot kg^{-1} \cdot K^{-1}$). The considered value of emissivity E was 0.88 (selected based on our previous experience [77]). Also, to simulate the hot compression test and reliably predict the results, it is crucial to properly set the boundary conditions related to the experiment. The hot compression test was characterized by two main conditions. The first one was that the temperature during the test had to be kept constant, with the only allowable temperature change being related to the development of deformation heat. Therefore, the thermal exchange conditions between the anvils and the test specimen were defined as adiabatic, i.e., no heat transfer was considered. The second condition was that during the compression test, the displacement of the anvil was controlled to ensure a constant strain rate. The anvil feed rate during the simulations was therefore defined by Equation (2) [78],

$$v = h_0 \dot{\varepsilon} \exp(-\dot{\varepsilon} t) \quad (2)$$

where v (mm·s^{-1}) is the anvil feed rate, h_0 (mm) is the initial sample height, and t (s) is time.

Regarding the friction conditions on the anvil–billet interface, the nickel-based lubricant applied during the real compression test featured a friction coefficient μ of 0.15. This coefficient was thus set in the simulation, too, along with a friction factor m of 0.3 (based on the combined Coulomb–Tresca friction law). The friction condition was selected based on our previous experience, with the values used corresponding to the nickel-based high-temperature grease, which was applied during the experimental hot compression test.

3. Results and Discussion

3.1. Microstructure

3.1.1. AM-Prepared State

The original microstructure of the AM-prepared workpiece was characterized via the orientation image map (OIM) depicted in Figure 3a and the grain size chart depicted in Figure 3b. The grain size was evaluated using the max. feret diameter parameter in μm [79,80]. The OIM image shows that the original as-printed microstructure featured a characteristic grains morphology (melt pools), and that the grains' orientations were more or less random, i.e., no prevailing texture was formed during the 3D printing (as was expected, see e.g., [81]). The microstructure also exhibited the presence of localized pores. The mean grain size of the AM-prepared workpiece was 32 μm; however, the sizes of the largest grains exceeded 130 μm. Figure 3a also shows that the microstructure was characterized by a dominant fraction of HAGBs, which is typical for additively manufactured steels, see e.g., [82].

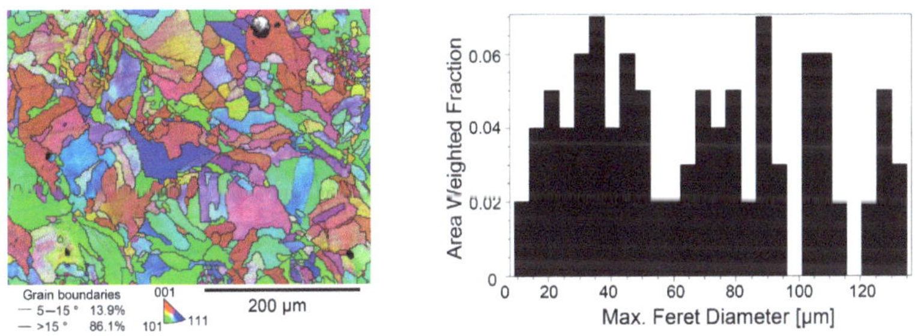

(a) (b)

Figure 3. Results of microstructure analysis of AM-prepared workpiece: OIM (**a**), grain size (**b**).

3.1.2. Deformed States

The results of the microstructure examinations of the samples subjected to hot compression testing are presented via OIM images and grain size charts, as follows: Figure 4a,b depict the OIM and grain size chart for the sample subjected to testing at 1250 °C and 0.1 s^{-1}, Figure 4c,d depict the OIM and grain size chart for the sample subjected to testing at 1250 °C and 100 s^{-1}, Figure 4e,f show the OIM and grain size chart for the sample subjected to testing at 900 °C and 0.1 s^{-1}, and, finally, Figure 4g,h depict the OIM and grain size chart for the sample subjected to testing at 900 °C and 100 s^{-1}. As can be seen from the OIM images, the majority of the grains within all the deformed samples exhibited random orientations. In other words, the deformed microstructures did not exhibit any dominant textures. All the examined samples also generally featured higher fractions of HAGBs than LAGBs. However, higher fractions of HAGBs were observed within the samples deformed at the higher temperature of 1250 °C, especially for the highest strain rate of 100 s^{-1}, as shown in Figure 4c. For the highest strain rate, the fraction of HAGBs within the sample deformed at 1250 °C was almost 20% higher than the HAGB fraction within the one deformed at 900 °C.

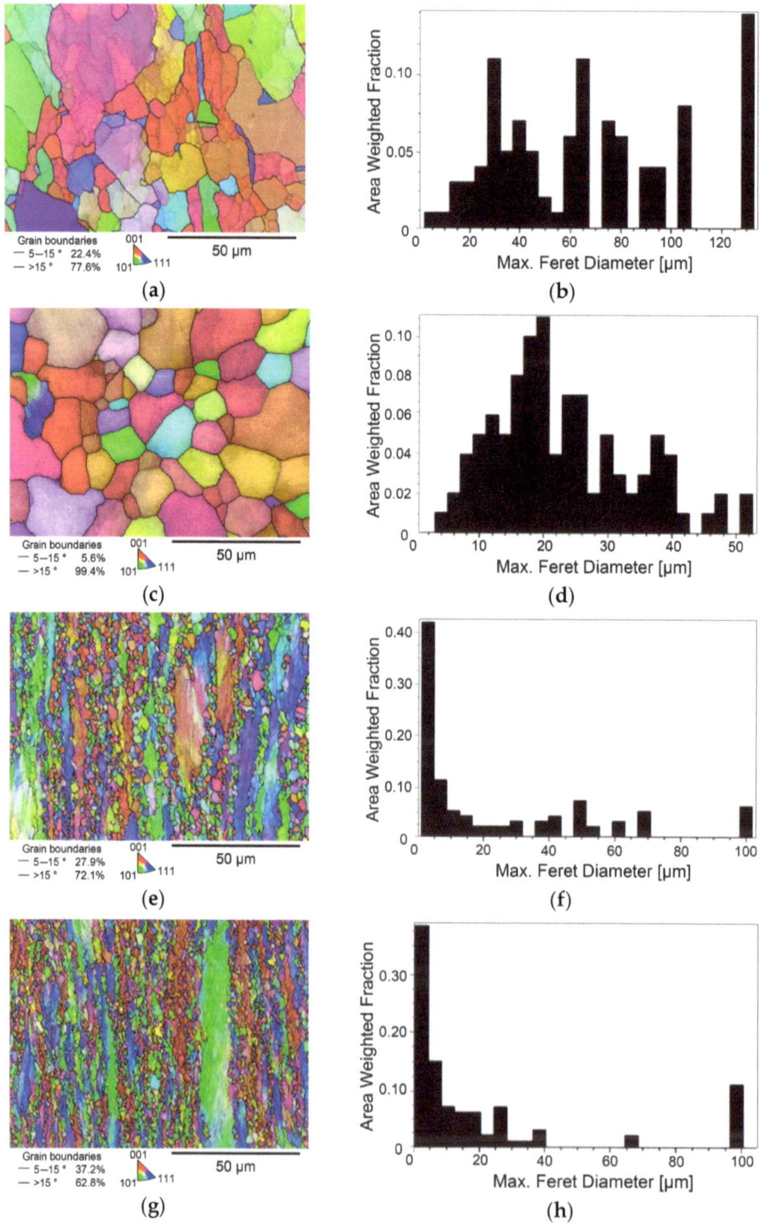

Figure 4. Results of microstructure analyses for samples subjected to hot compression test: 1250 °C and 0.1 s^{-1}, OIM (**a**), grain size chart (**b**); 1250 °C and 100 s^{-1}, OIM (**c**), grain size chart (**d**); 900 °C and 0.1 s^{-1}, OIM (**e**), grain size chart (**f**); 900 °C and 100 s^{-1}, OIM (**g**), grain size chart (**h**).

The results of the grain size analyses (max. feret diameter in μm) showed that the microstructure of the sample subjected to the highest deformation temperature and lowest strain rate of 0.1 s^{-1} featured the largest mean grain size, at 21.5 μm. This finding shows that even the processing conditions allowing for the most substantial development of recrystallization, i.e., a high temperature and low strain rate [83], resulted in grain refinement compared to the original microstructure of the AM-prepared steel. Nevertheless, the sizes of the largest grains within the deformed sample still exceeded 120 μm, as shown in Figure 4b. For the sample deformed at a temperature of 1250 °C and the highest strain rate of 100 s^{-1}, the grain refinement was even more significant; the mean grain size decreased to 10.5 μm. The largest grain size for this sample also decreased, to approximately 50 μm, as shown in Figure 4d. The grain refinement occurred due to the increased strain rate, which did not allow sufficient time for the grains to recrystallize. On the other hand, the (very) high temperature caused grain growth from the start of the test, which negatively affected the final grain refinement.

The lower deformation temperature of 900 °C led to significant grain refinement for all the applied strain rates. The mean grain size for the sample deformed at a temperature of 900 °C and strain rate of 0.1 s^{-1} was 2.6 μm, as shown in Figure 4f. The strain rate of 100 s^{-1} applied at 900 °C then resulted in the most substantial grain refinement, reducing the mean grain size down to the value of 1.3 μm, as shown in Figure 4h. However, deviations from the mean grain size were higher for the samples deformed at 900 °C than for those deformed at 1250 °C. The grains' morphology was evidently different for the samples deformed at 900 °C and 1250 °C, too. The samples deformed at 1250 °C evidently featured great fractions of (recrystallized and) grown grains, whereas the samples deformed at 900 °C featured more developed substructures with frequent occurrences of fine recrystallized grains. While the grains within the samples deformed at 1250 °C were more or less equiaxed, the samples deformed at 900 °C featured evident necklace-like structures consisting of combinations of the original deformed grains and small recrystallized grains. A higher processing temperature of 1250 °C provided the microstructures with increased (accumulated) energy, which not only initiated and supported the development of dynamic recrystallization, but enabled also subsequent grain growth [83,84]. A lower temperature of 900 °C, on the other hand, was sufficient to initiate recrystallization, but the accumulated energy was not sufficient for subsequent grain growth, and thus the microstructures of the samples deformed at 900 °C exhibited combinations of compressed elongated original grains, as well as small refined recrystallized grains. This hypothesis also corresponds to the fact that (i) the mean grain size for the samples deformed at 900 °C was substantially smaller than that for those deformed at 1250 °C, and (ii) the fractions of HAGBs were lower for the samples deformed at 900 °C than for those deformed at 1250 °C.

3.2. Microhardness

The results of the measurement of the Vickers microhardness for the samples subjected to testing at temperatures of 900 °C and 1250 °C and strain rates of 0.1 s^{-1} and 100 s^{-1} are summarized in Figure 5a. Figure 5b then shows a sample with a line drawn across its cross-section, along which the microhardness measurements were carried out. The average Vickers microhardness for the AM-prepared workpiece was 218.1 HV1. The HV1 values did not exhibit any decreasing/increasing trend across the workpiece cross-section, only local decreases which could be attributed to the localized porosity, as seen in Figure 3a. Figure 5a then shows that both the samples deformed at 1250 °C exhibited comparable microhardness values, with no visible increasing/decreasing trend across the cross sections; the average microhardness value for both the samples was 171.4 HV1. For the sample deformed at a deformation temperature of 900 °C and strain rate of 0.1 s^{-1}, the average microhardness increased and reached the value of 243.5 HV1, and the highest microhardness of 269.7 HV1 was acquired for the sample deformed at a temperature of 900 °C and strain rate of 100 s^{-1}. The microhardness ranged from 140 HV1 to 190 HV1 at a deformation temperature of

1250 °C, and from 230 HV1 to 280 HV1 at a deformation temperature of 900 °C. Forming at lower temperatures thus obviously resulted in higher microhardness values. It is also evident that the strain rate significantly affected the microhardness at a deformation temperature of 900 °C, while its effect was negligible at a temperature of 1250 °C. The acquired results thus confirm the results of the microstructure observations discussed above. Both the samples deformed at a temperature of 900 °C exhibited increased microhardness, which corresponds to the observed substantial grain refinement and presence of ultra-fine grains within the microstructure [85–87]. Processing at 1250 °C substantially supported softening via dynamic recrystallization, the formation of relaxed grains and consequent grain growth, which contributed to decreased microhardness. Despite the fact that the grains were refined after processing at 1250 °C, the grain size was not small enough and the dislocation density was not high enough to increase the microhardness. The highest value of microhardness was acquired for the sample deformed at 900 °C and the highest strain rate of 100 s^{-1}, which corresponds to the hypothesis that combinations of lower deformation temperatures and higher strain rates tend to lead to substantial substructure development and work hardening [84].

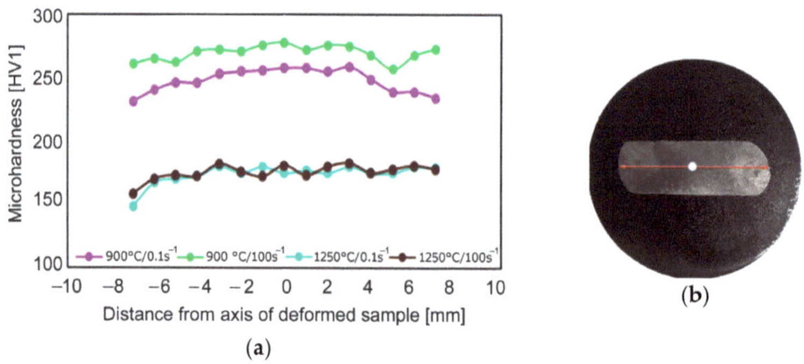

Figure 5. Vickers microhardness of examined samples (**a**); sample with depiction of location of microhardness measurement (red line across diameter) (**b**).

3.3. Flow Stress

Figure 6a–d depict the flow stress curves characterizing the development of material flow stress in the AM-prepared steel through the entire range of applied temperatures at strain rates of 0.1 s^{-1}, 1 s^{-1}, 10 s^{-1}, and 100 s^{-1}, respectively. The figures show that the achieved maximum flow stress generally increased with decreasing temperature and increasing strain rate, which is rather typical, see e.g., [88–90]. A higher strain rate in combination with a higher deformation temperature generally favorably influenced the formability. This finding was supported by the fact that the stress–strain curves acquired during testing at the highest strain rate, as shown in Figure 6d, did not exhibit any signs of dynamic recrystallization occurring. The initiation of dynamic recrystallization could be most notably seen at the curves acquired for the lowest deformation temperature. Figure 6a–c clearly show that the critical (peak) imposed strain necessary to initiate dynamic recrystallization decreased with an increase in the strain rate applied. In other words, a lower applied strain rate supported the development of dynamic recrystallization, when considering identical deformation temperatures. The maximum acquired flow stress, achieved for the sample deformed at 900 °C and 100 s^{-1}, was approximately 380 MPa. The acquired results can be compared, for example, with those acquired by Liu et al. [71] and Khodabakhshi et al. [72], similarly investigating AM-prepared AISI 316L steel samples. The researchers observed similar trends regarding the flow stress behavior, and reported that dynamic recrystallization within the AM-prepared AISI 316L steel occurred to higher extents at high processing temperatures, while lower processing temperatures decreased formability and led to the increased occurrence of cracks. Evidently, high strain rates combined with high deformation temperatures minimize surface cracking during processing due to the occurrence of dynamic recrystallization to a sufficient extent.

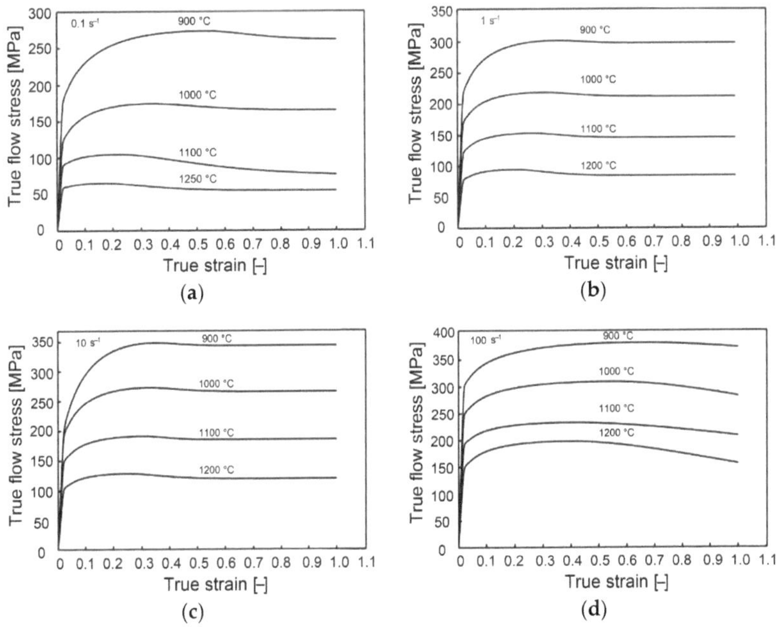

Figure 6. Experimental flow stress curves for four strain rates combined with each examined temperature: 0.1 s^{-1} (**a**); 1 s^{-1} (**b**); 10 s^{-1} (**c**); 100 s^{-1} (**d**).

Considering that the original AM-prepared AISI 316L steel is subjected to post-processing via hot rotary swaging, which is characterized by (very) high strain rates, optimized combinations of processing conditions based on this research will most probably offer a satisfactory solution to the reduction/elimination of the aforementioned disadvantages of AM technologies and result in the fabrication of AISI 316L steel components with enhanced performance.

3.4. Numerical Prediction

3.4.1. Deformation Forces

Figure 7 shows a comparison of the experimentally acquired and predicted dependencies of deformation force on the testing time for the examined samples (dashed curves depict the experimental results, while solid curves depict the predicted ones). Considering the maximum force, the predicted curves obviously exhibited trends comparable to the experimentally acquired ones, i.e., the maximum force increased as the testing temperature decreased. The highest force, both experimental and numerically predicted, was acquired when testing at 900 °C; the maximum experimentally measured force at 900 °C was approximately 65 kN. Nevertheless, the experimentally acquired curves generally featured higher force values compared to the predicted ones, and the difference between these two curves decreased towards an increasing processing temperature.

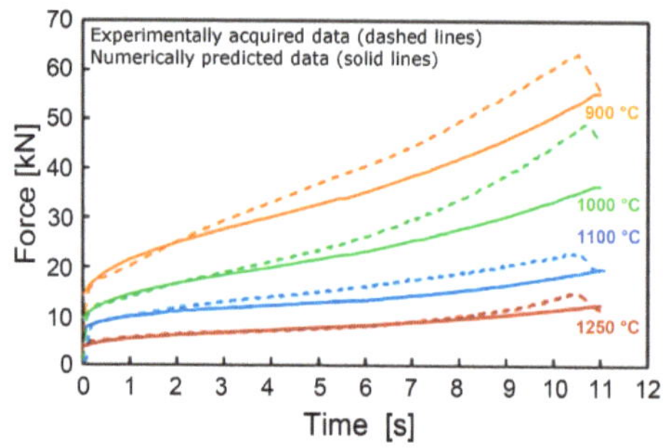

Figure 7. Comparison of experimental and numerically predicted forces for examined samples at strain rate of 0.1 s^{-1}.

The differences observed between the experimentally acquired and numerically predicted force values can most probably be attributed to the defined boundary conditions (e.g., differences in the friction conditions on the contact surfaces between the anvils and sample, etc.). The numerical simulation considered the constant condition of perfect lubrication during the entire test. In the experimental conditions, on the other hand, the lubricant was gradually squeezed out and, in addition, partially lost by evaporation due to high testing temperatures, which could result in higher friction forces, i.e., higher overall deformation forces measured in the experimental conditions. Another important factor influencing the deformation force is the homogeneity of temperature distribution, which evidently played a non-negligible role during testing at all the deformation temperatures. As can be seen from Figure 7 (and stated above), the difference between the experimentally acquired and predicted curves, especially towards the end of the test, decreased with increasing processing temperature. Regarding the influence of strain rate on the mechanical properties, there are two aspects. From the viewpoint of the mutual fit between the experimentally obtained and predicted curves for compression force under an identical temperature, there are only negligible differences with increasing strain. On the other hand, the curves gained an increasing character with an increasing strain rate. This made the individual curves steeper, i.e., there was a faster increase in the deformation force and overall higher values at all the testing temperatures. The reasons for the similar behavior is the shorter time needed for the activation and development of dynamic softening processes and the dynamic work hardening being faster, based on the very intensive accumulation of dislocations.

3.4.2. Equivalent Strain

Last but not least, the results of the predicted distributions of equivalent strain for the testing temperatures ranging from 900 °C to 1250 °C and strain rate of 0.1 s^{-1} are shown in Figure 8. For clarity, three distinct locations, characterized as location 1 (top surface of a sample), location 2 (axial region of a sample), and location 3 (lateral edge surface of a sample), are used for the samples. As can be seen in Figure 8, the distribution and magnitudes of the imposed strain were comparable for all the testing temperatures. The highest imposed strain within the bulk of all the samples (approximately 1.8) was observed in location 2, while location 1 generally exhibited the lowest strain values ($\leq$0.4). This corresponds to the effect of friction forces hindering the material plastic flow in the vicinity of the anvil-sample contact surfaces. In location 3, the imposed strain reached a value of approximately 1 for all the samples. The figure also shows that the temperature did not have any major effect on the distribution and magnitude of the equivalent strain. However, the maximum strain at higher temperatures was slightly lower, which was most probably related to the different nature of the material plastic flow (as discussed in the previous sections). The lower strain value can most probably be directly related to the previously mentioned occurring dynamic softening of the AM-prepared AISI 316L.

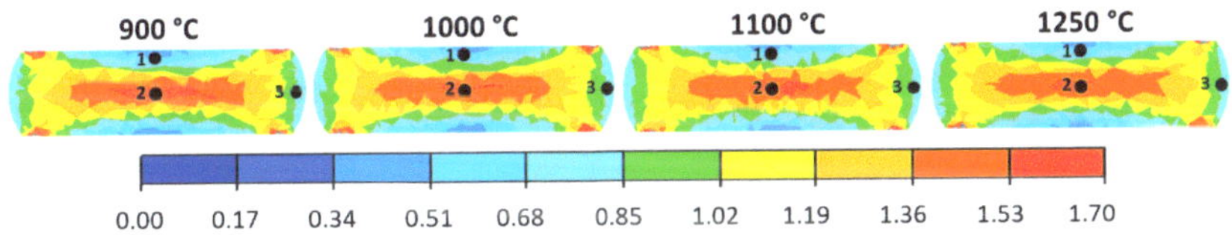

Figure 8. Equivalent strain at four temperatures at a strain rate of 0.1 s^{-1}, numbers 1, 2, and 3 correspond to location 1, location 2, and location 3.

4. Conclusions

This study focused on the experimental and numerical characterization of the deformation behavior of AM-prepared AISI 316L stainless steel in order to find the most suitable conditions for prospective thermomechanical post-processing. The most prominent results are the following:

- Higher deformation temperatures, especially in combination with lower strain rates, supported recrystallization and grain growth, while lower deformation temperatures supported grain refinement and the development of a substructure, resulting in work hardening;
- The deformation force and flow stress increased with a decreasing processing temperature, with maximum values of approximately 65 kN and 380 MPa, respectively, being acquired at 900 °C;
- Deformation at 900 °C and 100 s^{-1} resulted in the smallest mean grain size of 1.3 μm and highest average microhardness of 270 HV1;
- Despite a decrease in microhardness due to recrystallization and subsequent grain growth, deformation at 1250 °C resulted in grain refinement to a mean size of 21.5 μm.

The primary novelty of the presented research lies in the fact that the most favorable conditions for the prospective deformation (thermomechanical) post-processing of AM-prepared 316L stainless steel were given, and that it considered various viewpoints. Such post-processing will be the most favorable when performed at a temperature of 900 °C, from the viewpoint of mechanical properties, and higher strain rates, from the viewpoint of formability. The proposed combinations of higher temperatures and strain rates seem to be optimal for forming processes featuring high strain rates, such as rotary swaging. Nevertheless, the performed study confirmed that processing at the relatively low forming temperature of 900 °C is still feasible, and can provide favorable results. Future research, during which the acquired results will be applied, will be focused on the experimental post-processing of AM-prepared stainless steel workpieces using industrial processing methods and the verification of the acquired results under real conditions.

Author Contributions: Conceptualization, data curation, investigation, writing—original draft preparation: L.K.; investigation, project administration, funding acquisition, writing—review and editing: R.K.; resources, methodology: M.P. All authors have read and agreed to the published version of the manuscript.

Funding: Statement: This research was supported by SP2024/089 project by VŠB Technical Univerzity of Ostrava.

Data Availability Statement: The original contributions presented in this study are included in the article. Further inquiries can be directed to the corresponding author.

Acknowledgments: The assistance of Petr Opěla, Marek Benč, and Josef Walek (all Department of Metallurgical Technologies, Faculty of Materials Science and Technology, VŠB-Technical University of Ostrava) during the experimental works is appreciated.

Conflicts of Interest: The authors declare no conflicts of interest.

References

1. Sun, J.; Ye, D.; Zou, J.; Chen, X.; Wang, Y.; Yuan, J.; Liang, H.; Qu, H.; Binner, J.; Bai, J. A Review on Additive Manufacturing of Ceramic Matrix Composites. *J. Mater. Sci. Technol.* **2023**, *138*, 1–16. [CrossRef]
2. Vaezi, M.; Seitz, H.; Yang, S. A Review on 3D Micro-Additive Manufacturing Technologies. *Int. J. Adv. Manuf. Technol.* **2012**, *67*, 1721–1754. [CrossRef]
3. Das, T.; Mukherjee, M.; Chatterjee, D.; Samanta, S.K.; Lohar, A.K. A Comparative Evaluation of the Microstructural Characteristics of L-DED and W-DED Processed 316L Stainless Steel. *CIRP J. Manuf. Sci. Technol.* **2023**, *40*, 114–128. [CrossRef]
4. Gorunov, A.I. Additive Manufacturing of Ti6Al4V Parts Using Ultrasonic Assisted Direct Energy Deposition. *J. Manuf. Process.* **2020**, *59*, 545–556. [CrossRef]
5. Kocich, R.; Kunčická, L.; Benč, M.; Weiser, A.; Németh, G. Corrosion Behavior of Selective Laser Melting-Manufactured Bio-Applicable 316L Stainless Steel in Ionized Simulated Body Fluid. *Int. J. Bioprint.* **2024**, *10*, 1416. [CrossRef]
6. Puga, B.; Lomello, F.; Boussac, E.; Chniouel, A.; Fouchereau, A.; Laghoutaris, P.; Maskrot, H. Influence of Laser Powder Bed Fusion Processing Parameters on Corrosion Behaviour of 316L Stainless Steel in Nitric Acid. *Metall. Res. Technol.* **2022**, *119*, 523. [CrossRef]
7. Kazantseva, N.; Krakhmalev, P.; Yadroitsev, I.; Fefelov, A.; Merkushev, A.; Ilyinikh, M.; Vinogradova, N.; Ezhov, I.; Kurennykh, T. Oxygen and Nitrogen Concentrations in the Ti-6Al-4V Alloy Manufactured by Direct Metal Laser Sintering (DMLS) Process. *Mater. Lett.* **2017**, *209*, 311–314. [CrossRef]
8. Gong, H.; Rafi, K.; Gu, H.; Janaki Ram, G.D.; Starr, T.; Stucker, B. Influence of Defects on Mechanical Properties of Ti–6Al–4V Components Produced by Selective Laser Melting and Electron Beam Melting. *Mater. Des.* **2015**, *86*, 545–554. [CrossRef]
9. Rajaguru, K.; Karthikeyan, T.; Vijayan, V. Additive Manufacturing–State of Art. *Mater. Today Proc.* **2020**, *21*, 628–633. [CrossRef]
10. Savrai, R.A.; Toporova, D.V.; Bykova, T.M. Improving the Quality of AISI H13 Tool Steel Produced by Selective Laser Melting. *Opt. Laser Technol.* **2022**, *152*, 108128. [CrossRef]
11. Bakradze, M.M.; Rogalev, A.M.; Sukhov, D.I.; Aslanyan, G.G. Special Features of Formation of Surface by Selective Laser Melting. *Met. Sci. Heat Treat.* **2022**, *64*, 108–116. [CrossRef]
12. Aboulkhair, N.T.; Simonelli, M.; Parry, L.; Ashcroft, I.; Tuck, C.; Hague, R. 3D Printing of Aluminium Alloys: Additive Manufacturing of Aluminium Alloys Using Selective Laser Melting. *Prog. Mater. Sci.* **2019**, *106*, 100578. [CrossRef]
13. Tarasova, T.; Gvozdeva, G.; Ableyeva, R. Aluminium Matrix Composites Produced by Laser Based Additive Manufacturing. *Mater. Today Proc.* **2019**, *11*, 305–310. [CrossRef]
14. Karlsson, J.; Sjogren, T.; Snis, A.; Engqvist, H.; Lausmaa, J. Digital Image Correlation Analysis of Local Strain Fields on Ti6Al4V Manufactured by Electron Beam Melting. *Mater. Sci. Eng. A* **2014**, *618*, 456–461. [CrossRef]
15. Nocivin, A.; Raducanu, D.; Vasile, B.; Trisca-Rusu, C.; Cojocaru, E.M.; Dan, A.; Irimescu, R.; Cojocaru, V.D. Tailoring a Low Young Modulus for a Beta Titanium Alloy by Combining Severe Plastic Deformation with Solution Treatment. *Materials* **2021**, *14*, 3467. [CrossRef] [PubMed]
16. Cherry, J.A.; Davies, H.M.; Mehmood, S.; Lavery, N.P.; Brown, S.G.R.; Sienz, J. Investigation into the Effect of Process Parameters on Microstructural and Physical Properties of 316L Stainless Steel Parts by Selective Laser Melting. *Int. J. Adv. Manuf. Technol.* **2015**, *76*, 869–879. [CrossRef]
17. Shchitsyn, Y.D.; Terentev, S.A.; Neulybin, S.D.; Artyomov, A.O. Formation of 04Cr18Ni9 Steel Structure and Properties during Additive Manufacturing of Blanks. *Int. J. Adv. Manuf. Technol.* **2019**, *102*, 3719–3723. [CrossRef]
18. Moskvina, V.A.; Melnikov, E.V.; Astafurov, S.V.; Panchenko, M.Y.; Reunova, K.A.; Kolubaev, E.A.; Astafurova, E.G. Stable High-Nickel Austenitic Steel Produced by Electron Beam Additive Manufacturing Using Dual Wire-Feed System. *Mater. Lett.* **2021**, *305*, 130863. [CrossRef]
19. Astafurov, S.; Astafurova, E. Phase Composition of Austenitic Stainless Steels in Additive Manufacturing: A Review. *Metals* **2021**, *11*, 1052. [CrossRef]
20. Kuzminova, Y.; Shibalova, A.; Evlashin, S.; Shishkovsky, I.; Krakhmalev, P. Structural Effect of Low Al Content in the In-Situ Additive Manufactured CrFeCoNiAlx High-Entropy Alloy. *Mater. Lett.* **2021**, *303*, 130487. [CrossRef]
21. Kuzminova, Y.O.; Kudryavtsev, E.A.; Han, J.-K.; Kawasaki, M.; Evlashin, S.A. Phase and Structural Changes during Heat Treatment of Additive Manufactured CrFeCoNi High-Entropy Alloy. *J. Alloys Compd.* **2021**, *889*, 161495. [CrossRef]
22. Krinitcyn, M.; Kopytov, G.; Ryumin, E. Additive Manufacturing of Ti_3AlC_2/TiC and Ti_3AlC_2/SiC Ceramics Using the Fused Granules Fabrication Technique. *J. Manuf. Mater. Process.* **2024**, *8*, 123. [CrossRef]

23. Tamburrino, F.; Barone, S.; Paoli, A.; Razionale, A.V. Post-Processing Treatments to Enhance Additively Manufactured Polymeric Parts: A Review. *Virtual Phys. Prototyp.* **2021**, *16*, 221–254. [CrossRef]
24. Nguyen, H.D.; Pramanik, A.; Basak, A.K.; Dong, Y.; Prakash, C.; Debnath, S.; Shankar, S.; Jawahir, I.S.; Dixit, S.; Buddhi, D. A Critical Review on Additive Manufacturing of Ti-6Al-4V Alloy: Microstructure and Mechanical Properties. *J. Mater. Res. Technol.* **2022**, *18*, 4641–4661. [CrossRef]
25. Utyaganova, V.R.; Filippov, A.V.; Shamarin, N.N.; Vorontsov, A.V.; Savchenko, N.L.; Fortuna, S.V.; Gurianov, D.A.; Chumaevskii, A.V.; Rubtsov, V.E.; Tarasov, S.Y. Controlling the Porosity Using Exponential Decay Heat Input Regimes during Electron Beam Wire-Feed Additive Manufacturing of Al-Mg Alloy. *Int. J. Adv. Manuf. Technol.* **2020**, *108*, 2823–2838. [CrossRef]
26. Levin, V.A.; Zingerman, K.M.; Vershinin, A.V.; Konovalov, D.A. Numerical Modeling of Residual Stresses in Additive Manufacturing Products Using the Theory of Repeatedly Superimposed Finite Strains. *Math. Mech. Solids* **2022**, *27*, 2099–2109. [CrossRef]
27. Babaev, A.; Promakhov, V.; Schulz, N.; Semenov, A.; Bakhmat, V.; Vorozhtsov, A. Processes of Physical Treatment of Stainless Steels Obtained by Additive Manufacturing. *Metals* **2022**, *12*, 1449. [CrossRef]
28. Aboulkhair, N.T.; Maskery, I.; Tuck, C.; Ashcroft, I.; Everitt, N.M. Improving the Fatigue Behaviour of a Selectively Laser Melted Aluminium Alloy: Influence of Heat Treatment and Surface Quality. *Mater. Des.* **2016**, *104*, 174–182. [CrossRef]
29. Popovich, V.A.; Borisov, E.V.; Popovich, A.A.; Sufiiarov, V.S.; Masaylo, D.V.; Alzina, L. Impact of Heat Treatment on Mechanical Behaviour of Inconel 718 Processed with Tailored Microstructure by Selective Laser Melting. *Mater. Des.* **2017**, *131*, 12–22. [CrossRef]
30. Teixeira, Ó.; Silva, F.J.G.; Ferreira, L.P.; Atzeni, E. A Review of Heat Treatments on Improving the Quality and Residual Stresses of the Ti–6Al–4V Parts Produced by Additive Manufacturing. *Metals* **2020**, *10*, 1006. [CrossRef]
31. Peng, X.; Kong, L.; Fuh, J.Y.H.; Wang, H. A Review of Post-Processing Technologies in Additive Manufacturing. *J. Manuf. Mater. Process.* **2021**, *5*, 38. [CrossRef]
32. Mclean, N.; Bermingham, M.J.; Colegrove, P.; Sales, A.; Soro, N.; Ng, C.H.; Dargusch, M.S. Effect of Hot Isostatic Pressing and Heat Treatments on Porosity of Wire Arc Additive Manufactured Al 2319. *J. Mater. Process. Technol.* **2022**, *310*, 117769. [CrossRef]
33. Petrovskiy, P.; Travyanov, A.; Cheverikin, V.V.; Chereshneva, A.A.; Sova, A.; Smurov, I. Effect of Encapsulated Hot Isostatic Pressing on Properties of Ti6Al4V Deposits Produced by Cold Spray. *Int. J. Adv. Manuf. Technol.* **2020**, *107*, 437–449. [CrossRef]
34. Krakhmalev, P.; Fredriksson, G.; Thuvander, M.; Åsberg, M.; Vilardell, A.M.; Oikonomou, C.; Maistro, G.; Medvedeva, A.; Kazantseva, N. Influence of Heat Treatment under Hot Isostatic Pressing (HIP) on Microstructure of Intermetallic-Reinforced Tool Steel Manufactured by Laser Powder Bed Fusion. *Mater. Sci. Eng. A* **2020**, *772*, 138699. [CrossRef]
35. Wang, Z.; Chen, J.; Kocich, R.; Tardif, S.; Dolbnya, I.P.; Kunčická, L.; Micha, J.-S.; Liogas, K.; Magdysyuk, O.V.; Szurman, I.; et al. Grain Structure Engineering of NiTi Shape Memory Alloys by Intensive Plastic Deformation. *ACS Appl. Mater. Interfaces* **2022**, *14*, 31396–31410. [CrossRef]
36. Strunz, P.; Kunčická, L.; Beran, P.; Kocich, R.; Hervoches, C. Correlating Microstrain and Activated Slip Systems with Mechanical Properties within Rotary Swaged WNiCo Pseudoalloy. *Materials* **2020**, *13*, 208. [CrossRef]
37. Kunčická, L.; Kocich, R. Effect of Activated Slip Systems on Dynamic Recrystallization during Rotary Swaging of Electro-Conductive Al-Cu Composites. *Mater. Lett.* **2022**, *321*, 10–13. [CrossRef]
38. Shi, X.; Hussain, G.; Zha, G.; Wu, M.; Kong, F. Study on Formability of Vertical Parts Formed by Multi-Stage Incremental Forming. *Int. J. Adv. Manuf. Technol.* **2014**, *75*, 1049–1053. [CrossRef]
39. Vihtonen, L.; Puzik, A.; Katajarinne, T. Comparing Two Robot Assisted Incremental Forming Methods: Incremental Forming by Pressing and Incremental Hammering. *Int. J. Mater. Form.* **2008**, *1*, 1207–1210. [CrossRef]
40. Kunčická, L.; Kocich, R.; Németh, G.; Dvořák, K.; Pagáč, M. Effect of Post Process Shear Straining on Structure and Mechanical Properties of 316 L Stainless Steel Manufactured via Powder Bed Fusion. *Addit. Manuf.* **2022**, *59*, 103128. [CrossRef]
41. Canelo-Yubero, D.; Kocich, R.; Hervoches, C.; Strunz, P.; Kunčická, L.; Krátká, L. Neutron Diffraction Study of Residual Stresses in a W–Ni–Co Heavy Alloy Processed by Rotary Swaging at Room and High Temperatures. *Met. Mater. Int.* **2022**, *28*, 919–930. [CrossRef]
42. Lukáč, P.; Kocich, R.; Greger, M.; Padalka, O.; Szaraz, Z. Microstructure of AZ31 and AZ61 Mg Alloys Prepared by Rolling and ECAP. *Kov. Mater. Mater.* **2007**, *45*, 115–120.
43. Volokitina, I.E. Evolution of the Microstructure and Mechanical Properties of Copper under ECAP with Intense Cooling. *Met. Sci. Heat Treat.* **2020**, *62*, 253–258. [CrossRef]
44. Segal, V. Review: Modes and Processes of Severe Plastic Deformation (SPD). *Materials* **2018**, *11*, 1175. [CrossRef]
45. Kunčická, L.; Kocich, R.; Drápala, J.; Andreyachshenko, V.A. FEM Simulations and Comparison of the Ecap and ECAP-PBP Influence on Ti6Al4V Alloy's Deformation Behaviour. In Proceedings of the METAL 2013-22nd International Conference on Metallurgy and Materials, Conference Proceedings, Brno, Czech Republic, 15–17 May 2013; pp. 391–396.
46. Xia, K.; Wu, X. Back Pressure Equal Channel Angular Consolidation of Pure Al Particles. *Scr. Mater.* **2005**, *53*, 1225–1229. [CrossRef]

47. Kunčická, L.; Kocich, R.; Ryukhtin, V.; Cullen, J.C.T.; Lavery, N.P. Study of Structure of Naturally Aged Aluminium after Twist Channel Angular Pressing. *Mater. Charact.* **2019**, *152*, 94–100. [CrossRef]
48. Kocich, R.; Kunčická, L. Development of Structure and Properties in Bimetallic Al/Cu Sandwich Composite during Cumulative Severe Plastic Deformation. *J. Sandw. Struct. Mater.* **2021**, *23*, 4252–4275. [CrossRef]
49. Kocich, R.; Kunčická, L. Optimizing Structure and Properties of Al/Cu Laminated Conductors via Severe Shear Strain. *J. Alloys Compd.* **2023**, *953*, 170124. [CrossRef]
50. Fu, Z.; Gao, B.; Li, X.; Li, C.; Pan, H.; Niu, H.; Zhu, Y.; Zhou, H.; Zhu, X.; Wu, H.; et al. Improved Strength-Ductility Combination of Pure Zr by Multi-Scale Heterostructured Effects via Rotary Swaging and Annealing. *Mater. Sci. Eng. A* **2023**, *864*, 144584. [CrossRef]
51. Rybalchenko, O.; Torganchuk, V.; Rybalchenko, G.; Martynenko, N.; Lukyanova, E.; Tokar, A.; Prosvirnin, D.; Yusupov, V.; Dobatkin, S. Effect of Rotary Swaging on Microstructure and Properties of Cr-Ni-Ti Austenitic Stainless Steel. *Metals* **2023**, *13*, 1760. [CrossRef]
52. Xie, X.; Jiang, W.; Chen, J.; Zhang, X.; Tu, S.-T. Cyclic Hardening/Softening Behavior of 316L Stainless Steel at Elevated Temperature Including Strain-Rate and Strain-Range Dependence: Experimental and Damage-Coupled Constitutive Modeling. *Int. J. Plast.* **2019**, *114*, 196–214. [CrossRef]
53. Lee, S.; Ghiaasiaan, R.; Gradl, P.R.; Shao, S.; Shamsaei, N. Additively Manufactured 316L Stainless Steel via Laser Powder Directed Energy Deposition (LP-DED): Mechanical Properties at Cryogenic and Elevated Temperatures. *Int. J. Fatigue* **2024**, *182*, 108197. [CrossRef]
54. Behvar, A.; Aghayar, Y.; Avateffazeli, M.; Tridello, A.; Benelli, A.; Paolino, D.S.; Mohammadi, M.; Haghshenas, M. Synergistic Impact of Corrosion Pitting on the Rotating Bending Fatigue of Additively Manufactured 316L Stainless Steel: Integrated Experimental and Modeling Analyses. *Int. J. Fatigue* **2024**, *188*, 108491. [CrossRef]
55. Abramov, A.V.; Polovov, I.B.; Volkovich, V.A.; Rebrin, O.I.; Denisov, E.I.; Griffiths, T.R. Corrosion of Austenitic Steels and Their Components in Vanadium-Containing Chloride Melts. *ECS Trans.* **2013**, *50*, 685–698. [CrossRef]
56. Zach, L.; Kunčická, L.; Růžička, P.; Kocich, R. Design, Analysis and Verification of a Knee Joint Oncological Prosthesis Finite Element Model. *Comput. Biol. Med.* **2014**, *54*, 53–60. [CrossRef]
57. Kunčická, L.; Kocich, R.; Lowe, T.C. Advances in Metals and Alloys for Joint Replacement. *Prog. Mater. Sci.* **2017**, *88*, 232–280. [CrossRef]
58. Macháčková, A.; Kocich, R.; Bojko, M.; Kunčická, L.; Polko, K. Numerical and Experimental Investigation of Flue Gases Heat Recovery via Condensing Heat Exchanger. *Int. J. Heat Mass Transf.* **2018**, *124*, 1321–1333. [CrossRef]
59. Cao, X.; Xiao, D.; Li, Y.; Wen, W.; Zhao, T.; Chen, Z.; Jiang, Y.; Fang, D. Dynamic Compressive Behavior of a Modified Additively Manufactured Rhombic Dodecahedron 316L Stainless Steel Lattice Structure. *Thin-Walled Struct.* **2020**, *148*, 106586. [CrossRef]
60. Arteaga-Hernandez, L.A.; Cuao-Moreu, C.A.; Gonzalez-Rivera, C.E.; Alvarez-Vera, M.; Ortega-Saenz, J.A.; Hernandez-Rodriguez, M.A.L. Study of Boriding Surface Treatment in the Tribological Behavior of an AISI 316L Stainless Steel. *Wear* **2021**, *477*, 203825. [CrossRef]
61. D'Andrea, D. Additive Manufacturing of AISI 316L Stainless Steel: A Review. *Metals* **2023**, *13*, 1370. [CrossRef]
62. Shakil, S.I.I.; Smith, N.R.R.; Yoder, S.P.P.; Ross, B.E.E.; Alvarado, D.J.J.; Hadadzadeh, A.; Haghshenas, M. Post Fabrication Thermomechanical Processing of Additive Manufactured Metals: A Review. *J. Manuf. Process.* **2022**, *73*, 757–790. [CrossRef]
63. Bankong, B.D.; Abioye, T.E.; Olugbade, T.O.; Zuhailawati, H.; Gbadeyan, O.O.; Ogedengbe, T.I. Review of Post-Processing Methods for High-Quality Wire Arc Additive Manufacturing. *Mater. Sci. Technol.* **2023**, *39*, 129–146. [CrossRef]
64. Maamoun, A.H.; Veldhuis, S.C.; Elbestawi, M. Friction Stir Processing of AlSi10Mg Parts Produced by Selective Laser Melting. *J. Mater. Process. Technol.* **2019**, *263*, 308–320. [CrossRef]
65. Rubino, F.; Scherillo, F.; Franchitti, S.; Squillace, A.; Astarita, A.; Carlone, P. Microstructure and Surface Analysis of Friction Stir Processed Ti-6Al-4V Plates Manufactured by Electron Beam Melting. *J. Manuf. Process.* **2019**, *37*, 392–401. [CrossRef]
66. Zhao, L.; Santos Macías, J.G.; Dolimont, A.; Simar, A.; Rivière-Lorphèvre, E. Comparison of Residual Stresses Obtained by the Crack Compliance Method for Parts Produced by Different Metal Additive Manufacturing Techniques and after Friction Stir Processing. *Addit. Manuf.* **2020**, *36*, 101499. [CrossRef]
67. Hönnige, J.R.; Colegrove, P.A.; Ganguly, S.; Eimer, E.; Kabra, S.; Williams, S. Control of Residual Stress and Distortion in Aluminium Wire + Arc Additive Manufacture with Rolling. *Addit. Manuf.* **2018**, *22*, 775–783. [CrossRef]
68. Xu, X.; Ganguly, S.; Ding, J.; Seow, C.E.; Williams, S. Enhancing Mechanical Properties of Wire + Arc Additively Manufactured INCONEL 718 Superalloy through In-Process Thermomechanical Processing. *Mater. Des.* **2018**, *160*, 1042–1051. [CrossRef]
69. Shiyas, K.A.; Ramanujam, R. A Review on Post Processing Techniques of Additively Manufactured Metal Parts for Improving the Material Properties. *Mater. Today Proc.* **2021**, *46*, 1429–1436. [CrossRef]
70. Kunčická, L.; Kocich, R. High Strain Rate Induced Shear Banding within Additively Manufactured AISI 316L. *Mater. Lett.* **2024**, *363*, 136342. [CrossRef]

71. Liu, Y.; Zhang, C.; Wang, Y.; Xu, X.; Zhu, H.; Jiang, J. Reveal the Hot Deformation Behaviour and Microstructure Evolution in Additively Manufactured 316L Stainless Steel. *Mater. Sci. Eng. A* **2022**, *861*, 144290. [CrossRef]
72. Khodabakhshi, F.; Hasani, N.; Kalaie, M.R.; Hadadzadeh, A.; Wells, M.A.; Mohammadi, M. Dynamic Recrystallization under Hot Deformation of Additively Manufactured 316 L Stainless Steel. *Mater. Charact.* **2023**, *202*, 113055. [CrossRef]
73. Jin, M.; Piglione, A.; Dovgyy, B.; Hosseini, E.; Hooper, P.A.; Holdsworth, S.R.; Pham, M.-S. Cyclic Plasticity and Fatigue Damage of CrMnFeCoNi High Entropy Alloy Fabricated by Laser Powder-Bed Fusion. *Addit. Manuf.* **2020**, *36*, 101584. [CrossRef]
74. Hammami, S.; La Barbera-Sosa, J.G.; Chaari, F.; Sadat, T.; Zouari, B.; Dubar, L.; Elleuch, R. CuZn40Pb2 Brass Hot Deformation Behaviour Modelling Using Hansel Spittel Constitutive Model. *Adv. Mater. Process. Technol.* **2024**, 1–18. [CrossRef]
75. Niu, L.; Zhang, Q.; Wang, B.; Han, B.; Li, H.; Mei, T. A Modified Hansel-Spittel Constitutive Equation of Ti-6Al-4V during Cogging Process. *J. Alloys Compd.* **2022**, *894*, 162387. [CrossRef]
76. Wang, J.C.; Langlois, L.; Rafiq, M.; Bigot, R.; Lu, H. Experimental & Numerical Study of the Hot Upsetting of Weld Cladded Billets. *Key Eng. Mater.* **2013**, *554–557*, 287–299. [CrossRef]
77. Kawulok, P.; Opěla, P.; Schindler, I.; Kawulok, R.; Rusz, S. *Studium Pevnostních a Plastických Vlastností Kovových Materiálů Pomocí Simulátoru HDS-20*; Vysoká škola Báňská–Technická Univerzita Ostrava: Ostrava, Czech Republic, 2019.
78. Lv, J.; Ren, H.; Gao, K. Artificial Neural Network-Based Constitutive Relationship of Inconel 718 Superalloy Construction and Its Application in Accuracy Improvement of Numerical Simulation. *Appl. Sci.* **2017**, *7*, 124. [CrossRef]
79. Isobe, A.; Akaji, M.; Kurokawa, S. Proposal of New Polishing Mechanism Based on Feret's Diameter of Contact Area between Polishing Pad and Wafer. *Jpn. J. Appl. Phys.* **2013**, *52*, 126503. [CrossRef]
80. Dražić, S.; Sladoje, N.; Lindblad, J. Estimation of Feret's Diameter from Pixel Coverage Representation of a Shape. *Pattern Recognit. Lett.* **2016**, *80*, 37–45. [CrossRef]
81. Alsalla, H.; Hao, L.; Smith, C. Fracture Toughness and Tensile Strength of 316L Stainless Steel Cellular Lattice Structures Manufactured Using the Selective Laser Melting Technique. *Mater. Sci. Eng. A* **2016**, *669*, 1–6. [CrossRef]
82. Ziętala, M.; Durejko, T.; Panowicz, R.; Konarzewski, M. Microstructure Evolution of 316L Steel Prepared with the Use of Additive and Conventional Methods and Subjected to Dynamic Loads: A Comparative Study. *Materials* **2020**, *13*, 4893. [CrossRef] [PubMed]
83. Rollett, A.; Humphreys, F.; Rohrer, G.S.; Hatherly, M. *Recrystallization and Related Annealing Phenomena: Second Edition*; Elsevier Ltd: Amsterdam, The Netherlands, 2004; ISBN 9780080441641.
84. Verlinden, B.; Driver, J.; Samajdar, I.; Doherty, R.D. *Thermo-Mechanical Processing of Metallic Materials*; Elsevier: Amsterdam, The Netherlands, 2007; ISBN 9780080444970.
85. Purcek, G.; Saray, O.; Nagimov, M.I.; Nazarov, A.A.; Safarov, I.M.; Danilenko, V.N.; Valiakhmetov, O.R.; Mulyukov, R.R. Microstructure and Mechanical Behavior of UFG Copper Processed by ECAP Following Different Processing Regimes. *Philos. Mag.* **2012**, *92*, 690–704. [CrossRef]
86. Chinh, N.Q.; Olasz, D.; Ahmed, A.Q.; Bobruk, E.V.; Valiev, R.Z. Review on Grain Size- and Grain Boundary Phenomenon in Unusual Mechanical Behavior of Ultrafine-Grained Al Alloys. *Mater. Trans.* **2023**, *64*, MT-MF2022020. [CrossRef]
87. Filippov, A.V.; Shamarin, N.N.; Semenchuk, N.V.; Filippova, E.O. Formation of Ultra Fine Grained Microstructure in Aluminum Bronzes of the Cu-Al, Cu-Al-Si, and Cu-Al-Si-Mn Systems After Electron Beam Additive Manufacturing. *Russ. Phys. J.* **2024**, *67*, 1364–1372. [CrossRef]
88. Das, J.; Sarkar, R.; Rao, G.A.; Sankaranarayana, M.; Nandy, T.K.; Pabi, S.K. Flow Behaviour of a Heat Treated Tungsten Heavy Alloy. *Mater. Sci. Eng. A* **2012**, *553*, 119–127. [CrossRef]
89. Xue, Q.; Gray, G.T. Development of Adiabatic Shear Bands in Annealed 316L Stainless Steel: Part I. Correlation between Evolving Microstructure and Mechanical Behavior. *Metall. Mater. Trans. A* **2006**, *37*, 2435–2446. [CrossRef]
90. Huh, H.; Lee, H.J.; Song, J.H. Dynamic Hardening Equation of the Auto-Body Steel Sheet with the Variation of Temperature. *Int. J. Automot. Technol.* **2012**, *13*, 43–60. [CrossRef]

MDPI AG
Grosspeteranlage 5
4052 Basel
Switzerland
Tel.: +41 61 683 77 34

Metals Editorial Office
E-mail: metals@mdpi.com
www.mdpi.com/journal/metals

www.ingramcontent.com/pod-product-compliance
Lightning Source LLC
Chambersburg PA
CBHW041356210926
51850CB00013B/3096
* 9 7 8 3 7 2 5 8 4 2 2 9 2 *